Forest Trees

A Guide to the Eastern United States

Lisa J. Samuelson
Professor
School of Forestry and Wildlife Sciences
Auburn University

Michael E. Hogan
Photographer

Upper Saddle River, New Jersey
Columbus, Ohio

Library of Congress Cataloging in Publication Data

Samuelson, Lisa J.
  Forest trees : a guide to the eastern United States / Lisa J. Samuelson, Michael E. Hogan.
     p. cm.
  Includes bibliographical references.
  ISBN 0-13-113894-4
  1. Trees--East (U.S.)—Identification. 2. Forest plants—East (U.S.)—Identification.
I. Hogan, Michael E. II. Title.
  QK115.S26 2006
  582.16'0974--dc22

                                                                    2005022469

Executive Editor: Debbie Yarnell
Assistant Editor: Maria Rego
Production Editor: Alexandrina Benedicto Wolf
Production Coordination: Carlisle Publishers Services
Design Coordinator: Diane Ernsberger
Cover Designer: Ali Mohrman
Cover Image: Michael E. Hogan
Production Manager: Matt Ottenweller
Marketing Manager: Jimmy Stephens

All photos except where otherwise noted, by Michael E. Hogan; © Dorling Kindersley, pp. 61, 79, 147, 163, 185, 356, 362, 393, 425, 429, 441, 442, 450, 453, 459

This book was set in Galliard by Carlisle Communications, Ltd. It was printed and bound by Courier Kendallville, Inc. The cover was printed by Coral Graphic Services, Inc.

Copyright © 2006 by Pearson Education, Inc., Upper Saddle River, New Jersey 07458. Pearson Prentice Hall. All rights reserved. Printed in the United States of America. This publication is protected by Copyright and permission should be obtained from the publisher prior to any prohibited reproduction, storage in a retrieval system, or transmission in any form or by any means, electronic, mechanical, photocopying, recording, or likewise. For information regarding permission(s), write to: Rights and Permissions Department.

Pearson Prentice Hall™ is a trademark of Pearson Education, Inc.
Pearson® is a registered trademark of Pearson plc
Prentice Hall® is a registered trademark of Pearson Education, Inc.

Pearson Education Ltd.
Pearson Education Singapore Pte. Ltd.
Pearson Education Canada, Ltd.
Pearson Education–Japan

Pearson Education Australia Pty. Limited
Pearson Education North Asia Ltd.
Pearson Educación de Mexico, S.A. de C.V.
Pearson Education Malaysia Pte. Ltd.

10 9 8 7 6 5 4 3 2 1
ISBN 0-13-113894-4

For Our Mothers, Barbara and Jean

Contents

Preface *xiii*
About the Authors *xiv*
How to Identify Trees *xv*
Identification Features *xvi*

Guide to the Gymnosperm Families *1*

Cupressaceae Guide *2*
    Chamaecyparis thyoides (L.) B. S. P. 4
    Juniperus virginiana L. 6
    Taxodium distichum var. *distichum* (L.) Rich. 8
    Taxodium distichum var. *imbricarium* (Nutt.) Croom 10
    Thuja occidentalis L. 12

Pinaceae Guide *14*
    Abies balsamea (L.) Mill. 16
    Abies fraseri (Pursh) Poir. 18
    Larix laricina (Du Roi) K. Koch 20
    Picea glauca (Moench) Voss 22
    Picea mariana (Mill.) B. S. P. 24
    Picea rubens Sarg. 26
    Pinus banksiana Lamb. 28
    Pinus clausa (Chapm. ex Engelm.) Vasey ex Sarg. 30
    Pinus echinata Mill. 32
    Pinus elliottii Engelm. 34
    Pinus glabra Walt. 36
    Pinus palustris Mill. 38
    Pinus pungens Lamb. 40
    Pinus resinosa Soland. 42
    Pinus rigida Mill. 44
    Pinus serotina Michx. 46
    Pinus strobus L. 48
    Pinus sylvestris L. 50
    Pinus taeda L. 52
    Pinus virginiana Mill. 54
    Tsuga canadensis (L.) Carr. 56
    Tsuga caroliniana Engelm. 58

Taxaceae *60*
    Taxus floridana Nutt. ex Chapm. 60
    Torreya taxifolia Arn. 62

Guide to the Angiosperm Families   65

Aceraceae Guide   70

Acer barbatum Michx.   72
Acer leucoderme Small   74
Acer negundo L.   76
Acer nigrum Michx.   78
Acer pensylvanicum L.   80
Acer rubrum L.   82
Acer saccharinum L.   84
Acer saccharum Marsh.   86
Acer spicatum Lam.   88

Anacardiaceae Guide   90

Cotinus obovatus Raf.   92
Rhus copallina L.   94
Rhus glabra L.   96
Rhus typhina L.   98
Toxicodendron vernix (L.) Kuntze   100

Annonaceae   102

Asimina triloba (L.) Dunal   102

Aquifoliaceae Guide   104

Ilex amelanchier Curtis ex Chapm.   106
Ilex cassine L.   108
Ilex coriacea (Pursh) Chapm.   110
Ilex decidua Walt.   112
Ilex montana Torr. & Gray ex Gray.   114
Ilex myrtifolia Walt.   116
Ilex opaca Ait.   118
Ilex vomitoria Ait.   120

Araliaceae   122

Aralia spinosa L.   122

Betulaceae Guide   124

Alnus rugosa (Du Roi) Spreng.   126
Alnus serrulata (Ait.) Willd.   128
Betula alleghaniensis Britt.   130
Betula lenta L.   132
Betula nigra L.   134
Betula papyrifera Marsh.   136
Betula populifolia Marsh.   138
Carpinus caroliniana Walt.   140
Ostrya virginiana (Mill.) K. Koch   142

Bignoniaceae   144

Catalpa bignonioides Walt.   144
Catalpa speciosa Warder ex Engelm.   146

Caesalpiniaceae Guide   148
Cercis canadensis L.   150
Gleditsia aquatica Marsh.   152
Gleditsia triacanthos L.   154
Gymnocladus dioicus (L.) K. Koch   156

Caprifoliaceae Guide   158
Sambucus canadensis L.   160
Viburnum lentago L.   162
Viburnum nudum L.   164
Viburnum prunifolium L.   166
Viburnum rufidulum Raf.   168
Other *Viburnums*   170

Celastraceae   172
Euonymus atropurpureus Jacq.   172

Cornaceae Guide   174
Cornus alternifolia L. f.   176
Cornus amomum Mill.   178
Cornus drummondii C. A. Mey.   180
Cornus florida L.   182
Cornus racemosa Lam.   184
Cornus stricta Lam.   186
Nyssa aquatica L.   188
Nyssa biflora Walt.   190
Nyssa ogeche Bartr. ex Marsh.   192
Nyssa sylvatica Marsh.   194

Cyrillaceae   196
Cliftonia monophylla (Lam.) Britt. ex Sarg.   196
Cyrilla racemiflora L.   198

Ebenaceae   200
Diospyros virginiana L.   200

Ericaceae Guide   202
Kalmia latifolia L.   204
Oxydendrum arboreum (L.) DC.   206
Rhododendron maximum L.   208
Vaccinium arboreum Marsh.   210

Euphorbiaceae   212
Sapium sebiferum (L.) Roxb.   212

Fabaceae   214
Cladrastis kentukea (Dum.-Cours.) Rudd   214
Robinia pseudoacacia L.   216

Fagaceae Guide   218
Castanea dentata (Marsh.) Borkh.   222
Castanea pumila (L.) P. Mill.   224

Fagus grandifolia Ehrh.   226
Quercus acutissima Carr.   228
Quercus alba L.   230
Quercus austrina Small.   232
Quercus bicolor Willd.   234
Quercus chapmanii Sarg.   236
Quercus coccinea Muenchh.   238
Quercus durandii Buckl.   240
Quercus ellipsoidalis E. J. Hill   242
Quercus falcata Michx.   244
Quercus geminata Small   246
Quercus georgiana Curtis   248
Quercus hemisphaerica Bartr. ex Willd.   250
Quercus ilicifolia Wangenh.   252
Quercus imbricaria Michx.   254
Quercus incana Bartr.   256
Quercus laevis Walt.   258
Quercus laurifolia Michx.   260
Quercus lyrata Walt.   262
Quercus macrocarpa Michx.   264
Quercus margaretta Ashe   266
Quercus marilandica Muenchh.   268
Quercus michauxii Nutt.   270
Quercus muehlenbergii Engelm.   272
Quercus myrtifolia Willd.   274
Quercus nigra L.   276
Quercus oglethorpensis Duncan   278
Quercus pagoda Raf.   280
Quercus palustris Muenchh.   282
Quercus phellos L.   284
Quercus prinus L.   286
Quercus rubra L.   288
Quercus shumardii Buckl.   290
Quercus stellata Wangenh.   292
Quercus texana Buckl.   294
Quercus velutina Lam.   296
Quercus virginiana P. Mill.   298

Hamamelidaceae   300

Hamamelis virginiana L.   300
Liquidambar styraciflua L.   302

Hippocastanaceae Guide   304

Aesculus flava Ait.   306
Aesculus glabra Willd.   308
Aesculus pavia L.   310

Illiciaceae   312

Illicium floridanum Ellis   312

Juglandaceae Guide   314

Carya aquatica (Michx. f.) Nutt.   316
Carya cordiformis (Wangenh.) K. Koch   318
Carya glabra (Mill.) Sweet   320
Carya illinoinensis (Wangenh.) K. Koch   322
Carya laciniosa (Michx. f.) G. Don   324
Carya myristiciformis (Michx. f.) Nutt.   326
Carya ovalis (Wangenh.) Sarg.   328
Carya ovata (P. Mill.) K. Koch   330
Carya pallida (Ashe) Engelm. & Graebn.   332
Carya texana Buckl.   334
Carya tomentosa (Poir. ex Lam.) Nutt.   336
Juglans cinerea L.   338
Juglans nigra L.   340

Lauraceae Guide   342

Persea borbonia (L.) Spreng.   344
Persea palustris (Raf.) Sarg.   346
Sassafras albidum (Nutt.) Nees   348

Magnoliaceae Guide   350

Liriodendron tulipifera L.   352
Magnolia acuminata (L.) L.   354
Magnolia fraseri Walt.   356
Magnolia grandiflora L.   358
Magnolia macrophylla Michx.   360
Magnolia tripetala (L.) L.   362
Magnolia virginiana L.   364

Meliaceae   366

Melia azedarach L.   366

Mimosaceae   368

Albizia julibrissin Durazz.   368

Moraceae Guide   370

Broussonetia papyrifera (L.) L'Her. ex Vent.   372
Maclura pomifera (Raf.) Schneid.   374
Morus alba L.   376
Morus rubra L.   378

Myricaceae Guide   380

Myrica cerifera L.   382
Myrica heterophylla Raf.   384
Myrica pensylvanica Loisel.   386

Oleaceae Guide   388

Chionanthus virginicus L.   390
Forestiera acuminata (Mich.) Poir.   392
Fraxinus americana L.   394
Fraxinus caroliniana P. Mill.   396

Fraxinus nigra Marsh.   398
Fraxinus pennsylvanica Marsh.   400
Fraxinus profunda (Bush) Bush   402
Fraxinus quadrangulata Michx.   404
Ligustrum sinense Lour.   406
Osmanthus americanus (L.) Benth. & Hook. f. ex Gray   408

Platanaceae   410

Platanus occidentalis L.   410

Rhamnaceae   412

Rhamnus caroliniana Walt.   412

Rosaceae Guide   414

Amelanchier arborea (Michx. f.) Fern.   416
Amelanchier laevis Wieg.   418
Crataegus spp.   420
Malus angustifolia (Ait.) Michx.   422
Malus coronaria (L.) P. Mill.   424
Malus pumila P. Mill.   426
Prunus americana Marsh.   428
Prunus angustifolia Marsh.   430
Prunus caroliniana (P. Mill.) Ait.   432
Prunus mexicana S. Wats.   434
Prunus pensylvanica L. f.   436
Prunus serotina Ehrh.   438
Prunus virginiana L.   440
Sorbus americana Marsh.   442

Rubiaceae   444

Cephalanthus occidentalis L.   444
Pinckneya bracteata (Bartr.) Raf.   446

Rutaceae Guide   448

Ptelea trifoliata L.   450
Zanthoxylum americanum P. Mill.   452
Zanthoxylum clava-herculis L.   454

Salicaceae Guide   456

Populus balsamifera L.   458
Populus deltoides Bartr. ex Marsh.   460
Populus grandidentata Michx.   462
Populus tremuloides Michx.   464
Salix nigra Marsh.   466
Other *Willows*   468

Sapotaceae Guide   470

Bumelia lanuginosa (Michx.) Pers.   472
Bumelia lycioides (L.) Pers.   474
Bumelia tenax (L.) Willd.   476

Scrophulariaceae   478

Paulownia tomentosa (Thumb.) Sieb. & Zucc. ex Steud.   478

Simaroubaceae 480
Ailanthus altissima (P. Mill.) Swingle 480
Staphyleaceae 482
Staphylea trifolia L. 482
Styracaceae Guide 484
Halesia diptera Ellis 486
Halesia tetraptera Ellis 488
Styrax americanus Lam. 490
Styrax grandifolius Ait. 492
Symplocaceae 494
Symplocos tinctoria (L.) L'Her. 494
Theaceae 496
Gordonia lasianthus (L.) Ellis 496
Stewartia ovata (Cav.) Weatherby 498
Tiliaceae 500
Tilia americana L. 500
Ulmaceae Guide 502
Celtis laevigata Willd. 504
Celtis occidentalis L. 506
Celtis tenuifolia Nutt. 508
Planera aquatica J.F. Gmel. 510
Ulmus alata Michx. 512
Ulmus americana L. 514
Ulmus crassifolia Nutt. 516
Ulmus rubra Muhl. 518
Ulmus serotina Sarg. 520
Ulmus thomasii Sarg. 522

Glossary 524
Bibliography 527
Index 529

Preface

Our goal in writing this book is to provide students and tree enthusiasts with a guide using minimal technical language and excellent photography for identification of tree species common to forests of the eastern United States.

The text includes over 1,000 photographs with useful leaf, twig, bud, flower, fruit, and form characteristics, habitat and ecology, multiple common names, similar species, derivation of the binomial name, wildlife uses, landscape uses, wood uses, and interesting historic uses. A "Quick Guide" is given for each species to highlight key features. The average sizes of leaves, flowers, fruits, cones, and buds are listed, but beware that sizes and shapes change with site quality, microclimate, genetics, tree age and size, and even within the same tree.

Families and species within families are arranged alphabetically. Range maps are presented only as general guides. We define *tree* as a woody plant that commonly reaches 20 feet or more in height. Because taxonomy and nomenclature varies to some degree between texts, we follow for the most part *Harlow and Harrar's Textbook of Dendrology* (Hardin et al., 2001). For example, some authors keep *Nyssa* in Nyssaceae and *Taxodium* in Taxodiaceae and do not recognize Mimosaceae or Caesalpiniaceae. *Rhamnus* is also known as *Frangulata*.

Acknowledgments

We wish to acknowledge the support of the School of Forestry and Wildlife Sciences at Auburn University. The support of Thomas Stokes of Auburn University in "holding down the fort" was invaluable. Cameragraphics in Auburn, Alabama, provided excellent film development and photographic support. Range maps were drawn based on the *Atlas of the United States Trees*, Vols. 1 and 4, by Elbert Little, USDA Forest Service. *Silvics of North America*, Vols. 1 and 2, edited by Burns and Honkala, served as an important reference for describing shade tolerance, forest associates, and tree sizes.

Finally, we thank Debbie Yarnell and Alex Wolf at Prentice Hall for sharing our vision.

Lisa Samuelson
Michael Hogan

About the Authors

Lisa J. Samuelson

Lisa Samuelson is Professor of Tree Physiology in the School of Forestry and Wildlife Sciences at Auburn University. She received her B.S. and M.S. in Forestry from the School of Forest Resources at the University of Georgia, and her Ph.D. in Forestry in 1992 from Virginia Tech. She has co-authored *Forest Trees: A Guide to the Southeastern and Mid-Atlantic Regions of the United States* and *Guide and Key to Alabama Trees*. In addition to University teaching, she gives lectures on trees identification to nature societies, 4-H clubs, garden clubs, arboreta and botanical gardens, forestry groups, and community organizations.

Michael E. Hogan

Michael Hogan is a fine woodworker and freelance photographer who shares his wife's love of trees. His tree photos have been on the cover of *Functional Ecology* and *New Phytologist* and used in educational, extension, and outreach publications from universities and government and nonprofit organizations.

How to Identify Trees

These guidelines will help you use the Identification Features and Guides for identifying your trees. Always be sure to use identification features that are stable rather than some minute trait that may not appear in all trees of that species.

Gymnosperms

Let's start with the Gymnosperms, including the pines, firs, spruces, junipers, cedars, larches, cypresses, hemlocks, and yews. First, take a look at the type of leaf and determine if it is needlelike or scalelike and evergreen or deciduous. The majority of species are evergreen; therefore, if your tree is deciduous, you have a cypress or larch. See if the needles are bundled in a sheath (fascicle) as in the pines, or unbundled as in the spruces, firs, larches, cypresses, cedars, hemlocks, and yews. To separate the pines, determine how many needles are bundled in a fascicle, as well as the length of the needle. For unbundled leaves, determine if the leaves are flat, fragrant, and with white stripes (firs, hemlocks); angled and often sharp tipped (spruces); featherlike (cypresses); or in clusters from knobby small shoots (larches). Next take a look at the cone. Determine if it is woody, papery, or fleshy. Members of the yew family have seeds in a fleshy sac or cup. Look at the cone scales and determine if they are spirally arranged (Pinaceae) or with peltate scales and a round or fleshy appearance (Cupressaceae). Look at the size and color of a cone and whether a prickle exists. See if the cones are pendant (pines, spruces, hemlocks) or upright (firs, larches) on the branch.

Angiosperms

When first learning to identify Angiosperm trees in summer, it is best to use a series of identification features beginning with the most general and ending with the most specific features. First, determine whether your specimen has an opposite, whorled, or alternate leaf arrangement. Most species are alternate, so observing an opposite or whorled arrangement narrows your options. Some of the most common opposite species are maples, ashes, dogwoods, viburnums, and buckeyes. The catalpas are a common whorled species. When determining the leaf arrangement, make sure that you are looking at where the petiole inserts on the branch, especially when dealing with a compound species. Next, determine if your specimen has simple or compound leaves. Most species are simple, so a compound arrangement again narrows down your options. An opposite and compound arrangement is the least common and includes the ashes, boxelder, elderberries, buckeyes (palmately compound), and bladdernut. Next, take a look at the bark. Is it grooved, flaky, or smooth, and what is its color? It's best to use bark features on large trees because for most species the bark is smooth on small trees. Do you see any flowers or fruit? Look for thorns or spines on the twigs. Next look for more specific leaf features. Are there any obvious or unusual leaf features such as a distinctive leaf shape or a unique odor when the leaf is crushed? Look at the margin of the leaf and see if there are any teeth, glands, or lobes. Examine the shape of the leaf base and apex. Look for distinctive textures on the upper and lower surfaces, such as sandpapery, waxy, hairy, or glandular.

Here are a few general hints for winter identification of Angiosperms. You must use bark and twig features and, sometimes, fruit if the fruit persists over winter. In some ways winter identification is easier because twig characteristics are unique for most genera and it's easier to see up in the canopy. A hand lens is very helpful for seeing bud characteristics. First, examine the arrangement of small branches or leaf scars to determine if your tree is opposite or alternate. Be sure to examine the portion of a young twig that grew the past summer to see the best identification features. The leaf scar is the scar left behind when the petiole senesced. The dots within the leaf scar are the bundle scars, which were made by the xylem and phloem that transported sugar and water in and out of the leaf. The shape of the leaf scar and arrangement of bundle scars are unique. Take a look at the terminal bud. Do you see a clump of buds, such as in the oaks, or is there only one terminal bud? Are the buds round or pointed, small or large, smooth, hairy, or glandular? Look at the size and color of the buds and whether the bud scales are overlapping (imbricate) or not overlapping (valvate). If you can't see any bud scales, usually because the bud is too hairy, the bud is described as naked.

Examine the texture and color of the hair on the buds. Examine the twig for hair, wax, glands, lenticels, thorns, and any odor when broken.

To help separate families and species, guides to families and to species within families with more than two members are included. The guides work by selecting one choice from numerous choices. First, work through the Gymnosperm or Angiosperm Guide to select a family. Once you have selected a family, use the family guide to select a species or go directly to the species pages. When using a guide, select the most appropriate heading and subheading (if given), and then the description that best describes your specimen. If a characteristic is listed for one species but not another, you can assume that only the species with the description possesses that characteristic.

Identification Features

Leaf Margins

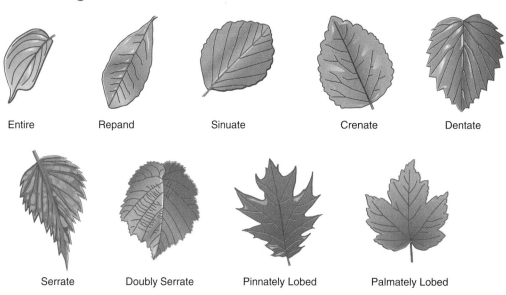

Twig Features

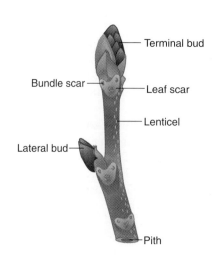

Leaf Arrangements

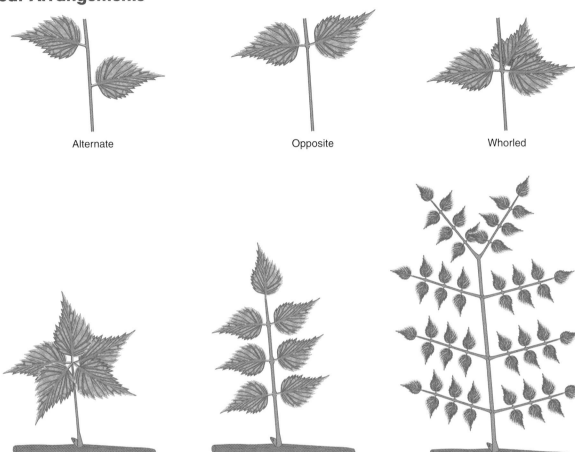

Leaf Apices

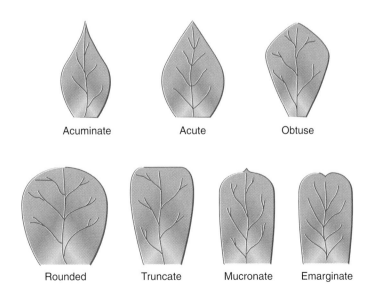

Leaf Bases

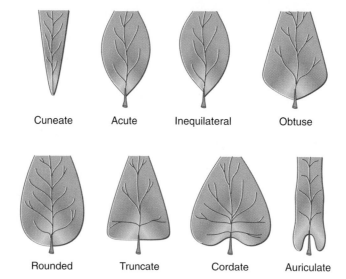

Leaf Shapes

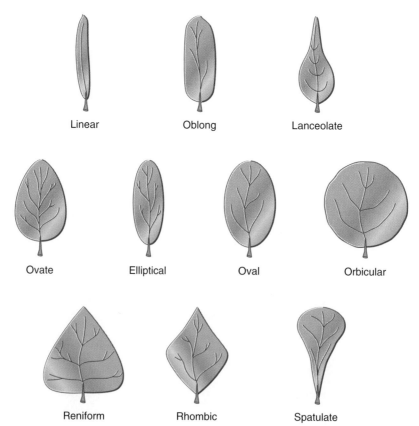

Guide to the Gymnosperm Families

Most species with needle or scalelike leaves, seeds not enclosed in an ovary
1. Leaves needlelike or linear, in bundles or solitary, 2–4 sided; cone woody or papery with spiral scales. *Pinaceae* (pine family)
2. Leaves not in bundles, linear, flat, apex pointed or sharp; seeds in a green sac or red fleshy cup. *Taxaceae* (yew family)
3. Leaves not in bundles, linear or awl- or scalelike; cone berrylike or with woody peltate scales or leathery scales. *Cupressaceae* (cypress, cedar family)

Cupressaceae Guide

1. Leaves scalelike; cone woody with small peltate scales when mature. *Chamaecyparis thyoides*
2. Leaves scalelike (awl-like on seedlings), cone blue and berrylike. *Juniperus virginiana*
3. Leaves scalelike, cone ovoid, upright, with 5–6 pairs of prickle-tipped leathery scales. *Thuja occidentalis*
4. Leaves linear, cone with large peltate scales. *Taxodium distichum* var. *distichum*
5. Leaves awl-like, overlapping; cone with large peltate scales. *Taxodium distichum* var. *imbricarium*

Cupressaceae

Juniperus communis bark

Taxodium distichum var. *imbricarium* bark

Juniperus communis leaves

Chamaecyparis thyoides (L.) B. S. P.
Atlantic white-cedar, southern white-cedar, swamp-cedar, white-cedar

Quick Guide: *Needles* scalelike in bright green, flat, graceful sprays; *Cones* small, scales woody and peltate when mature; *Bark* ash-gray with fibrous interlacing ridges, often twisting around the trunk; *Habitat* swamps and bogs.

Leaves: Up to 5 mm long, scalelike, rich green to blue-green, glandular on the underside, in flattened sprays, persisting for several years.

Twigs: Brown-gray, scaly.

Buds: Small, hidden.

Bark: Gray-brown and smooth on very small trees; large trees ash-gray with red-brown inner bark and fibrous, often stringy, interlacing ridges; bark may twist around the trunk.

Seed Cone: 6 mm wide, yellow-green when immature, nearly round, ridged; scales brown, woody, and peltate after opening; maturing in one growing season.

Form: Up to 35 m (115 ft) in height and 1.5 m (5 ft) in diameter but usually smaller.

Habitat and Ecology: Intermediate shade tolerance; found in acidic freshwater swamps and bogs. Forest associates include *Acer rubrum*, *Cliftonia monophylla*, *Cyrilla racemiflora*, *Magnolia grandiflora*, *Magnolia virginiana*, *Nyssa biflora*, *Pinus palustris* on swamp edges, *Quercus virginiana*, and *Taxodium distichum* var. *imbricarium*.

Uses: Wood pale to pinkish, soft, straight grained, decay resistant; used for siding, shingles, boxes, fencing, small-boat construction, and telephone poles. The blue-green foliage against the ash-gray bark makes for an attractive landscape tree on moist to wet soils.

Botanical Name: *Chamaecyparis* means "low-growing cypress," perhaps referring to the low habitat; *thyoides* means "similar to *Thuja*," referring to the needles.

Chamaecyparis thyoides leaves

Chamaecyparis thyoides bark

Chamaecyparis thyoides bark

Chamaecyparis thyoides seed cones

Juniperus virginiana L.

eastern redcedar, redcedar, red juniper, pencil cedar

Quick Guide: *Needles* dimorphic (awl- or scalelike); *Cones* smooth, berrylike, waxy blue; *Bark* red-brown, fibrous, and stringy; *Habitat* dry uplands to swamp edges.

Leaves: Green to yellow-green, persisting 2 years, needles dimorphic being either awl-like about 8 mm long and very sharp or scalelike up to 2 mm long and glandular. Awl-like needles typical on seedlings.

Twigs: Brown-gray, angled, scaly.

Buds: Hidden, green.

Bark: Red-brown to gray, peeling into long fibrous strips.

Seed Cone: 5 mm wide, round, berrylike, white to dark blue, waxy, maturing in one growing season; species is dioecious.

Form: Generally up to 15 m (50 ft) in height and 61 cm (2 ft) in diameter.

Habitat and Ecology: Shade intolerant; the most widely distributed conifer in the eastern United States, found on a wide variety of sites such as abandoned fields, swamp edges, and upland forests. Forest associates include *Carya* spp., *Diospyros virginiana, Liquidambar styraciflua, Pinus echinata, Pinus taeda, Pinus virginiana, Quercus alba, Quercus coccinea, Quercus falcata, Quercus rubra, Quercus stellata, Rhus* spp., *Sassafras albidum,* and *Ulmus alata.*

Uses: Wood with yellow-white sapwood and red-maroon heartwood, fragrant, close textured, durable; used for fence posts, railroad ties, furniture, paneling, cedar chests, pencils, woodenware, and cedarwood oil. Also used in Christmas tree production and in windbreaks. Browsed by white-tailed deer and seed cones eaten by a wide variety of game birds, song-birds (such as cedar waxwings), and small to midsize mammals such as opossum and raccoon. Dense foliage also provides excellent

nesting, roosting, and foraging substrate for birds. Older trees are interesting because of the bark and low, wide, evergreen crown. Cultivars with better form and foliage color have been developed for landscaping. This species is an intermediate host for apple-rust disease, however.

Botanical Name: *Juniperus* is Latin for "juniper tree"; *virginiana* refers to the geographic range.

Similar Species: Southern red-cedar [*Juniperus virginiana* var. *silicicola* (Small) E. Murray] is found on coastal sites such as dunes, sandy ridges, and brackish sites from North Carolina southward and reported to have smaller cones and more drooping needlelike foliage. Common juniper (*Juniperus communis* L.) is a small tree or shrub found on well-drained soils from northern Georgia to New York and identified by linear sharp-pointed needles (but not scalelike) in whorls of three (see page 3). Ashe juniper (*Juniperus ashei* Buchh.) is found in Arkansas on limestone outcrops and identified by scalelike leaves with minute teeth, needlelike leaves on new growth, and crooked form.

Juniperus virginiana leaves and seed cones

Juniperus virginiana bark

Juniperus virginiana bark

Juniperus virginiana form

Taxodium distichum var. *distichum* (L.) Rich.

baldcypress, southern-cypress, gulf-cypress, white-cypress, yellow-cypress, red-cypress, tidewater cypress

Quick Guide: *Needles* flat, 2 ranked, feathery, deciduous; *Branchlets* deciduous; *Bark* red-brown, fibrous, "knees" often seen near the base of the tree; *Habitat* swamps and bottoms.

Leaves: Needles 1 cm long, linear, flat, 2 ranked, green to yellow-green, soft, feathery, deciduous; autumn color red-brown.

Twigs: Red-brown to gray-brown, rough, branchlets deciduous.

Buds: Green, round, buried in peglike branches.

Bark: Gray to red-brown, fibrous, and stringy; large trees with loose, narrow strips.

Seed Cone: Up to 3.5 cm wide, round, green when immature, brown and woody with peltate scales when mature; maturing in one growing season.

Form: Old growth trees up to 46 m (150 ft) in height and 4 m (12 ft) in diameter; base of the trunk swollen and fluted, knees often seen near the base.

Habitat and Ecology: Intermediate shade tolerance; found in swamps, ponds, and bottomlands; can withstand prolonged flooding. Common forest associates include *Nyssa aquatica*, *Nyssa ogeche*, *Nyssa biflora*, and *Taxodium distichum* var. *imbricarium*.

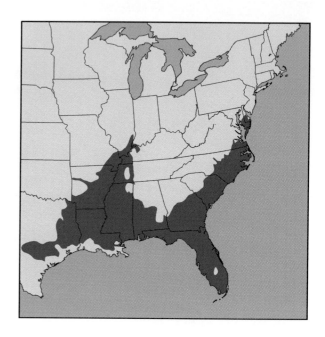

Uses: Wood yellow-white with reddish latewood, light, soft, decay resistant, easily worked; used for greenhouses, fencing, shingles, caskets, railroad ties, small-boat construction, and boxes. Pecky cypress, caused by a fungus, is a popular decorative wood. Seeds eaten by waterfowl and small mammals. Serves as a nesting tree for osprey and bald eagles. Planted as an ornamental on a variety of sites.

Botanical Name: *Taxodium* is Latin for "yewlike," referring to the needles; *distichum* refers to the 2 ranked needles. Also named *Taxodium distichum* (L.) Rich and grouped in Taxodiaceae.

Taxodium distichum var. *distichum* leaves

Taxodium distichum var. *distichum* bark

Taxodium distichum var. *distichum* bark

Taxodium distichum var. *distichum* seed cone

Cupressaceae

Taxodium distichum var. *imbricarium* (Nutt.) Croom

pondcypress, black-cypress

Quick Guide: *Needles* awl-like, overlapping, deciduous; *Branchlets* curving upward and deciduous; *Bark* red-brown, narrowly plated, "knees" often seen near the base of the tree; *Habitat* swamps.

Leaves: Needles awl-like or linear, appressed and often appearing as a pine needle from a distance, green to yellow-green, deciduous; autumn color red-brown.

Twigs: Red-brown to gray-brown and rough; branchlets curving upward and deciduous.

Buds: Green, round, buried in peglike branches.

Bark: Gray to red-brown, fibrous and stringy; large trees becoming ridged with long narrow plates. Large trees more plated than in var. *distichum*.

Seed Cone: Up to 3.5 cm wide, round, green when immature, brown and woody with peltate scales when mature; maturing in one growing season.

Form: Old growth trees up to 40 m (130 ft) in height and 2 m (7 ft) in diameter; base of the trunk swollen and fluted, knees often seen near the base.

Habitat and Ecology: Intermediate shade tolerance; found in swamps, ponds, and blackwater river swamps. Although both species can be found

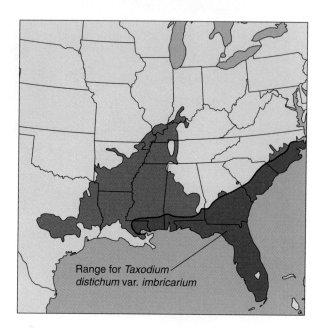

Range for *Taxodium distichum* var. *imbricarium*

together, *T. distichum* var. *imbricarium* is more common on nutrient-poor soils than var. *distichum*. Forest associates include *Chamaecyparis thyoides*, *Nyssa ogeche*, *Nyssa biflora*, and *Taxodium distichum* var. *distichum*.

Uses: Same as for *T. distichum* var. *distichum*.

Botanical Name: *Taxodium* is Latin for "yewlike," referring to the needles; *imbricarium* refers to the overlapping (imbricate) needles. Also named *Taxodium ascendens* Brongn and grouped in Taxodiaceae.

Taxodium distichum var. *imbricarium* leaves

Taxodium distichum var. *imbricarium* bark

Taxodium distichum var. *imbricarium* bark

Taxodium distichum var. *imbricarium* seed cone

Thuja occidentalis L.

northern white-cedar, swamp-cedar, American arborvitae, eastern white-cedar

Quick Guide: *Needles* scalelike, glandular, aromatic, forming flattened sprays; *Cones* small, ovoid, upright, with 5–6 pairs of prickle-tipped leathery scales; *Bark* red-brown to gray, fibrous; *Habitat* swamps to uplands.

Leaves: Scalelike, up to 3 mm long, green to yellow-green, glandular, aromatic, persisting for several growing seasons.

Twigs: In flattened fanlike sprays.

Buds: Minute.

Bark: Red-brown to gray, with stringy fibrous ridges.

Seed Cone: Up to 1.2 cm long, ovoid, with 5–6 pairs of prickle-tipped leathery scales, in upright clusters; maturing in one growing season.

Form: Up to 15 m (50 ft) in height and 61 cm (2 ft) in diameter with short branches and a pyramidal crown. Lower branches may root in moss and produce seedlings.

Habitat and Ecology: Shade tolerant; found on wide range of sites from nutrient-rich swamps to limestone uplands but common on moist fertile soils. Forest associates include *Abies balsamea, Acer rubrum, Betula alleghaniensis, Betula papyrifera, Larix laricina, Picea mariana, Picea rubens, Pinus strobus, Populus grandidentata, Populus tremuloides,* and *Tsuga canadensis*.

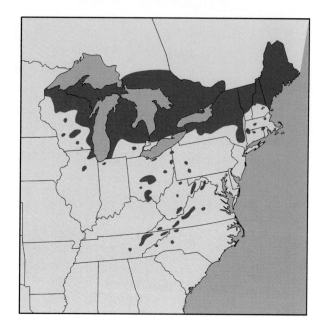

Uses: Wood with yellow-white sapwood and yellow-brown heartwood, fragrant, durable; used for pulp, particleboard, fence posts, railroad ties, furniture, shingles, and woodenware; once used for telegraph poles. Leaves are used to make cedar oil that is used in soaps, perfumes, and insect repellents. Leaves and bark were used in medicinal teas (for example, to heal scurvy) and skin salves. Browsed by white-tailed deer, hares, and small mammals; provides cover for a variety of wildlife. A popular landscape tree and cultivars have been developed to enhance stress tolerance, form, and foliage color.

Botanical Name: *Thuja* means "aromatic cedar"; *occidentalis* means "of the Western Hemisphere."

Cupressaceae

Thuja occidentalis leaves

Thuja occidentalis bark

Thuja occidentalis bark

Thuja occidentalis seed cones

Pinaceae Guide

I. Needles not in bundles

1. Needles 4 sided, sharp pointed, shiny, born on pegs; cone pendent, scales with round only slightly erose margins lacking prickles; bark with gray-brown scaly plates. *Picea rubens*
2. Needles 4 sided, blunt, dull, born on pegs; cone pendent, scales with round erose margins lacking prickles; bark with gray scaly plates. *Picea mariana*
3. Needles 4 sided, blue-green, waxy, with a fetid odor when crushed, born on pegs; cone pendent, scales with smooth truncate margins lacking prickles; bark gray and scaly with red-brown inner bark. *Picea glauca*
4. Similar to *Picea glauca* but branches drooping and cone larger. *Picea abies*
5. Needles 2 sided, somewhat 2 ranked, with white bands below, born on pegs; cone very small, pendent, lacking prickles; bark red-brown and deeply grooved. *Tsuga canadensis*
6. Needles 2 sided, born on pegs spirally around the twig with white bands below; cone larger than *Tsuga canadensis;* bark red-brown and deeply grooved. *Tsuga caroliniana*
7. Needles blue-green, 2 sided, with white bands below, fragrant, leaving round leaf scars; cone upright with reflexed and exserted bracts; bark with resin blisters. *Abies fraseri*
8. Needles dark green, 2 sided, with white bands below, fragrant, leaving round leaf scars; cone upright with bracts not visible above the cone scale; bark with resin blisters. *Abies balsamea*
9. Needles angled, in clusters from knobby spur shoots, deciduous; cone nearly upright, small, lacking prickles; bark gray and scaly with red-brown inner bark. *Larix laricina*

II. Needles in bundles of two

1. Needles not twisted, flexible, 6–13 cm long; cone flexible with small sharp prickles; bark plates with resin holes. *Pinus echinata*
2. Needles not twisted, brittle and snapping when bent, 10–16 cm long; cone lacking prickles; bark gray with red-brown plates. *Pinus resinosa*
3. Needles often twisted, stiff, sharp pointed, 3–7 cm long; cone serotinous, misshapen when open with large clawlike prickles and the base often embedded in the branch or trunk; bark thin with gray scales. *Pinus pungens*
4. Needles sometimes twisted, flexible, 5–10 cm long; cone with weak deciduous prickles; bark on large trees deeply grooved rather than plated. *Pinus glabra*
5. Needles twisted, stiff, blue-green, 4–8 cm long; cone lacking prickles and scales 4 sided; bark with orange-gray scaly plates. *Pinus sylvestris*
6. Needles twisted or forked, stiff, 2–4 cm long; cone serotinous, misshapen, in clusters pointing forward on the branch; bark thin and brown-gray. *Pinus banksiana*
7. Needles twisted, stiff, 3–8 cm long; cone with sharp prickles; bark orange-gray and scaly. *Pinus virginiana*
8. Needles twisted, flexible, 5–9 cm long; cone with sharp prickles, sometimes serotinous; bark scaly and orange or red-gray. *Pinus clausa*

III. Needles in bundles of more than two

1. Needles in groups of five, blue-green, 8–14 cm long; cone slim and lacking prickles; branching in whorls. *Pinus strobus*

2. Needles in groups of three, 20–46 cm long, clustered in tufts at the end of stout branches; cone large; bark red plated. *Pinus palustris*
3. Needles in groups of three, 13–21 cm long; cone serotinous; bark may show tufts of needles or sprouts; found on wet sites. *Pinus serotina*
4. Needles in groups of three, 12–23 cm long; cone gray-brown with sharp prickles; bark red-brown plated. *Pinus taeda*
5. Needles in groups of two and three, 15–28 cm long; cone chocolate brown and shiny with weak prickles; bark plates possibly tan flecked. *Pinus elliottii*
6. Needles in groups of three or sometimes two, 6–15 cm long, stiff; cone serotinous, nearly globose, with rigid prickles, the base often embedded in the branch; bark may show tufts of needles. *Pinus rigida*

Pinus virginiana pollen cones

Pinus palustris pollen cones

Pinus taeda pollen cones

Pinus glabra pollen cones

Abies balsamea (L.) Mill.
balsam fir, eastern fir, Canada balsam

Quick Guide: *Needles* flat, blunt, dark green with two white bands below, very fragrant; *Cones* upright with smooth scales and inserted bracts; *Bark* with resin blisters; *Habitat* moist to poorly drained soils.

Leaves: Needles 1–2.5 cm long, flat, blunt, lying horizontal or curving upwards, shiny dark green above with two white bands below, very fragrant, persisting up to 10 years.

Twigs: Stout, gray-green, glabrous or pubescent; leaf scar round and smooth.

Buds: Round, green-orange, resinous.

Bark: Green-gray to brown-gray, smooth with resin blisters; large trees more scaly.

Seed Cone: 5–10 cm long, upright, oblong, purple-green to gray-brown; scales without prickles, margins rounded and smooth, dehiscing at maturity; bracts not visible; maturing in one growing season.

Form: Up to 18 m (60 ft) tall and 46 cm (18 in) in diameter with a tapering, steeplelike crown.

Habitat and Ecology: Very shade tolerant; found on a variety of sites but common on moist to poorly drained soils in association with *Acer rubrum, Acer saccharum, Betula alleghaniensis, Betula papyrifera, Larix laricina, Picea glauca, Picea mariana, Picea rubens, Pinus strobus, Populus grandidentata, Populus tremuloides, Thuja occidentalis,* and *Tsuga canadensis.*

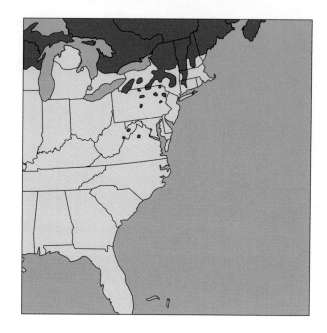

Uses: Wood white to light brown, soft, light; used for pulpwood, construction, and crates. Planted in Christmas tree plantations and branches used for Christmas wreaths. Resin drained from bark blisters ("Canada balsam" and "oil of fir") used in mounting microscopic specimens and as a folk remedy. Seeds and buds eaten by birds and small mammals; foliage provides browse and cover for many animals including deer and moose.

Botanical Name: *Abies* is Latin for "fir"; *balsamea* refers to the fragrant resin.

Abies balsamea seed cones

Abies balsamea seed cone

Abies balsamea bark

Abies balsamea leaves

Abies balsamea bark

Pinaceae

Abies fraseri (Pursh) Poir.

Fraser fir, southern balsam fir

Quick Guide: *Needles* flat, blunt, bushy, blue-green to dark green with two white bands on the underside; *Cones* upright with erose scales and reflexed bracts; *Bark* gray-green with resin blisters; *Habitat* southern Appalachians.

Leaves: Needles 1–2.5 cm long, flat, dark green to blue-green and shiny above, apex blunt, underside with two white bands, highly fragrant, persisting up to 10 years.

Twigs: Stout, gray-green with round leaf scars.

Buds: Round, orange-red, resinous.

Bark: Gray-green to brown and smooth with resin blisters, becoming more scaly on large trees.

Seed Cone: 5–9 cm long, upright, oblong, yellow-brown to purple-brown; scales without prickles, margins round and erose, dehiscing at maturity; bracts yellow-green to brown, exserted and reflexed; maturing in one growing season.

Form: Up to 24 m (80 ft) in height and 61 cm (2 ft) in diameter.

Habitat and Ecology: Shade tolerant; found at high elevations usually above 1300 m on moist cool sites in isolated populations in the southern Appalachians with *Acer spicatum, Betula alleghaniensis, Betula lenta, Picea rubens, Prunus*

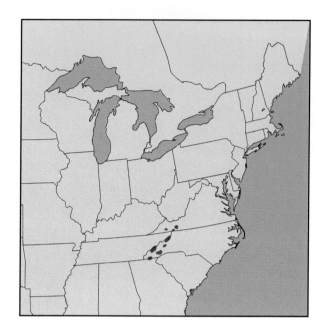

pensylvanica, and *Sorbus americana*. Natural distribution of *Abies fraseri* greatly affected by the balsam wooly adelgid, an insect that damages the cambium and vegetative tissues.

Uses: Wood white to light brown, soft, and light. Planted in Christmas tree plantations. A beautiful tree with blue-green foliage and upright cones. Seeds and buds eaten by birds and small mammals.

Botanical Name: *Abies* is Latin for "fir"; *fraseri* refers to the discoverer John Fraser from Scotland.

Abies fraseri seed cone

Abies fraseri bark

Abies fraseri bark

Abies fraseri seed cones and leaves

Larix laricina (Du Roi) K. Koch

tamarack, eastern larch, American larch

Quick Guide: *Needles* angled, soft, in clusters from knobby spur shoots, deciduous; *Cones* nearly upright, small, lacking prickles; *Bark* with gray scales and red-brown inner bark; *Habitat* moist to poorly drained soils.

Leaves: Needles 2–3 cm long, angled, blue-green or light-green, apex blunt, born in clusters from spur shoots on older twigs, deciduous; autumn color yellow.

Twigs: Orange-brown to gray, glabrous, older twigs with knobby spur shoots.

Buds: Round, maroon, mostly glabrous.

Bark: Gray-brown and smooth on small trees; large trees with loose gray scales and red-brown to orange inner bark.

Seed Cone: 1–2 cm long, nearly upright, ovoid to oblong, brown; scales without prickles, margins somewhat erose; maturing in one growing season.

Form: Up to 23 m (75 ft) in height and 51 cm (20 in) in diameter with a pyramidal crown and low branches.

Habitat and Ecology: Shade intolerant; found on moist to wet organic soils of peatlands, swamps, bogs, and edges of streams and lakes with *Abies balsamea, Acer rubrum, Fraxinus nigra, Picea mariana, Populus balsamifera, Thuja occidentalis,*

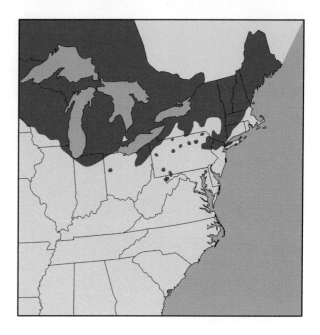

and *Ulmus americana*. May be found on well-drained uplands in its northern range.

Uses: Wood with white sapwood and yellow to red-brown heartwood, moderately heavy, moderately hard, slightly oily, durable; used for pulpwood, posts, construction, boxes, telephone poles and crates. Roots were used to sew birch bark canoes by Native Americans. Browsed by porcupine and hares; seeds eaten by birds and small mammals. Planted as an ornamental in cool climates.

Botanical Name: *Larix* means "fat" and refers to the oily wood; *laricina* is Latin for "larch."

Larix laricina leaves and seed cones

Larix laricina bark

Larix laricina bark

Larix laricina leaves

Pinaceae

Picea glauca (Moench) Voss

white spruce, skunk spruce, cat spruce, Canadian spruce

Quick Guide: *Needles* 4 sided, pointed, blue-green, waxy, smelling fetid when crushed; *Cones* pendent, papery, scales without prickles, margins smooth; *Bark* gray-brown and scaly with red-brown inner bark; *Habitat* a wide range of soils and climates.

Leaves: Needles 1.2–2.0 cm long, stout, 4 sided, blue-green, waxy, with bands of white on all sides, apex pointed, base pegged, with a fetid odor when crushed, persisting up to 10 years.

Twigs: Orange-brown to gray-brown, glaucous.

Buds: Ovoid, blunt, brown-gray, mostly glabrous.

Bark: Gray-brown and scaly revealing red-brown inner bark.

Seed Cone: 3–6 cm long, pendent, papery, brown, flexible; scales without prickles and with smooth, truncate margins; maturing in one growing season.

Form: Up to 30 m tall (100 ft) and 91 cm (36 in) in diameter with a conical, irregular crown.

Habitat and Ecology: Intermediate shade tolerance; found on a wide variety of soils and adapted to a wide range of climates. In the Northeast, associated with *Abies balsamea, Acer saccharum, Betula alleghaniensis, Betula papyrifera, Picea mariana, Picea rubens, Pinus resinosa, Pinus strobus, Populus tremuloides, Thuja occidentalis,* and *Tsuga canadensis.*

Uses: Wood yellow-white to red-brown, soft, light, resilient; used for pulpwood, plywood, residential

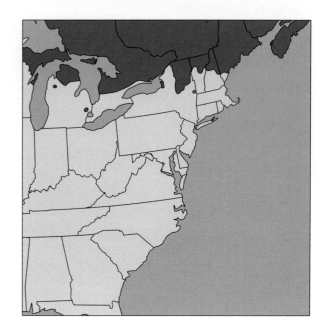

framing, boxes, crates, furniture, and musical instruments. Planted as an ornamental for its hardiness and used in hedges and windbreaks. A dwarf cultivar is widely planted. Browsed by squirrel and grouse; provides cover for moose, rabbit, and other large and medium size mammals.

Botanical Name: *Picea* is Latin for "pitchy"; *glauca* means "waxy," referring to the needles.

Similar Species: Norway spruce (*Picea abies* (L.) Karst.) is from Europe but planted in plantations and as an ornamental, and has naturalized in some areas of the Northeast. Distinguished by drooping branches and a much larger cone (15 cm long).

Picea abies seed cones

Picea glauca seed cones

Picea glauca bark

Picea glauca leaves

Picea glauca bark

Picea mariana (Mill.) B. S. P.
black spruce, swamp spruce, bog spruce

Quick Guide: *Needles* 4 sided, blunt pointed, dull green; *Cones* pendent, papery, scales without prickles, margins round and erose; *Bark* brown-gray and scaly; *Habitat* wet, organic soils.

Leaves: Needles 6–15 mm long, 4 sided, dull green to blue-green, somewhat waxy, with bands of white on all sides, apex blunt pointed, base pegged, persisting up to 10 years.

Twigs: Brown to yellow-brown, rough, pubescent.

Buds: Ovoid, red-brown to gray, with hair.

Bark: Red-brown to gray and scaly with green-yellow inner layers.

Seed Cone: 1.3–4 cm long and on average smaller than the other spruces, pendent, papery, purplish to brown, nearly round when open; scales unarmed and brittle with rough margins; maturing in one growing season but may remain on tree for decades.

Form: Up to 20 m (66 ft) tall and 23 cm (9 in) in diameter, characterized by short branches and an irregularly cylindrical crown often misshapen at the top with clusters of cones. Lower branches may root in organic soils and produce seedlings.

Habitat and Ecology: Intermediate shade tolerance; slow growing, especially on poorly drained sites. Found on a variety of sites from dry to wet but more commonly on poorly drained sites such as floating mats, acid bogs, and swamps. Forest associates include *Abies balsamea*, *Picea glauca*, *Picea rubens*, *Larix laricina*, and *Thuja occidentalis*.

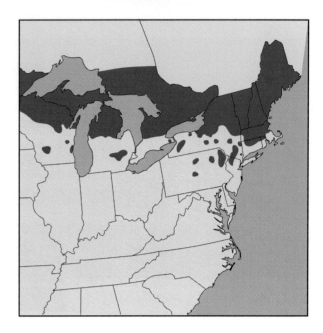

Uses: Similar to *Picea glauca* but used primarily for pulp. Twigs boiled to make spruce beer and resin was used in chewing gum. Planted in Christmas tree plantations. Seeds and buds eaten by many songbirds and small mammals; needles eaten by grouse.

Botanical Name: *Picea* is Latin for "pitchy"; *mariana* refers to the New World.

Picea mariana seed cone

Picea mariana leaves

Picea mariana bark

Picea mariana bark

Picea rubens Sarg.

red spruce, eastern spruce, yellow spruce

Quick Guide: *Needles* 4 sided, sharp pointed; *Cones* pendent, papery, scales without prickles, margins round and slightly erose; *Bark* gray-brown and scaly; *Habitat* a variety of sites.

Leaves: Needles 1.0–1.5 cm long, 4 sided, dark green to yellow-green and shiny, apex sharp pointed, base pegged, white bands on all sides, persisting up to 10 years.

Twigs: Brown-gray to orange-brown, pubescent.

Buds: Conical, red-brown.

Bark: Red-brown to gray-brown with scaly plates.

Seed Cone: 3–6 cm long, red-brown, pendent, papery; scales without prickles, margins round and only slightly erose; maturing in one growing season.

Form: Up to 34 m (110 ft) in height and 1 m (3 ft) in diameter.

Habitat and Ecology: Shade tolerant; found on a variety of sites including mountain slopes, uplands, rocky slopes, and poorly drained soils with *Abies balsamea, Abies fraseri, Acer pensylvanicum, Acer saccharum, Acer spicatum, Betula alleghaniensis, Betula lenta, Betula papyrifera, Picea mariana, Pinus strobus, Prunus pensylvanica, Thuja occidentalis,* and *Tsuga canadensis.*

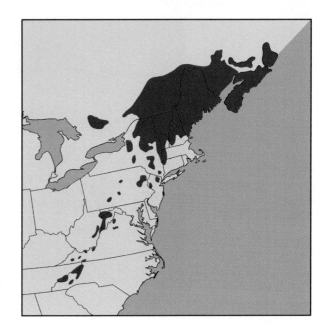

Uses: Wood yellow-white to red-brown, soft, light; used for pulpwood, plywood, residential framing, boxes, crates, and furniture. Twigs boiled to make spruce beer and resin was used in chewing gum. Planted in Christmas tree plantations. Seeds and buds eaten by songbirds and small mammals; needles eaten by blue and spruce grouse.

Botanical Name: *Picea* is Latin for "pitchy"; *rubens* means "reddish," referring to the red-brown cones, buds or bark.

Picea rubens leaves

Picea rubens bark

Picea rubens bark

Picea rubens seed cones

Pinus banksiana Lamb.

jack pine, scrub pine, gray pine, banksian pine

Quick Guide: *Needles* fascicled in bundles of two, short, forked or twisted, thick; *Cones* serotinous, misshapen, in clusters often appressed to the branch; *Bark* rough with gray or reddish scaly plates; *Habitat* common on dry, rocky soils.

Leaves: Needles in bundles of two, 2–4 cm long, stout, yellow-green, may twist or fork, persisting up to three growing seasons.

Twigs: Brown-gray, rough.

Buds: Ovoid, red-brown, resinous.

Bark: Red-brown to dark gray or black, thin, scaly.

Seed Cone: 3–6 cm long, light brown to yellow-brown, often serotinous, variable in shape, uneven and often curved, prickle small and deciduous, found in clusters pointing toward the tip of the branch, remaining on the tree for many years and sometimes embedded in the branch.

Form: Up to 20 m (66 ft) tall and 25 cm (10 in) in diameter, with an irregular, thin crown.

Habitat and Ecology: Shade intolerant; forming pure stands after fire, found on a variety of sites including bogs but common on sterile, dry soils and rocky ridges. Forest associates include *Betula papyrifera*, *Pinus resinosa*, *Pinus strobus*, *Populus balsamifera*, *Populus tremuloides*, and *Quercus rubra*.

Uses: Wood with orange-brown heartwood, sapwood pale with prominent latewood, moderately heavy, often knotty; used for pulpwood, railroad ties, posts, crates, fuel, and was used by Native Americans for canoe frames. Young stands are habitat for the endangered Kirtland's warbler. Not an attractive ornamental but planted in windbreaks and shelterbelts.

Botanical Name: *Pinus* is Latin for "pine tree"; *banksiana* is for Joseph Banks.

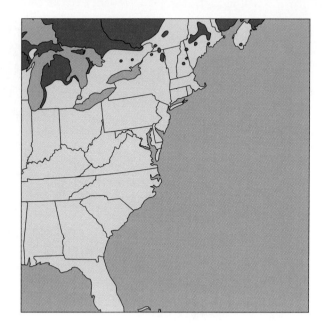

Pinus banksiana seed cones

Pinus banksiana leaves

Pinus banksiana bark

Pinus banksiana bark

Pinus banksiana seed cone

Pinus clausa (Chapm. ex Engelm.) Vasey ex Sarg.

sand pine, scrub pine, spruce pine

Quick Guide: *Needles* fascicled in bundles of two, short, moderately twisted, flexible; *Cones* sometimes serotinous, scales with sharp prickles; *Bark* orange-gray and scaly; *Habitat* dry, sandy soils.

Leaves: Needles in bundles of two, 5–9 cm long, light green, thin, flexible, often twisted, persisting 2–3 years.

Twigs: Gray-green, smooth, older twigs may show embedded cones.

Buds: Conical, gray-brown.

Bark: Orange-gray to red-brown or gray with scaly plates.

Seed Cone: 4–9 cm long, conical, brown-gray, armed with sharp stiff prickles, open or serotinous.

Form: Usually only up to 15 m (50 ft) in height and 61 cm (2 ft) in diameter.

Habitat and Ecology: Intermediate shade tolerance; found on sandy, acidic, infertile soils with *Quercus incana*, *Quercus laevis*, *Quercus margaretta*, *Pinus elliottii*, and *Pinus palustris*. The Ocala (var. *clausa*) geographic race is found in northeast to south Florida and has serotinous cones. The Choctawhatchee (var. *immuginata*) geographic race is found in northwest Florida and extreme southwest Alabama, and has open cones at maturity.

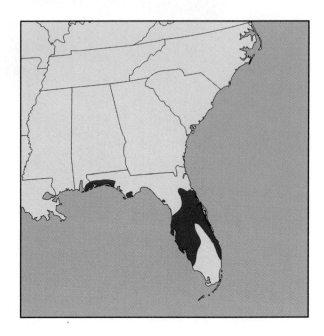

Uses: Wood with red-brown heartwood and yellow sapwood with prominent red-brown latewood, resinous, moderately heavy; used for pulpwood, construction lumber, and fuel. Planted in Christmas tree plantations. Same wildlife uses as for *Pinus echinata*.

Botanical Name: *Pinus* is Latin for "pine tree"; *clausa* means "closed," referring to the serotinous cones.

Pinus clausa seed cone

Pinus clausa bark

Pinus clausa bark

Pinus clausa leaves

Pinaceae

Pinus echinata Mill.

shortleaf pine, southern yellow pine, yellow pine, Arkansas pine, shortstraw pine

Quick Guide: *Needles* fascicled in bundles of two, short but not twisted, flexible; *Cones* gray, with small sharp prickles; *Bark* red-brown plated, plates often with small resin holes; *Habitat* floodplains to dry, upland forests.

Leaves: Needles in bundles of two or sometimes three, 6–13 cm long, green to yellow-green, flexible, persisting 3–5 years.

Twigs: Red-brown, rough, resinous.

Buds: Conical, with red-brown scales.

Bark: Red-brown and rough on small trees; large trees with orange-brown to red-brown plates and pin-size resin holes in the plates.

Seed Cone: 4–7 cm long, ovoid, flexible, red-brown to gray, armed with small sharp prickles; maturing in two growing seasons.

Form: Up to 30 m (100 ft) in height and 1 m (3 ft) in diameter.

Habitat and Ecology: Shade intolerant; found on a range of sites of poor to medium quality including floodplains, rocky upland forests, and up to 900 m in the southern Appalachians. Forest associates include *Juniperus virginiana, Liquidambar styraciflua, Pinus palustris, Pinus rigida, Pinus strobus, Pinus taeda, Pinus virginiana, Quercus alba, Quercus marilandica, Quercus prinus, Quercus rubra, Quercus stellata,* and *Quercus velutina*.

Uses: Wood with red-brown heartwood and yellow sapwood with prominent red-brown latewood, resinous, moderately heavy; an important commercial species used for pulpwood, plywood, and construction lumber. Seeds eaten by birds and small mammals, and buds eaten by small mammals such as fox squirrel. Evergreen foliage provides excellent nesting, roosting, and loafing cover for a variety of bird species.

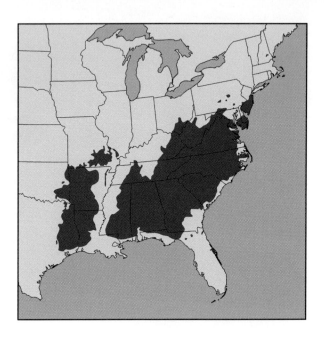

Botanical Name: *Pinus* is Latin for "pine tree"; *echinata* means "prickly," referring to the cone.

Pinus echinata seed cones

Pinus echinata leaves

Pinus echinata bark

Pinus echinata bark

Pinus elliottii Engelm.

slash pine, swamp pine, pitch pine, southern yellow pine, southern pine, Cuban pine

Quick Guide: *Needles* fascicled in bundles of two or three, long, flexible, moderately stout; *Cones* shiny chocolate brown with weak prickles; *Bark* with reddish or white-tan flecking on the plates; *Habitat* edges of streams, swamps, and bays as well as drier sites.

Leaves: Needles in bundles of two or three, 15–28 cm long, green to dark green and shiny, moderately stout, flexible, persisting 2 years.

Twigs: Stout but not as stout as *Pinus palustris,* orange-brown, rough, resinous.

Buds: Conical, scales red-brown, fringed with white hair.

Bark: Brown and rough on small trees; large trees with red-brown plates and may have white-tan flecking on the plates.

Seed Cone: 7–18 cm long, ovoid, shiny, chocolate brown with weak prickles; maturing in two growing seasons.

Form: Up to 30 m (100 ft) in height and 1 m (3 ft) in diameter.

Habitat and Ecology: Shade intolerant; found on low sites near ponds and streams, in bays and depressions, and on drier sites with *Chamaecyparis thyoides, Gordonia lasianthus, Nyssa* spp., *Pinus palustris, Pinus serotina,* and *Pinus taeda*. Widely planted on a range of sites in the lower Coastal Plain of Alabama, Georgia, and Florida.

Uses: Wood with red-brown heartwood and yellow sapwood with prominent red-brown latewood, resinous, moderately heavy; an important commercial species used for pulpwood, plywood, and construction lumber; also a source of turpentine and resin. Needles were once used to make baskets, pine wool, and rough textiles. Same wildlife uses as for *Pinus echinata*.

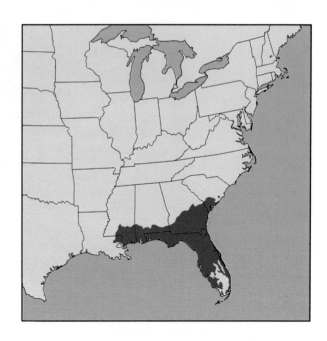

Botanical Name: *Pinus* is Latin for "pine tree"; *elliottii* is for Stephen Elliott, a botanist from South Carolina.

Similar Species: South Florida or Florida slash pine (*Pinus elliottii* var. *densa* Little & Dorman) is found in pure stands on wet to dry sites in southern Florida and is distinguished by longer needles clumped at the ends of stouter branches, needles in mostly twos and rarely in threes, and a seedling grass stage.

Pinus elliottii seed cones

Pinus elliottii leaves

Pinus elliottii bark

Pinus elliottii bark

Pinus glabra Walt.

spruce pine, cedar pine, bottom white pine, Walter pine

Quick Guide: *Needles* fascicled in bundles of two, short, slender, flexible, blue-green, sometimes twisted; *Cones* with weak deciduous prickles; *Bark* smooth on branches and on stems of young trees, becoming furrowed rather than plated on large trees; *Habitat* stream banks and bottoms.

Leaves: Needles in bundles of two, 5–10 cm long, blue-green, sometimes twisted, thin, flexible, fragrant when crushed, persisting 2–3 years.

Twigs: Slender, gray-green to brown, smooth.

Buds: Conical, gray-brown.

Bark: Gray-green to brown and smooth or shallowly ridged on small trees; large trees brown to dark gray and deeply furrowed and may show scaly ridges.

Seed Cone: 4–8 cm long, ovoid, red-brown to gray, prickles weak and deciduous; maturing in two growing seasons.

Form: Up to 37 m (121 ft) in height and 1 m (3 ft) in diameter.

Habitat and Ecology: Shade tolerant; found on damp to poorly drained soils of stream banks, flatwoods, bottomlands, and hummocks in association with *Acer rubrum, Carya aquatica, Liquidambar styraciflua, Magnolia virginiana, Nyssa* spp., *Pinus elliottii, Pinus taeda, Quercus laurifolia, Quercus lyrata, Quercus michauxii, Quercus virginiana,* and *Taxodium distichum* var. *distichum* and var. *imbricarium*.

Uses: Wood light, soft, brittle, sometimes used for lumber and pulpwood. Same wildlife uses as for *Pinus echinata* plus bark eaten by beaver.

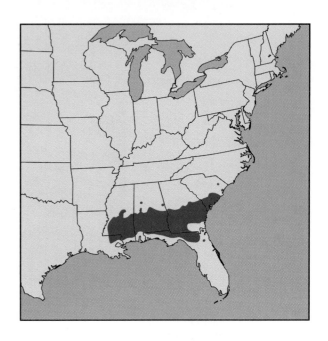

Botanical Name: *Pinus* is Latin for "pine tree"; *glabra* means "smooth."

Pinus glabra seed cones

Pinus glabra leaves

Pinus glabra bark

Pinus glabra bark

Pinus palustris Mill.

longleaf pine, heart pine, swamp pine, southern yellow pine, yellow pine, pitch pine, hard pine, longstraw pine

Quick Guide: *Needles* fascicled in bundles of three, very long, moderately stout, flexible, clustered at the ends of stout branches; *Terminal bud* large and silvery white; *Cones* large; *Seedling* grass stage; *Habitat* a variety of sites.

Leaves: Needles in bundles of three, 20–46 cm long, dark green and shiny, moderately stout, flexible, densely clustered in tufts on ends of branches, persisting 2 years.

Twigs: Stout, red-brown, rough.

Buds: Stout, large (6 cm long), scales silvery white.

Bark: Gray-brown to orange-gray and rough on small trees; large trees with red-brown plates.

Seed Cone: 15–25 cm long, cylindrical, red-brown to gray, prickles slender and moderately sharp; maturing in two growing seasons.

Form: Up to 37 m (121 ft) in height and 1 m (3 ft) in diameter. Seedlings may exist in a "grass stage" for many years to aid root development.

Habitat and Ecology: Shade intolerant; found on dry sandy soils, clay soils, in flatwoods, and on sand hills, swamp edges, and mountain ridges in Alabama. Forest associates include *Acer rubrum, Carya glabra, Carya pallida, Carya tomentosa, Chamaecyparis thyoides, Diospyros virginiana, Nyssa sylvatica, Oxydendrum arboreum, Pinus clausa, Pinus elliottii, Pinus taeda, Quercus incana, Quercus falcata, Quercus laevis, Quercus nigra, Quercus margaretta, Quercus marilandica, Quercus prinus,* and *Quercus stellata.*

Uses: Wood with red-brown heartwood (heart pine) and yellow sapwood with prominent red-brown latewood, resinous, moderately heavy; used for pulpwood, plywood, and construction lumber; was the most important commercial southern

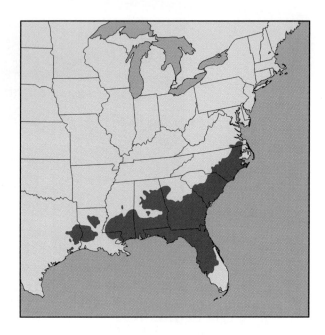

species before extensive logging; a source of turpentine and resin; foliage used for pine straw. Needles were once used for baskets, pine wool, upholstery stuffing, and antiseptic dressings. Same wildlife uses as for *Pinus echinata*, but larger seed more preferred than others. Once infected with red heart fungus, a nesting tree for the red-cockaded woodpecker.

Botanical Name: *Pinus* is Latin for "pine tree"; *palustris* means "swamp."

Pinus palustris seed cone

Pinus palustris leaves

Pinus palustris bark

Pinus palustris grass stage seeding

Pinus pungens Lamb.

Table Mountain pine, prickly pine, mountain pine, hickory pine, poverty pine

Quick Guide: *Needles* fascicled in bundles of two, short, very stiff, sharp pointed, twisted; *Cones* large and often misshapen, heavy, with stout prickles, closed cones persistent on branches; *Habitat* dry, rocky slopes.

Leaves: Needles in bundles of two, 3–7 cm long, twisted, apex sharp pointed, very stiff, persisting 3 years.

Twigs: Stout, orange-brown, scaly, resinous.

Buds: Conical, scaly, brown.

Bark: Thin, breaking into scaly gray-brown plates.

Seed Cone: 5–10 cm long, serotinous, ovoid, gray-brown, heavy, often misshapen when open, prickles clawlike and stout, closed cones may persist on branches for many years.

Form: Usually only up to 9 m (30 ft) in height and 61 cm (2 ft) in diameter, often leaning with gnarly form.

Habitat and Ecology: Shade intolerant; found on dry rocky ridges and slopes of the Appalachians. Pure stands are common in areas frequented by fire. Forest associates include *Carya glabra, Carya tomentosa, Kalmia latifolia, Nyssa sylvatica, Oxydendrum arboreum, Pinus echinata, Pinus rigida, Pinus virginiana, Quercus coccinea, Quercus falcata, Quercus prinus, Quercus stellata,* and *Quercus velutina.*

Uses: Sapwood yellow with prominent red-brown latewood, heartwood red-brown; used for pulpwood and fuel. Same wildlife uses as for *Pinus echinata.*

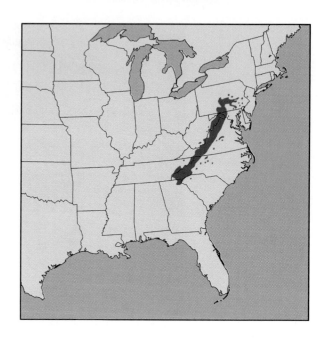

Botanical Name: *Pinus* is Latin for "pine tree"; *pungens* means "prickly or sharp pointed," referring to the cone.

Pinus pungens seed cones

Pinus pungens leaves

Pinus pungens bark

Pinus pungens seed cone

Pinus resinosa Soland.

red pine, Norway pine

Quick Guide: *Needles* fascicled in bundles of two, moderately long, sharp pointed, breaking when bent; *Cones* lacking prickles; *Bark* gray with red-brown plates; *Habitat* rocky ridges and sandy plains.

Leaves: Needles in bundles of two, 10–16 cm long, thick, dark green, brittle and snapping when bent, sharp pointed, persisting up to five growing seasons.

Twigs: Orange-brown.

Buds: Acute, red-brown, resinous.

Bark: Gray-orange and scaly on small trees; large trees with red-brown plates.

Seed Cone: 4–7 cm long, ovoid, brown to red-brown, cone scales without prickles and with a thickened lip; maturing in two growing seasons.

Form: Up to 24 m (80 ft) tall and 1 m (3 ft) in diameter.

Habitat and Ecology: Shade intolerant; found on a variety of sites including swamp and lake margins but more commonly on sandy soils. Associated with a variety of species including *Acer rubrum*, *Pinus banksiana*, *Pinus rigida*, *Pinus strobus*, *Pinus virginiana*, *Populus grandidentata*, *Populus tremuloides*, and *Quercus prinus*.

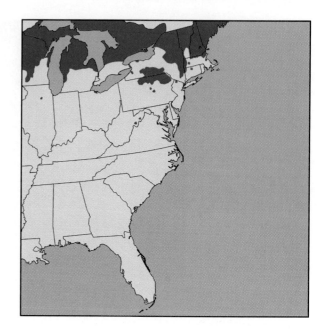

Uses: Wood with red-brown heartwood and white-yellow sapwood with prominent red-brown late-wood, oily, resinous, moderately heavy; used for pulpwood and cabin logs, planted in Christmas tree plantations, and was used for bridges, docks, and masts; a good windbreak for poor sites. Same wildlife uses as for *Pinus echinata*.

Botanical Name: *Pinus* is Latin for "pine tree"; *resinosa* means "resinous", referring to the wood.

Pinus resinosa leaves

Pinus resinosa bark

Pinus resinosa bark

Pinus resinosa seed cone

Pinus rigida Mill.
pitch pine, knotty pine

Quick Guide: *Needles* fascicled in bundles of three, moderately long, stiff; *Cones* serotinous, nearly globose, with rigid prickles; *Bark* gray with rough, red-brown plates and sometimes with tufts of needles; *Habitat* rocky, sandy soils.

Leaves: Needles in bundles of three or sometimes two, 6–15 cm long, green to yellow-green, stiff, sharp-pointed, sometimes twisted, persisting 2–3 years.

Twigs: Brown, scaly, resinous.

Buds: Conical, scales red-brown, fringed, resinous.

Bark: Brown-gray and rough with scaly, gray-brown, yellowish, or red-brown plates that may show tufts of needles. More deeply grooved on large trees.

Seed Cone: 3–8 cm long, ovoid to nearly globose, serotinous, base may be imbedded in the branch, milk chocolate brown to gray, prickles rigid.

Form: Up to 24 m (80 ft) in height and 61 cm (2 ft) in diameter, but usually smaller.

Habitat and Ecology: Shade intolerant; found on a range of sites but mostly on sandy plains, rocky shallow slopes, and dry ridges. Occasionally found in cold swamps. Forest associates include *Carya glabra*, *Carya tomentosa*, *Nyssa sylvatica*, *Oxydendrum arboreum*, *Pinus echinata*, *Pinus pungens*, *Pinus strobus*, *Pinus virginiana*, *Quercus coccinea*, *Quercus prinus*, *Quercus rubra*, and *Quercus velutina*.

Uses: Wood with red-brown heartwood and yellow sapwood with prominent red-brown late-wood, coarse, very resinous (especially the knots), durable; used for pulpwood and railroad ties, was tapped for turpentine. Same wildlife uses as for *Pinus echinata*.

Botanical Name: *Pinus* is Latin for "pine tree"; *rigida* means "rigid."

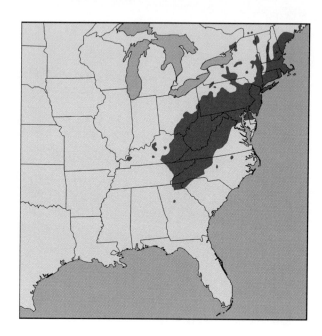

Pinus rigida seed cones

Pinus rigida leaves

Pinus rigida bark

Pinus rigida bark

Pinus serotina Michx.

pond pine, bay pine, marsh pine, pocosin pine

Quick Guide: *Needles* fascicled in bundles of three, long, flexible; *Cones* nearly globose and serotinous; *Bark* may show tufts of needles; *Habitat* wet sites.

Leaves: *Needles* in bundles of three or sometimes four, 13–21 cm long, flexible, persisting 3–4 years.

Twigs: Brown, scaly.

Buds: Conical, scales red-brown.

Bark: Gray-brown and rough on small trees; large trees with irregular red-brown and gray plates. May show clumps of needles or sprouts sticking out of the bark.

Seed Cone: 5–7 cm long, ovoid to nearly globose, serotinous, milk chocolate brown to gray, with small prickles, base may be embedded in the branch.

Form: Up to 21 m (70 ft) in height and 61 cm (2 ft) in diameter.

Habitat and Ecology: Shade intolerant; found in swamps, ponds, flatwoods, pocosins, and bays with *Acer rubrum, Chamaecyparis thyoides, Gordonia lasianthus, Liquidambar styraciflua, Magnolia virginiana, Nyssa aquatica, Nyssa biflora, Pinus elliottii, Pinus taeda, Quercus virginiana,* and *Taxodium distichum* var. *distichum* and var. *imbricarium*.

Uses: Wood with red-brown heartwood and yellow sapwood with prominent red-brown latewood, resinous, moderately heavy; used for pulpwood and construction lumber. Same wildlife uses as for *Pinus echinata*.

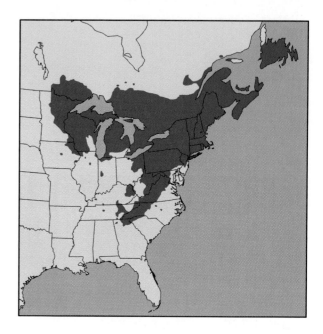

Botanical Name: *Pinus* is Latin for "pine tree"; *serotina* means "late," referring to the serotinous cones.

Pinus serotina seed cones

Pinus serotina seed cone

Pinus serotina bark

Pinus serotina bark

Pinus serotina leaves

Pinus strobus L.

eastern white pine, white pine, northern white pine, soft pine, Weymouth pine

Quick Guide: *Needles* fascicled in bundles of five (the only native eastern conifer with five needles); *Cones* long and slim, scales without prickles; *Bark* green-gray and smooth on young stems, becoming red-brown and grooved on large trees; *Habitat* moist, cool forests including drier soils in the North.

Leaves: Needles in bundles of five, 8–14 cm long, blue-green, slender, flexible, ethereal from a distance, persisting 2–3 years.

Twigs: Slender, green-brown.

Buds: Conical, with red-brown scales, clustered in groups of about five around the terminal.

Bark: Green-gray and smooth on small trees; large trees gray-brown to red-brown and more deeply furrowed and blocky.

Seed Cone: 8–20 cm long, oblong, slim, brown, often curved, scales without prickles and often with white resin; maturing in two growing seasons.

Form: Up to 46 m (150 ft) in height and 1 m (3 ft) in diameter; branches produced in whorls on the trunk. The largest pine in the eastern United States.

Habitat and Ecology: Intermediate shade tolerance; found in moist cool forests and near streams and rivers and on rocky, sandy sites in its northern range. Forest associates are numerous and include *Abies balsamea, Acer rubrum, Acer saccharum, Betula papyrifera, Fagus grandifolia, Halesia tetraptera, Liriodendron tulipifera, Magnolia acuminata, Picea rubens, Pinus rigida, Populus grandidentata, Populus tremuloides, Prunus serotina, Quercus alba, Quercus prinus, Quercus rubra, Tilia americana, Thuja occidentalis,* and *Tsuga canadensis*

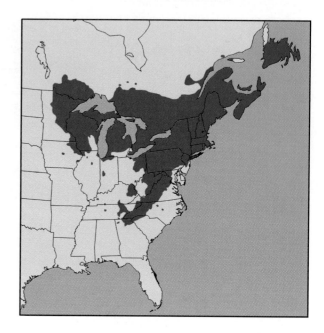

Uses: Wood cream colored, soft, light; used for boxes, crates, millwork, furniture, and matches. One of the most important eastern species before extensive logging. Planted as an ornamental on moist cool sites and in Christmas tree plantations. Same wildlife uses as for *Pinus echinata*.

Botanical Name: *Pinus* is Latin for "pine tree"; *strobus* refers to a fragrant, gum-yielding tree.

Pinus strobus seed cones

Pinus strobus bark

Pinus strobus bark

Pinus strobus leaves

Pinus strobus bark

Pinaceae 49

Pinus sylvestris L.

Scotch pine, Scots pine

Quick Guide: *Needles* fascicled in bundles of two, short, stiff, twisted; *Cone* scales thickened and 4 sided without a prickle; *Bark* with orange-gray, scaly plates; *Habitat* naturalized in New England on well-drained soils.

Leaves: Needles in bundles of two, 4–8 cm long, stiff, blue-green, twisted, persisting up to 4 years.

Twigs: Orange-red to brown-gray.

Buds: Ovoid, red-brown, resinous.

Bark: Gray-brown to bright orange and scaly.

Seed Cone: 3–6 cm long, light brown, often curved with an elongated asymmetrical base, pointing back towards the trunk; scales thickened and pyramidal (4 sided), lacking a prickle; maturing in two growing seasons.

Form: Up to 21 m (68 ft) tall and 1 m (3 ft) in diameter.

Habitat and Ecology: Shade intolerant; originally from Europe and an important timber species there, recognized as the most widely distributed pine in the world. Naturalized in New England in association with *Acer saccharum, Fagus grandifolia, Pinus strobus,* and *Prunus serotina.*

Uses: Widely planted for pulp and sawtimber, as landscape trees, and for Christmas trees.

Botanical Name: *Pinus* is Latin for "pine tree"; *sylvestris* means "of the forest."

Pinus sylvestris seed cones

Pinus sylvestris leaves

Pinus sylvestris bark

Pinus sylvestris bark

Pinus taeda L.

loblolly pine, old-field pine, meadow pine, torch pine

Quick Guide: *Needles* fascicled in bundles of three, long, flexible; *Cones* conical, brown-gray, with sharp prickles; *Bark* red-brown plated; *Habitat* a wide range of sites from dry to wet.

Leaves: Needles in bundles of mostly three or sometimes four, 12–23 cm long, green to dark green, flexible, persisting 3–4 years.

Twigs: Red-brown, scaly, resinous.

Buds: Conical, with red-brown scales.

Bark: Gray-black to red-brown and scaly on small trees; large trees with tight red-brown plates.

Seed Cone: 6–14 cm long, conical, brown-gray, scales with sharp prickles; maturing in two growing seasons.

Form: Up to 30 m (100 ft) tall and 1 m (3 ft) in diameter.

Habitat and Ecology: Shade intolerant; found on a wide variety of sites including ridges, swamp edges, upland forests, old fields, and bottomlands. Forest associates are numerous and site dependent. Many plantations exist in the valleys of the Ridge and Valley province of Alabama, Georgia, and Tennessee.

Uses: Wood with red-brown heartwood and yellow sapwood with prominent red-brown latewood, resinous, moderately heavy; the most important commercial species in the South, used for pulpwood, plywood, and construction lumber. Same wildlife uses as for *Pinus echinata*. Once infected with red heart fungus, mature trees also used as a nesting tree for the red-cockaded woodpecker.

Botanical Name: *Pinus* is Latin for "pine tree"; *taeda* means "resinous," referring to the wood.

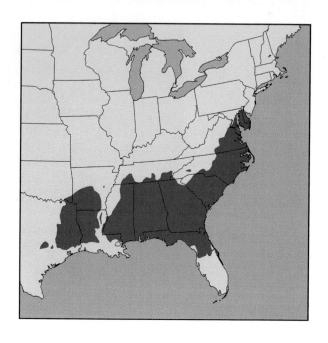

Pinus taeda seed cones

Gymnosperm Families

Pinus taeda leaves, pollen cones and seed cone

Pinus taeda bark

Pinus taeda bark

Pinus virginiana Mill.
Virginia pine, scrub pine, Jersey pine

Quick Guide: *Needles* fascicled in bundles of two, short, twisted, stiff; *Cone* scales with a thickened lip and slender sharp prickles; *Bark* red-brown and scaly; *Habitat* upland sites.

Leaves: Needles in bundles of two, 3–8 cm long, stiff, twisted, fragrant, persisting 3–4 years.

Twigs: Brown, glaucous, resinous.

Buds: Acute, red-brown, resinous.

Bark: Gray-brown to orange-brown and scaly on small trees; large trees with thin, scaly, red-brown plates.

Seed Cone: 4–7 cm long, ovoid, red-brown to gray, cone scales thin with a thick lip and slender sharp prickle; maturing in two growing seasons.

Form: Usually only up to 23 m (75 ft) in height.

Habitat and Ecology: Shade intolerant; found on a variety of upland sites. Forest associates include *Acer rubrum*, *Carya* spp., *Juniperus virginiana*, *Liquidambar styraciflua*, *Nyssa sylvatica*, *Pinus echinata*, *Pinus pungens*, *Pinus rigida*, *Pinus strobus*, *Pinus taeda*, *Quercus alba*, *Quercus marilandica*, *Quercus prinus*, *Quercus rubra*, *Quercus stellata*, and *Quercus velutina*.

Uses: Wood with red-brown heartwood and yellow sapwood with prominent red-brown latewood, resinous; used for pulpwood, construction lumber, and fuel. Planted in Christmas tree plantations and for privacy screens. Same wildlife uses as for *Pinus echinata*.

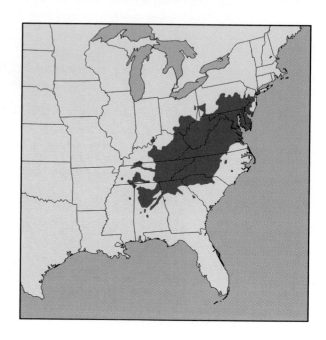

Botanical Name: *Pinus* is Latin for "pine tree"; *virginiana* refers to the geographic range.

Pinus virginiana seed cones

Pinus virginiana leaves, pollen cones and seed cone

Pinus virginiana bark

Pinus virginiana bark

Tsuga canadensis (L.) Carr.

eastern hemlock, hemlock spruce, northern hemlock, Canada hemlock

Quick Guide: *Needles* small, flat, pegged at the base, underside with double white bands; *Cones* small, scales without prickles; *Bark* dark red-brown and deeply furrowed; *Habitat* moist to damp soils.

Leaves: Needles 8–17 mm long, appearing 2 ranked, flat, dark shiny green or dark yellow-green, apex rounded or notched, base pegged, underside with two white narrow bands, persisting 3 years.

Twigs: Slender, yellow-brown to red-brown, pubescent when young.

Buds: Small, with pubescent scales.

Bark: Dark cinnamon brown and scaly or shallowly grooved on smaller trees; large trees deeply furrowed with purple inner bark.

Seed Cone: 1.3–2 cm long, ovoid to round, brown, without prickles, born on ends of twigs.

Form: Up to 37 m (121 ft) in height and 1.2 m (4 ft) in diameter, old growth trees (>500 years of age) may be larger. Crown with pendulous graceful branches.

Habitat and Ecology: Shade tolerant; found on moist to damp soils, such as edges of streams, lakes and ponds, and coves and northern slopes. Forest associates include *Abies balsamea, Abies fraseri, Acer pensylvanicum, Acer rubrum, Acer saccharum, Aesculus flava, Betula alleghaniensis, Betula lenta, Betula papyrifera, Halesia tetraptera, Liriodendron tulipifera, Magnolia acuminata, Picea rubens, Pinus strobus, Prunus serotina, Quercus alba, Quercus rubra,* and *Tilia americana*. The hemlock woolly adelgid is an introduced pest native to Asia and a serious problem for forest and urban hemlocks.

Uses: Wood pale red-brown, light, moderately hard; used for pulpwood, containers, construction lumber, and the bark was a major source of tannin.

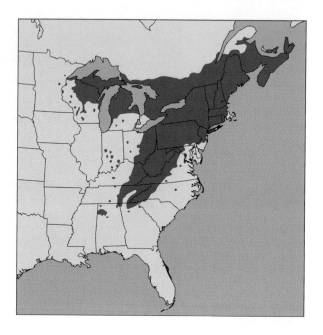

Planted as an ornamental but heat intolerant and susceptible to insects. Important cover for white-tailed deer, birds, and small mammals; seeds eaten by birds and small mammals; and foliage browsed by white-tailed deer, squirrel, and rabbit.

Botanical Name: *Tsuga* is Japanese for "hemlock"; *canadensis* refers to the geographic range.

Tsuga canadensis (smaller) and *Tsuga caroliniana* (larger) seed cones

Tsuga canadensis leaves and seed cones

Tsuga canadensis bark

Tsuga canadensis bark

Tsuga caroliniana Engelm.
Carolina hemlock

Quick Guide: Similar to *Tsuga canadensis* but *Needles* longer with entire margins (*Tsuga canadensis* with small teeth) arranged around the twig rather than 2 ranked; *Cones* larger (see page 56); *Habitat* drier sites in the southern Appalachians.

Leaves: Needles 1–2 cm long, arranged around the twig, flat, shiny dark green, apex rounded or notched, underside with two white bands, persisting 3 years.

Twigs: Similar to *Tsuga canadensis*.

Buds: Similar to *Tsuga canadensis*.

Bark: Similar to *Tsuga canadensis*.

Seed Cone: 2–4 cm long, ovoid, brown, lacking prickles, born on ends of twigs.

Form: Up to 18 m (60 ft) tall and 61 cm (2 ft) in diameter.

Habitat and Ecology: Shade tolerant; usually found on drier sites than *Tsuga canadensis* in the southern Appalachians at elevations between 700 and 1200 m.

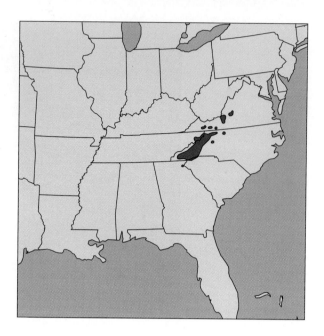

Uses: Sometimes planted as an ornamental.

Botanical Name: *Tsuga* is Japanense for "hemlock"; *caroliniana* refers to the geographic range.

Tsuga caroliniana leaves

Tsuga caroliniana bark

Tsuga caroliniana bark

Tsuga caroliniana leaves and seed cones

Taxus floridana Nutt. ex Chapm.
Florida yew

Quick Guide: *Needles* flat, shiny, dull-pointed; *Seeds* in a fleshy, red cup; *Bark* reddish and scaly; *Habitat* slopes of the Appalachicola River.

Leaves: Needles 2–2.5 cm long, flat, evergreen, appearing 2 ranked, shiny dark green, flexible, apex dull-pointed, base pegged, pale green below.

Twigs: Gray-brown, smooth, with warty lenticels.

Bark: Red-brown, thin, scaly.

Seed: A single seed enclosed in a 2 cm long, red, fleshy cup; maturing in one growing season; species is dioecious.

Form: Shrubby but up to 9 m (30 ft) in height.

Habitat and Ecology: Shade tolerant; rare, found on wooded slopes and ravines of the Appalachicola River in association with *Ilex opaca, Illicium floridanum, Fagus grandifolia, Magnolia grandiflora, Magnolia pyramidata, Persea borbonia,* and *Torreya taxifolia.* Listed as an Endangered Species.

Uses: Seed cones eaten by birds and bark eaten by beaver.

Botanical Name: *Taxus* means "yew"; *floridana* refers to the geographic range.

Taxus floridana leaves

Taxus floridana bark

Taxus floridana seeds

Taxus floridana leaves

Taxus floridana bark

Torreya taxifolia Arn.
Florida torreya, stinking-cedar, gopherwood

Quick Guide: *Needles* flat, shiny, stiff, very sharp pointed, with a disagreeable odor when bruised; *Seed* in a fleshy, green sac; *Habitat* limestone bluffs along the Apalachicola River.

Leaves: Needles 2.5–4.0 cm long, flat, appearing 2 ranked, dark green and shiny, apex with a spiny bristle tip, base with a short curved petiole, with a disagreeable odor when bruised, underside with two white lines, persisting several years.

Twigs: Whorled, stiff, green, with a disagreeable odor when cut.

Buds: With green-brown overlapping scales.

Bark: Brown and lightly ridged; large trees with scaly ridges.

Seed: A single seed enclosed in a green, fleshy, leathery sac 3 cm long; maturing in two growing seasons; species is dioecious.

Form: Up to 18 m (60 ft) in height.

Habitat and Ecology: Shade tolerant; rare because of fungal disease and habitat alteration, found on limestone bluffs and ravines of the Apalachicola River in southwest Georgia and northwest Florida

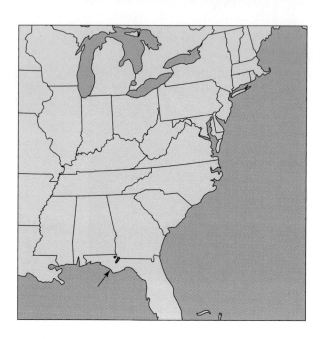

in association with *Taxus floridana*. Less than 250 reported in the wild and listed as an Endangered Species.

Uses: The durable wood was used for fence posts and cabinets.

Botanical Name: *Torreya* is for the botanist John Torrey; *taxifolia* is Latin for "yewlike leaves."

Torreya taxifolia seed

Torreya taxifolia leaves

Torreya taxifolia bark

Torreya taxifolia leaves

Guide to the Angiosperm Families

Most species broadleaved, seeds enclosed in an ovary

I. Leaves opposite or whorled and simple

1. Leaves whorled or opposite, heart shaped, margin entire; flowers white and trumpetlike; fruit long and beanlike; bark scaly or grooved. *Bignoniaceae* (trumpet creeper family)
2. Leaves opposite, heart shaped, margin entire, underside tomentose; flowers purple and trumpetlike; fruit a persistent nutlike capsule; bark smooth. *Scrophulariaceae* (figwort family)
3. Leaves opposite, margin serrate or entire, underside possibly glandular or rusty pubescent; flowers in white clusters, fruit a red-stalked drupe. *Caprifoliaceae* (honeysuckle family)
4. Leaves opposite or whorled, margin entire; twigs with stipule scars connecting the leaf scars, leaf scar with one bundle scar; fruit a bumpy capsule or a ball of nutlets. *Rubiaceae* (madder family)
5. Leaves opposite or whorled, margin entire, venation arcuate; flowers petal-like or in white flat-topped heads; fruit a red, white or blue drupe; twigs telescopic; fine hairs visible when the leaf is split. *Cornaceae* (dogwood family)
6. Leaves opposite or subopposite, margin entire or serrate; leaf scar with one bundle scar; flowers white and bell shaped, fringelike, or in small yellow clusters; fruit a blue or purplish drupe. *Oleaceae* (olive family)
7. Leaves opposite, palmately lobed, margin entire or serrate; fruit a double samara. *Aceraceae* (maple family)
8. Leaves opposite or subopposite, margin with small teeth; twigs bright green when young; flowers with maroon petals; fruit a 4 lobed pink to purple capsule with red seeds. *Celastraceae* (staff-tree family)

II. Leaves opposite and compound

1. Leaves palmately compound with mostly five leaflets. *Hippocastanaceae* (buckeye family)
2. Leaves trifoliately compound. *Staphyleaceae* (bladdernut family)
3. Leaves pinnately compound, with three to nine leaflets often shallowly lobed; twigs bright green. *Aceraceae* (maple family)
4. Leaves pinnately compound, leaflets five or more, margin serrate or entire; fruit a single samara; buds with dark suedelike scales; bark with a diamond pattern or scaly. *Oleaceae* (olive family)
5. Leaves pinnately compound, margin serrate, rachis grooved and with stiff hairs; fruit a juicy drupe in stalked clusters. *Caprifoliaceae* (honeysuckle family)

III. Leaves alternate, simple, and lobed

1. Leaves fan shaped and large. *Platanaceae* (sycamore family)
2. Leaves tulip shaped with a broadly notched apex. *Magnoliaceae* (magnolia family)
3. Leaves star shaped. *Hamamelidaceae* (witch hazel family)
4. Leaves pinnately lobed. *Fagaceae* (oak, beech family)
5. Leaves shallowly lobed at the apex. *Fagaceae* (oak, beech family)

IV. Leaves alternate, simple, and unlobed: fragrant when crushed

1. Leaves with an anise smell when crushed, leathery, with obscure lateral venation. *Illiciaceae* (anise tree family)
2. Leaves spicy aromatic when crushed and yellow glandular; fruit a waxy drupe. *Myricaceae* (wax myrtle family)

3. Leaves spicy aromatic when crushed, leaves can be unlobed, mittenlike or disfigured. *Lauraceae* (laurel family)

V. Leaves alternate, simple, and unlobed: twigs armed

1. Leaves with an acuminate apex and long petiole exuding milky sap when cut, margin entire; fruit a green, large, brainlike ball of drupes. *Moraceae* (mulberry family)
2. Leaves with a short petiole and entire margin; twigs with milky sap; fruit a black drupe. *Sapotaceae* (sapodilla family)
3. Margin entire, serrate or irregularly toothed; twigs may have a bitter almond smell when cut; some species with thorns, a pair of glands on the petiole, or basal leaf lobing; fruit a drupe or pome. *Rosaceae* (rose family)

VI. Leaves alternate, simple, and unlobed: leaf margin without teeth

1. Leaves triangular with a pair of glands on the long petiole; fruit popcornlike and waxy. *Euphorbiaceae* (spurge family)
2. Leaf lopsided, margin wavy; flowers yellow and stringy in autumn or winter; fruit a knobby capsule; young twigs and buds with pale scurfy pubescence. *Hamamelidaceae* (witch hazel family)
3. Leaves obovate to oblong, with a green tomato smell when crushed; leaf underside, twig and buds with velvety maroon hair; flowers with maroon petals in threes; fruit bananalike. *Annonaceae* (custard apple, pawpaw family)
4. Leaves heart shaped, petiole swollen at both ends; flowers pink in early spring; fruit a legume. *Caesalpiniaceae* (legume family)
5. Leaf deciduous or evergreen; twigs with stipule scars completely encircling the twig; flowers white or yellow; fruit red follicles in a conelike structure. *Magnoliaceae* (magnolia family)
6. Leaves leathery with obscure lateral venation; flowers white and fragrant; fruit buckwheatlike. *Cyrillaceae* (cyrilla family)
7. Leaves leathery with arcuate venation; twigs three-angled; flowers and fruit in drooping clusters. *Cyrillaceae* (cyrilla family)
8. Leaves leathery, sweet tasting, with a bright yellow midrib; flowers yellow in spring resembling pom-poms; young bark streaked. *Symplocaceae* (sweetleaf family)
9. Leaves with black spots in late summer; twigs with orange lenticels; buds black and triangular; fruit a large edible berry; bark alligatorlike. *Ebenaceae* (ebony family)
10. Leaf margin entire or with an occasional coarse tooth; fruit a juicy purple-black drupe with a ribbed or winged stone; base of trunk may be swollen in wet habitats. *Cornaceae* (dogwood family)
11. Leaves may show a bristle tip at the apex; buds clustered at twig ends; fruit an acorn; pith star shaped. *Fagaceae* (beech, oak family)
12. Leaves may appear opposite; leathery flowers very showy; leaf scar with one bundle scar; shrublike form, often forming thickets. *Ericaceae* (heath family)
13. Leaves obovate with a long purple-red petiole; leaf scar bluish; fruit a drupe on feathery stalks. *Anacardiaceae* (cashew family)
14. Leaves obovate to nearly round, margin may show irregular teeth; leaf scar with one bundle scar; flowers in white bells; fruit a wingless drupe. *Styracaceae* (storax family)

VII. Leaves alternate, simple, and unlobed: margin toothed

1. Margin serrate or irregularly toothed; twigs may have a bitter almond smell when cut; some species with thorns, a pair of glands on the petiole, or basal leaf lobing; fruit a drupe or pome. *Rosaceae* (rose family)

2. Leaf margin finely serrate, dentate, or irregularly toothed; leaf scar with one bundle scar; flowers in white bells; fruit a dry winged or dry wingless drupe. *Styracaceae* (storax family)

3. Leaf margin entire, crenate, serrate, irregularly serrate or with spiny teeth, leathery or thin; bark smooth even on large trees; fruit a round shiny red or black drupe. *Aquifoliaceae* (holly family)

4. Leaf margin with blunt teeth, or upwardly curved teeth with a bristle tip; fruit an acorn or a nut in a spiny bur. *Fagaceae* (beech, oak family)

5. Leaf triangular to nearly round, margin finely toothed or with rounded or dentate teeth; petiole very long and may show glands near the blade; mature bark deeply grooved. *Salicaceae* (willow, poplar family)

6. Leaf often lanceolate, margin entire or with small sometimes glandular teeth; buds with one scale; mature bark scaly or grooved. *Salicaceae* (willow, poplar family)

7. Leaf heart shaped, margin coarsely serrate, possibly sandpapery above or densely hairy, some leaves one to three lobed, petiole with milky sap. *Moraceae* (mulberry family)

8. Leaf heart shaped, margin with fine teeth, base inequilateral; fruit a nutlet attached to an unlobed leafy bract. *Tiliaceae* (linden family)

9. Leaves leathery and margin with fine sharp teeth or leaf thin with fine teeth; flowers with five, white, silky, hairy petals. *Theaceae* (camellia family)

10. Leaf margin serrate or doubly serrate, leaves 2 ranked; fruit a nutlet in a papery cone, hoplike sac or three-lobed leafy bract. *Betulaceae* (birch family)

11. Leaf margin serrate; leaves appearing 2 ranked; lateral buds stalked with two to three maroon valvate scales; fruit a nutlet in a woody cone. *Betulaceae* (birch family)

12. Leaf margin doubly serrate, serrate or irregularly serrate, base cordate or inequilateral, some species with three veins arising from the leaf base; fruit a round or elliptical samara, a berrylike drupe, or burlike drupe. *Ulmaceae* (elm family)

13. Leaf margin serrate, midrib with long stiff hairs, sour tasting; fruit in persistent drooping clusters; mature bark dark and deeply grooved with orange in the grooves. *Ericaceae* (heath family)

14. Leaf margin finely serrate, leaves glossy with prominent parallel venation; fruit a red to black drupe; bark streaked. *Rhamnaceae* (buckthorn family)

VIII. Leaves alternate and compound: twigs armed

1. Twigs with a pair of spines at the node; leaves pinnately compound; mature bark light brown and loosely ridged; flowers in dangling, fragrant, white clusters; fruit a legume. *Fabaceae* (legume family)

2. Twigs with branched thorns; leaves bipinnately compound; bark with clusters of thorns; fruit a long twisted legume. *Caesalpiniaceae* (legume family)

3. Twigs with spines or prickles; leaves pinnately compound; trunk smooth or with pyramidal growths. *Rutaceae* (rue family)

4. Twigs with spines or prickles; leaves very large and bi- or tripinnately compound; rachis and leaf scar spiny; flowers in large drooping clusters; fruit a purple drupe in heavy clusters; usually single stemmed. *Araliaceae* (ginseng family)

IX. Leaves alternate and compound: twigs unarmed

1. Leaves trifoliately compound. *Rutaceae* (rue family)

2. Leaves mostly bipinnately compound; leaflets scallop shaped, margin entire; mature bark smooth; pink flowers like pom-poms; fruit a legume. *Mimosaceae* (mimosa family)

3. Leaves large and mostly bipinnately compound; leaflets subopposite or alternate on the rachis and with an entire margin; twigs stout and mottled; bark with scaly ridges; fruit a thick leathery legume. *Caesalpiniaceae* (legume family)

4. Leaves mostly bipinnately compound; leaflets with a coarsely serrate margin, some basally lobed; young bark purplish; flowers like small purple firecrackers; fruit a yellow drupe in persistent clusters. *Meliaceae* (mahogany family)
5. Leaves pinnately compound; leaflets number seven to nine, sometimes alternate on the rachis, margin entire, petiole swollen and covering the bud; mature bark smooth; flowers in dangling, fragrant, white clusters; fruit a legume. *Fabaceae* (legume family)
6. Leaves pinnately compound, leaf margin entire, rachis bright red; twigs stout with large shield-shaped leaf scars; young bark with black patches; fruit a persistent white drupe; do not touch. *Anacardiaceae* (cashew family)
7. Leaves pinnately compound; leaflets basally lobed with black glands on the lobe underside; twigs stout and curving upward; mature bark smooth; fruit a single samara in persistent clusters. *Simaroubaceae* (bitterwood family)
8. Leaves pinnately compound, leaf margin serrate, lemon fragrant when crushed; twigs with heart-shaped or monkey-faced leaf scars; fruit a nut. *Juglandaceae* (walnut, hickory family)
9. Leaves pinnately compound, leaf margin coarsely serrate or entire, rachis red or winged; buds fuzzy; fruit a drupe in red, conical, upright terminal clusters. *Anacardiaceae* (cashew family)
10. Leaves pinnately compound, leaf margin coarsely serrate; fruit a small pome in bright red-orange, flat-topped clusters. *Rosaceae* (rose family)

Quercus nigra pistillate flowers

Acer barbatum pistillate flowers

Albizia julibrissin flowers

Cornus florida flowers

Aceraceae Guide

1. Leaves compound, with three to seven or occasionally nine leaflets. *Acer negundo*
2. Leaves simple, margin entire, palmately five-lobed, apices acuminate, underside pale and glabrous. *Acer saccharum*
3. Leaves simple, margin entire, palmately three to five lobed but mostly three-lobed, apices acuminate, underside densely hairy, leaves droopy. *Acer nigrum*
4. Leaves simple, margin entire, palmately three to five lobed, apices acute or blunt, underside white. *Acer barbatum*
5. Leaves simple, margin entire, palmately three to five lobed, apices acuminate, underside with yellow-green pubescence. *Acer leucoderme*
6. Leaves simple, margin serrate, palmately five-lobed, sinuses V-shaped, underside silvery white. *Acer saccharinum*
7. Leaves simple, margin finely doubly serrate, palmately three to five lobed, apices acuminate, sinuses shallow; young bark green-white streaked. *Acer pensylvanicum*
8. Leaves simple, margin coarsely serrate, palmately three to five lobed, apices acute to acuminate, sinuses shallow, underside with hair. *Acer spicatum*
9. Leaves simple, margin irregularly doubly serrate, palmately three to five lobed, apices acute, sinuses acute and shallow. *Acer rubrum*

Acer rubrum staminate flowers

Acer negundo pistillate flowers

Acer negundo staminate flowers

Acer rubrum pistillate flowers

Acer barbatum Michx.
Florida maple, southern sugar maple, sugar maple

Quick Guide: *Leaves* opposite, simple, deciduous, palmately three to five lobed with blunt sometimes squarish apices, margin entire, underside white; *Fruit* a double samara with divergent wings; *Bark* gray-black with scaly plates on large trees; *Habitat* moist soils.

Leaves: Simple, opposite, deciduous, nearly round, blade 3 to 9 cm long, lobes three to five and palmate, apices blunt to acute, margin entire, underside white and possibly pubescent; autumn color red, orange, or yellow.

Twigs: Red-brown, pubescent, with lenticels; leaf scar V-shaped with three bundle scars.

Buds: Terminal bud ovoid, 3 mm long; scales overlapping, red-brown, pubescent.

Bark: Gray-brown and smooth on small trees; large trees gray-black with scaly plates.

Flowers: Perfect and imperfect in long, dangling, green-yellow clusters before or with the leaves.

Fruit: Double samara, green to brown; wings up to 3 cm long, forming a 100-degree angle; maturing in early to midsummer.

Form: Usually an understory tree but sometimes up to 18 m (60 ft).

Habitat and Ecology: Shade tolerant; found on fertile moist soils with *Acer rubrum, Carpinus caroliniana, Celtis laevigata, Fagus grandifolia,*

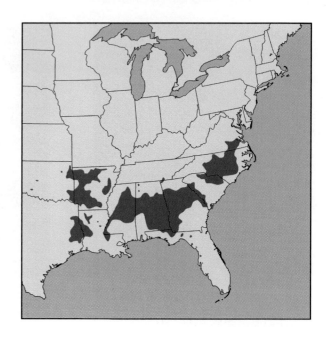

Fraxinus pennsylvanica, Liquidambar styraciflua, Liriodendron tulipifera, Magnolia macrophylla, Ostrya virginiana, Quercus alba, Quercus nigra, Quercus rubra, and *Tsuga canadensis.*

Uses: Wood of limited economic importance; large trees can be used for pulpwood, veneer, furniture, and flooring. Reported to have been tapped for syrup. A browse for white-tailed deer, and seed eaten by birds and small to midsize mammals. An attractive landscape tree in the South due to good form, heat tolerance, and attractive fall color.

Botanical Name: *Acer* is Latin for "maple tree" and refers to the hardness of the wood; *barbatum* means "bearded," referring to the hairy flowers.

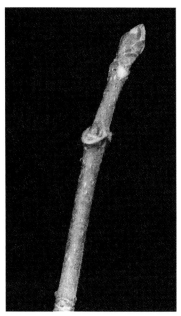

Acer barbatum twig

Acer barbatum leaves

Acer barbatum pistillate flowers

Acer barbatum staminate flowers

Acer barbatum fruit

Acer barbatum bark

Aceraceae

Acer leucoderme Small

chalk maple, white-bark maple

Quick Guide: *Leaves* opposite, simple, deciduous, palmately three to five lobed with attenuated apices, margin entire, underside yellow-green with pubescence; *Fruit* a double samara with widely divergent wings; *Bark* chalky white on large trees; *Habitat* moist soils.

Leaves: Simple, opposite, deciduous, nearly round, blade 3 to 9 cm long, lobes three to five but often only three and palmate, apices attenuated, margin entire, underside yellow-green and often with pubescence; autumn color red, orange, or yellow.

Twigs: Red-brown, glabrous, with lenticels; leaf scar V-shaped with three bundle scars.

Buds: Terminal bud ovoid, 3 mm long; scales overlapping, red-brown, pubescent.

Bark: Brown-gray and smooth on small trees; large trees chalky white and flaky but possibly dark at the base.

Flowers: Perfect and imperfect in long, dangling, green-yellow clusters before or with the leaves.

Fruit: Double samara, green to brown; wings up to 3 cm long, widely divergent, forming a 120-degree angle; maturing in early to midsummer.

Form: Similar to *Acer barbatum*.

Habitat and Ecology: Shade tolerant; found on fertile, moist soils and well-drained bottomlands

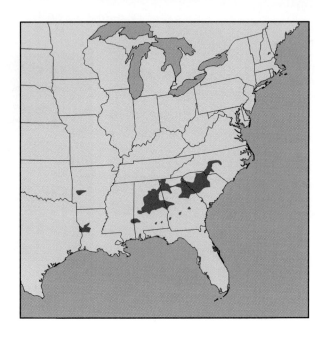

with *Acer rubrum, Carpinus caroliniana, Celtis laevigata, Fagus grandifolia, Fraxinus pennsylvanica, Liquidambar styraciflua, Liriodendron tulipifera, Ostrya virginiana, Populus deltoides, Quercus alba, Quercus nigra,* and *Quercus rubra*.

Uses: Similar to *Acer barbatum*.

Botanical Name: *Acer* is Latin for "maple tree" and refers to the hardness of the wood; *leucoderme* means "white skin," referring to the white bark.

Acer leucoderme staminate flowers

Acer leucoderme leaf

Acer leucoderme twig

Acer leucoderme fruit

Acer leucoderme bark

Acer negundo L.

boxelder, ashleaf maple, negundo maple, three-leaved maple, Manitoba maple

Quick Guide: *Leaves* opposite, pinnately compound, leaflets three to seven, margin coarsely toothed or shallowly lobed; *Twigs* bright green; *Fruit* a double samara with a flattened seed cavity; *Bark* gray-brown and grooved; *Habitat* floodplains and water edges.

Leaves: Pinnately compound, opposite, deciduous, up to 25 cm long and sometimes longer, petiole swollen at the base. Leaflets three to seven and occasionally nine, ovate to elliptical, 5 to 10 cm long, margin coarsely serrate and often shallowly lobed; autumn color yellow.

Twigs: Bright green, glabrous, possibly glaucous, with lenticels; leaf scar very narrowly V-shaped with three bundle scars, opposite leaf scars meet.

Buds: Terminal bud ovoid to acute, 5 mm long; scales overlapping, yellow-green, and pubescent; lateral buds yellow-white, pubescent.

Bark: Green-gray and smooth on small trees; large trees gray-brown and shallowly grooved.

Flowers: Species is dioecious; staminate flowers in long, dangling, yellow-green clusters; pistillate flowers in small clusters at shoot tips; appearing before or with the leaves (see page 71).

Fruit: Double samara, green to brown; wings 2.5 to 4 cm long, forming a 90-degree angle; seed cavity long, narrow, and flattened; maturing in mid to late summer.

Form: Up to 23 m (75 ft) in height and 1 m (3 ft) in diameter.

Habitat and Ecology: Shade intolerant; found commonly in floodplains and on edges of streams

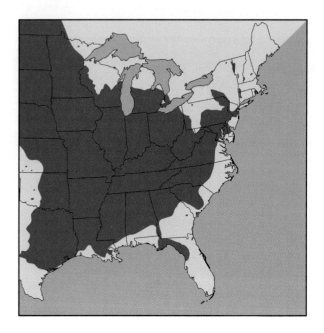

and swamps but sometimes on dry sites. Forest associates include *Acer rubrum, Acer saccharinum, Betula nigra, Celtis laevigata, Fraxinus* spp., *Liquidambar styraciflua, Platanus occidentalis, Populus deltoides, Quercus macrocarpa, Quercus lyrata, Quercus michauxii, Quercus nigra, Quercus phellos, Salix nigra,* and *Ulmus americana*.

Uses: Wood weak, brittle, soft; used for fuel, crates, woodenware, and pulpwood. Seed eaten by birds and small to midsize mammals. Because of fast growth, used for wind breaks and soil erosion control. A poor landscape tree due to excessive sprouting, poor form, and brittle branches.

Botanical Name: *Acer* is Latin for "maple tree" and refers to the hardness of the wood; *negundo* refers to similarities with *Vitex negundo*.

Acer negundo bark

Acer negundo leaf

Acer negundo twig

Acer negundo fruit

Acer negundo bark

Aceraceae

Acer nigrum Michx.

black maple, black sugar maple, hard maple

Quick Guide: Similar to *Acer saccharum* but *Leaves* mostly three-lobed, wilted or drooping, and hairy below.

Leaves: Simple, opposite, deciduous, nearly round, 8 to 20 cm long, lobes three to five but mostly three and palmate, dark green, appearing drooping or wilted, apices attenuated, sinuses shallow and rounded, margin entire, underside with dense, soft, pubescence; autumn color yellow.

Twigs: Green-brown to gray or red, glabrous or pubescent, with lenticels; leaf scar V-shaped with three bundle scars.

Buds: Terminal bud acute, 1 cm long; scales overlapping, red-brown to brown-black, with pale pubescence.

Bark: Brown-gray to gray-black and smooth on small trees; large trees more furrowed and scaly.

Flowers: Perfect and imperfect, yellow, on long pubescent stalks, in drooping clusters with the leaves.

Fruit: Double samara, green to brown; wings 2.5 to 4 cm long; seed cavity swollen; maturing in late summer.

Form: Up to 34 m (110 ft) in height and 1 m (3 ft) in diameter.

Habitat and Ecology: Shade tolerant; found on moist soils in hardwood forests and more tolerant of

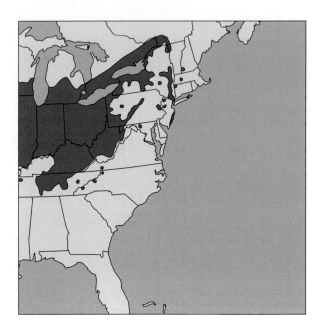

dry soils than *Acer saccharum*. Often hybridizes with *Acer saccharum*.

Uses: Wood similar to *Acer saccharum*, blond, close-grained, hard and heavy; used for pulpwood, furniture, flooring, cabinets, veneer, plywood, and firewood. Source of bird's-eye maple and tapped for maple syrup. A browse for white-tailed deer and rodents, and seed eaten by birds and small to medium-size mammals. Sometimes planted as an ornamental.

Botanical Name: *Acer* is Latin for "maple tree" and refers to the hardness of the wood; *nigrum* means "dark," possibly referring to the leaves relative to *Acer saccharum*.

Acer nigrum bark

Acer nigrum leaf

Acer nigrum leaf

Acer nigrum fruit

Acer nigrum twig

Acer nigrum bark

Aceraceae

Acer pensylvanicum L.

striped maple, moosewood, moose maple, whistlewood, snakebark maple

Quick Guide: *Leaves* opposite, simple, palmately three to five lobed, some apices attenuated, margin doubly serrate; *Flowers* yellow-green, bell shaped, in dangling clusters; *Fruit* a double samara with an indented seed cavity; *Bark* on twigs and young stems green with vertical white stripes; *Habitat* cool, moist sites.

Leaves: Simple, opposite, deciduous, nearly round, 10 to 18 cm long, palmately and shallowly three to five lobed, apices acute or acuminate, base cordate, margin finely doubly serrate, surface venation prominent, petiole red; autumn color yellow.

Twigs: Bright green or red-brown, glabrous; leaf scar crescent shaped with three bundle scars.

Buds: Terminal bud 1 cm long, conical, with two purple-green to red glabrous scales; lateral buds stalked.

Bark: Green and smooth with vertical white stripes on small trees; large trees red-brown to gray and rough with vertical stripes.

Flowers: Imperfect, bell-like with bright yellow-green and wide (1 cm) petals, in very long (up to 12 cm) drooping racemes after the leaves.

Fruit: Double samara, green to brown; wings 2 cm long, forming a 110-degree angle; seed cavity indented on one side; maturing in late summer.

Form: Up to 15 m (50 ft) in height.

Habitat and Ecology: Shade tolerant; found on north-facing slopes and in coves with *Abies*

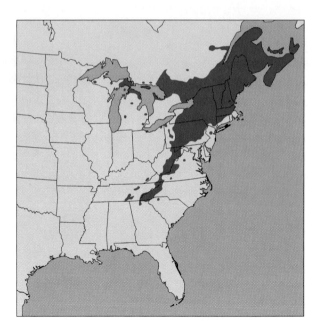

balsamea, Acer rubrum, Acer saccharum, Acer spicatum, Aesculus flava, Betula alleghaniensis, Betula lenta, Betula papyrifera, Halesia tetraptera, Pinus strobus, Picea rubens, Prunus pensylvanica, Prunus serotina, Tilia americana, and *Tsuga canadensis.*

Uses: Wood of little ecomic importance, used in specialty items. Browse for white-tailed deer, beaver, ruffed grouse, cottontail rabbits, and moose (hence "moosewood"). Seed eaten by birds and small to midsize mammals. In the spring, the tough bark of young twigs slips off easily and the twig can be used as a whistle (hence "whistlewood").

Botanical Name: *Acer* is Latin for "maple tree" and refers to the hardness of the wood; *pensylvanicum* refers to the geographic range.

Acer pensylvanicum leaf

Acer pensylvanicum fruit

Acer pensylvanicum bark

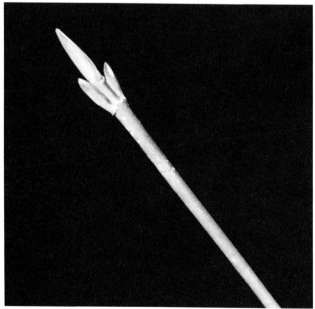

Acer pensylvanicum twig

Acer rubrum L.

red maple, scarlet maple, swamp maple, water maple, white maple, soft maple

Quick Guide: *Leaves* opposite, simple, palmately three to five lobed, sinuses shallow, margin irregularly doubly serrate; *Fruit* a bright red to brown double samara; *Bark* smooth, becoming darker, plated, and scaly on large trees; *Habitat* rocky uplands to swamps.

Leaves: Simple, opposite, deciduous, nearly round, 6 to 19 cm long, lobes three to five and palmate, apices mostly acute, sinuses acute and shallow, margin irregularly doubly serrate, underside pale and glabrous, petiole long and often red; autumn color yellow, orange, or scarlet.

Twigs: Shiny, red-brown, with lenticels; leaf scar crescent shaped with three bundle scars.

Buds: Terminal bud blunt, 4 mm long; scales overlapping, red-brown, mostly glabrous; flower buds round and plump.

Bark: Highly variable, brown-gray to white and smooth on small trees; large trees dark gray to brown, loosely plated or scaly.

Flowers: Imperfect, in red or yellowish clusters before the leaves (see page 71).

Fruit: Double samara, bright red to brown; wings 2.0 cm long, forming a 70-degree angle; maturing in spring.

Form: Up to 30 m (100 ft) in height and 1 m (3 ft) in diameter.

Habitat and Ecology: Shade tolerant; found on a variety of sites such as dry ridges, rocky uplands, stream borders, cove forests, bottomlands, and swamps. Because of its wide range and ecological amplitude, *Acer rubrum* is associated with more than 70 tree species (Hutnick and Yawney 1961).

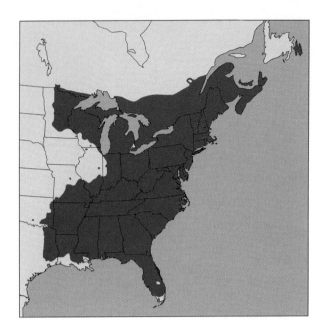

Uses: Wood white to gray-green, close grained, moderately hard; used for veneer, furniture, pallets, paneling, and boxes. Browse for white-tailed deer, and seed eaten by birds and small to midsize mammals. A common ornamental because of the attractive flowers, brilliant fall color, rapid growth, and tolerance of a variety of sites. Many cultivars have been developed to enhance form and fall color.

Botanical Name: *Acer* is Latin for "maple tree" and refers to the hardness of the wood; *rubrum* means "red," referring to the red buds, flowers, fruit, and autumn leaves.

Similar Species: Trident maple or Carolina red maple (*Acer rubrum* var. *trilobum* Torr. and Gray ex K. Koch) is found in wet areas and has smaller mostly 3-lobed leaves and smaller fruits. Drummond red maple (*Acer rubrum* var. *drummondii* (Hook. and Arn. ex Nutt.) Sarg.) is found in swamps and has leaves that are densely hairy below.

Acer rubrum leaves

Acer rubrum twig

Acer rubrum bark

Acer rubrum fruit

Acer rubrum bark

Acer saccharinum L.

silver maple, soft maple, white maple, river maple, silverleaf maple, water maple

Quick Guide: *Leaves* opposite, simple, palmately five lobed, margin coarsely serrate, sinuses deeply V-shaped, underside silvery white; *Fruit* a large double samara with widely divergent wings; *Bark* with loose silvery gray plates; *Habitat* moist to poorly drained soils.

Leaves: Simple, opposite, deciduous, nearly round, 7 to 20 cm long, lobes five and palmate, apices acute to acuminate, sinuses deeply V-shaped, margin coarsely serrate, underside prominently silvery white; autumn color yellow.

Twigs: Red-brown, shiny or glaucous, with lenticels, emitting an unpleasant odor when cut; leaf scar crescent shaped with three bundle scars.

Buds: Terminal bud blunt, 5 mm long; scales overlapping, red-brown, pubescent tipped; flower buds round and plump.

Bark: Light gray and smooth on small trees; large trees with long, loose, silvery gray plates.

Flowers: Perfect and imperfect on short stalks, in dense clusters before the leaves.

Fruit: Double samara, large (the largest of the native maples), green to brown, tomentose when young; wings 4 to 6 cm long, widely divergent, forming a 130-degree angle, often with only one wing producing seed; maturing in spring.

Form: Up to 30 m (98 ft) in height and 1 m (3 ft) in diameter.

Habitat and Ecology: Shade tolerance varies with site quality; found on well-drained to poorly drained alluvial soils of stream banks, lake edges, and floodplains. Forest associates include *Acer*

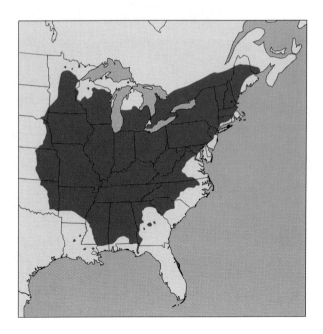

rubrum, Betula nigra, Fraxinus spp., *Liquidambar styraciflua, Platanus occidentalis, Populus deltoides, Quercus bicolor, Quercus lyrata, Quercus macrocarpa, Quercus nigra, Quercus palustris, Quercus phellos, Salix nigra, Ulmus americana,* and *Ulmus rubra.*

Uses: Wood white to gray-green, moderately hard; used for pulpwood, paneling, boxes, pallets, and furniture. A browse for beaver, and seed eaten by birds and small to midsize mammals. Planted as an ornamental for the fast growth, attractive leaves, and low crown, but often a poor landscape tree because of weak limbs, excessive sprouting, and an invasive shallow root system.

Botanical Name: *Acer* is Latin for "maple tree" and refers to the hardness of the wood; *saccharinum* means "sugary" and refers to the sap, which was occasionally boiled for sugar.

Acer saccharinum flowers

Acer saccharinum twig

Acer saccharinum leaf

Acer saccharinum bark

Acer saccharinum fruit

Aceraceae 85

Acer saccharum Marsh.

sugar maple, hard maple, rock maple, sweet maple

Quick Guide: *Leaves* opposite, simple, palmately five lobed, margin entire, apices long attenuated, underside pale and glabrous; *Fruit* a double samara with a swollen seed cavity; *Bark* gray-brown and smooth, becoming loose and flaky on large trees; *Habitat* moist, cool sites.

Leaves: Simple, opposite, deciduous, nearly round, 7 to 20 cm long, lobes mostly five and palmate, apices long attenuated, margin entire, underside pale and glabrous; autumn color brilliant red, orange, or yellow.

Twigs: Brown, shiny, with lenticels; leaf scar V-shaped with three bundle scars.

Buds: Terminal bud acute, 7 mm long; scales overlapping, dark maroon-brown, with pale pubescence.

Bark: Brown-gray and smooth or ridged on smaller trees; large trees scaly with loose curved plates.

Flowers: Perfect and imperfect, yellow, on long pubescent stalks, in drooping clusters with the leaves.

Fruit: Double samara, green to brown; wings 2.5 to 4.0 cm long, forming a 50-degree angle; seed cavity swollen; maturing in late summer.

Form: Up to 35 m (115 ft) in height and 1 m (3 ft) in diameter.

Habitat and Ecology: Shade tolerant; found primarily on moist, cool sites with *Abies balsamea, Acer rubrum, Aesculus flava, Betula alleghaniensis, Betula lenta, Betula papyrifera, Diospyros virginiana, Fagus grandifolia, Halesia tetraptera, Liriodendron tulipifera, Magnolia acuminata, Picea rubens, Prunus serotina, Pinus strobus, Quercus alba, Quercus rubra, Tilia americana,* and *Tsuga canadensis.*

Uses: Wood blond, close grained, hard, heavy; used for pulpwood, furniture, flooring, cabinets, veneer, plywood, and firewood. Source of bird's-eye maple and main source of maple syrup. A browse for white-tailed deer and rodents, and seed eaten by birds and small to midsize mammals. A popular

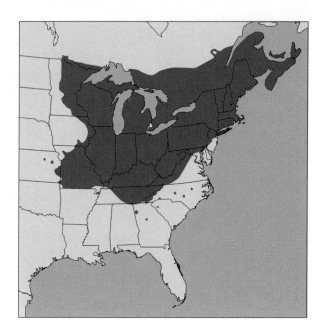

ornamental due to brilliant fall foliage, good form, and pest resistance but requires a cool site. Cultivars have been developed to increase stress tolerance.

Botanical Name: *Acer* is Latin for "maple tree" and refers to the hardness of the wood; *saccharum* means "sugary" and refers to the sap.

Similar Species: Norway maple (*Acer platanoides* L.) is from Europe and has naturalized in some areas. It is identified by five to seven lobed leaves and milky sap from the cut petiole.

Acer platanoides leaf

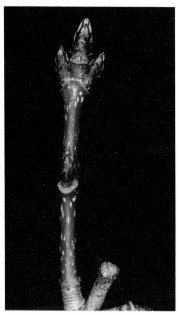

Acer saccharum twig

Acer saccharum bark

Acer saccharum fruit

Acer saccharum flowers

Acer saccharum leaves

Acer saccharum bark

Aceraceae

Acer spicatum Lam.

mountain maple, moose maple, low maple

Quick Guide: *Leaves* opposite, simple, shallowly three lobed, margin with coarse teeth; *Flowers* in erect yellow-green clusters after the leaves; *Fruit* a double samara with an indented seed cavity; *Bark* gray-brown; *Habitat* moist, cool sites.

Leaves: Simple, opposite, deciduous, nearly round, 6 to 20 cm long, lobes three to five but mostly three and palmate, sinuses shallow, apices blunt to acuminate, margin coarsely serrate, underside with white pubescence, surface venation prominent, petiole red and up to 10 cm long; autumn color red, orange, or yellow.

Twigs: Yellow-green to red-brown, with gray pubescence; leaf scar U-shaped with three bundle scars.

Buds: Terminal bud 6 mm long, conical, stalked; scales red and valvate.

Bark: Red-brown to gray-brown, shallowly ridged.

Flowers: Imperfect and perfect, appearing in upright, yellow-green, terminal clusters after the leaves.

Fruit: Double samara, red to brown; wings 1 to 2.5 cm long, forming a 90-degree angle; seed cavity indented on one side; maturing in late summer.

Form: A shrub or small understory tree up to 9 m (30 ft) in height.

Habitat and Ecology: Shade tolerant; found on moist, cool sites such as mountain slopes, stream edges, and coves. Forest associates include *Abies fraseri, Acer pensylvanicum, Betula alleghaniensis, Betula lenta, Picea rubens, Prunus pensylvanica,* and *Tsuga canadensis.*

Uses: Wood of little economic importance, used in specialty items. Browse for white-tailed deer, rabbit, and beaver; seed eaten by birds and small to midsize mammals. Buds eaten by ruffed grouse.

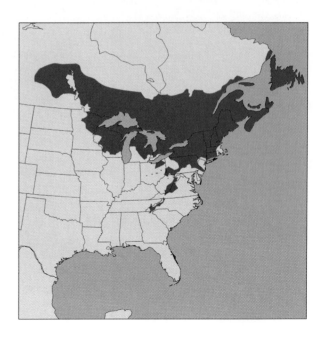

Botanical Name: *Acer* is Latin for "maple tree" and refers to the hardness of the wood; *spicatum* means "spikelike," referring to the flowers.

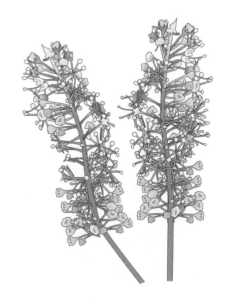

Acer spicatum leaves

Acer spicatum leaf

Acer spicatum bark

Acer spicatum fruit

Anacardiaceae Guide

1. Leaves simple, obovate to oval, petiole purple-red. *Cotinus obovatus*
2. Leaves compound, leaflets elliptical, margin entire, rachis bright red; twig glabrous with large shield-shaped leaf scars. *Toxicodendron vernix*
3. Leaves compound, leaflets lanceolate or falcate, margin mostly entire, rachis green winged; twig pubescent. *Rhus copallina*
4. Leaves compound, leaflets lanceolate or falcate, margin serrate, rachis red and glabrous; twig glaucous. *Rhus glabra*
5. Leaves compound, leaflets lanceolate or falcate, margin serrate; rachis, petiole, and twig pubescent or tomentose. *Rhus typhina*

Rhus glabra form

Rhus glabra twig

Rhus typhina twig

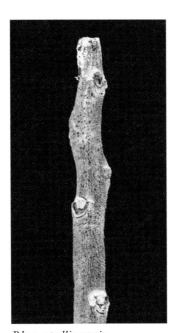

Rhus copallina twig

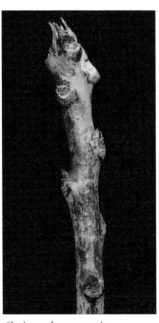

Cotinus obovatus twig

Anacardiaceae

Cotinus obovatus Raf.

smoketree, mist tree, common smoketree, American smoketree, chittamwood

Quick Guide: *Leaves* alternate, simple, obovate to oval, base and apex rounded, margin entire; *Fruit* stalks creating a smoky appearance in midsummer; *Bark* gray-brown and scaly; *Habitat* limestone soils.

Leaves: Simple, alternate, deciduous, obovate to oval, 10 to 15 cm long, apex rounded or notched, base rounded, margin entire or repand, underside with green pubescence on veins, petiole up to 5 cm long and purple-red; autumn color red or yellow.

Twigs: Purple-gray to red-brown, glaucous, with lenticels; leaf scar shield shaped to crescent shaped, bluish around the scar (see page 91).

Buds: Terminal bud 3 mm long, acute; scales overlapping, red-brown, pubescent.

Bark: Red-brown to gray-brown, thin, with loose scales.

Flowers: Small, clustered on long stalks, petals yellow-green, anthers bright yellow, in spring after the leaves; species is dioecious.

Fruit: Drupe, compact, kidney shaped, 5 mm wide, on feathery-hairy pinkish stalks, many stalks without fruit creating a smoky appearance in mid to late summer.

Form: Shrub or small tree up to 10 m (33 ft) in height, sometimes forming thickets.

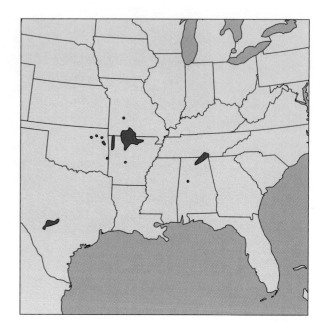

Habitat and Ecology: Uncommon; found on rocky limestone soils in northern Alabama, northern Arkansas, northwest Georgia, and Tennessee.

Uses: Durable wood was used for fenceposts, a dye extracted from the wood was used during the Civil War. Planted as an ornamental for the smoky fruit stalks, red fall foliage, tolerance of dry soils, and lack of insect and disease problems.

Botanical Name: *Cotinus* is derived from Greek for "wild-olive"; *obovatus* refers to the obovate leaves.

Cotinus obovatus leaves

Cotinus obovatus flowers

Cotinus obovatus fruit

Cotinus obovatus bark

Anacardiaceae

Rhus copallina L.

winged sumac, shining sumac, flameleaf sumac, dwarf sumac

Quick Guide: *Leaves* alternate, pinnately compound, leaflets from 9 to 21, shiny green, margin mostly entire, rachis green winged; *Twigs* velvety pubescent; *Fruit* a conical collection of dull red drupes in terminal clusters in late summer; *Habitat* old fields and disturbed areas.

Leaves: Pinnately compound, alternate, deciduous, up to 30 cm long. Leaflets from 9 to 21, lanceolate or falcate, shiny green, 4 to 7 cm long, margin mostly entire; rachis green winged; autumn color bright red.

Twigs: Slender with brown pubescence and raised lenticels; leaf scar horseshoe shaped with numerous bundle scars, nearly encircling the lateral bud (see page 91).

Buds: Lacking a true terminal bud; lateral buds round, 3 mm long, light brown, fuzzy.

Bark: Gray-brown and smooth with lenticels, becoming warty or scaly on large stems.

Flowers: Greenish white, clustered in dense panicles at the twig terminal in early summer; plants dioecious.

Fruit: Drupes, pubescent, dull red, 3 mm wide, in dense conical clusters up to 20 cm long at branch terminals, maturing in late summer and autumn, persisting over winter.

Form: Shrub or small tree up to 6 m (20 ft) in height with an open crown, often forming dense thickets due to root sprouting.

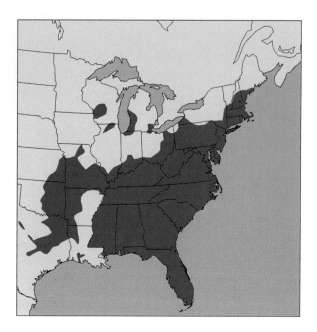

Habitat and Ecology: Shade intolerant; common in old fields, cutover areas, and wood edges.

Uses: Beaver, fox squirrels, and rabbits eat the bark, particularly in winter; fruit eaten by many songbirds and game birds including northern bobwhite, wild turkey, ruffed and sharp-tailed grouse, ring-necked pheasants, woodpeckers, thrushes, and vireos as well as white-tailed deer and opossum. The crushed fruit has a tart flavor and was mixed with sugar to make lemonade. Can be used to naturalize rough areas. Cultivars have been developed to enhance form and fall color.

Botanical Name: *Rhus* is derived from the Celtic word meaning "red"; *copallina* is Spanish for "varnishlike sap." Also named *Rhus copallinum* L.

Rhus copallina twig and fruit

Rhus copallina leaf

Rhus copallina bark

Rhus copallina flowers

Anacardiaceae

Rhus glabra L.
smooth sumac

Quick Guide: *Leaves* alternate, pinnately compound, leaflets from 11 to 31, margin serrate, rachis purple-red and glabrous; *Twigs* glaucous; *Fruit* a conical collection of bright red drupes in terminal clusters in fall; *Habitat* old fields and disturbed areas.

Leaves: Pinnately compound, alternate, deciduous, up to 40 cm long. Leaflets from 11 to 31, lanceolate to falcate, 5 to 14 cm long, margin serrate; rachis purple-red and glabrous; autumn color bright red.

Twigs: Stout, glaucous, purple-red, sticky when cut; leaf scar heart shaped with numerous bundle scars, nearly encircling the lateral bud (see page 91).

Buds: Lacking a true terminal bud; lateral buds round, 5 mm long, light brown, fuzzy.

Bark: Gray-brown and smooth with lenticels, becoming warty and scaly on large stems.

Flowers: Greenish white, clustered in dense panicles at the twig terminal in early summer; plants dioecious.

Fruit: Drupes, pubescent, bright red, 3 mm wide, in dense conical clusters up to 20 cm long at branch terminals, maturing in late summer and autumn, persisting over winter.

Form: Shrub or small tree up to 6 m (20 ft) in height with an open crown, often forming dense thickets due to root sprouting.

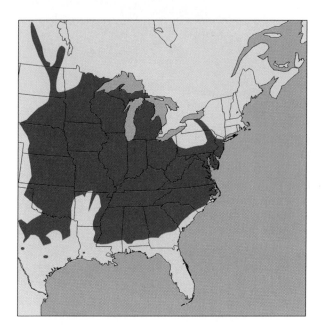

Habitat and Ecology: Shade intolerant; common in old fields, wood edges, and cutover areas.

Uses: Same wildlife uses as for *Rhus copallina*. The crushed fruit has a tart taste and was mixed with sugar to make lemonade. Limited landscape value because of a shallow root system and prolific root sprouting, but can be used for naturalizing rough areas.

Botanical Name: *Rhus* is derived from Celtic word meaning "red"; *glabra* means "hairless," referring to the hairless rachis and twig.

Rhus glabra fruit

Rhus glabra leaf

Rhus glabra bark

Rhus glabra flowers

Anacardiaceae

Rhus typhina L.
staghorn sumac, velvet sumac

Quick Guide: *Leaves* alternate, pinnately compound, leaflets from 11 to 31, margin serrate, rachis red and pubescent; *Twigs* velvety hairy; *Fruit* a conical collection of bright red velvety drupes in terminal clusters in late summer; *Habitat* old fields and disturbed areas.

Leaves: Pinnately compound, alternate, deciduous, up to 60 cm long. Leaflets from 11 to 31, lanceolate or falcate, 5 to 14 cm long, margin serrate; rachis purple-red and pubescent; autumn color bright red.

Twigs: Stout, with velvety brown hair; leaf scar U-shaped with numerous bundle scars, nearly encircling the lateral bud. Twigs likened to the velvet stage of antlers of white-tailed deer (see page 91).

Buds: Lacking a true terminal bud; lateral buds round, 6 mm long, yellow-brown, fuzzy.

Bark: Red-brown to gray-brown and smooth with lenticels, becoming warty or scaly on large stems.

Flowers: Greenish white, clustered in dense panicles at the twig terminal in early summer; plants dioecious.

Fruit: Drupes, bright red, velvety, 3 mm wide, in dense conical clusters up to 20 cm long at branch terminals, maturing in late summer and autumn, persisting over winter.

Form: Shrub or small tree up to 6 m (20 ft) in height with an open crown, often forming dense thickets.

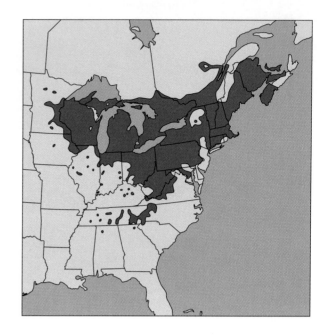

Habitat and Ecology: Shade intolerant; common in old fields, wood edges, and cutover areas.

Uses: Same wildlife uses as for *Rhus copallina*. The crushed fruit has a tart taste and has been mixed with sugar to make lemonade. Early settlers smoked the dried leaves, and the leaves and twigs were used for tanning leather. Limited landscape value because of a shallow root system and prolific root sprouting, but the velvet fruit and hairy stems are attractive. Cultivars have been developed for landscape use.

Botanical Name: *Rhus* is derived from the Celtic word meaning "red"; *typhina* means "like a cat's tail," refering to the densely hairy twig, which may look like a curved cat's tail. Also named *Rhus hirta* (L.) Sudworth.

Rhus typhina flowers

Rhus typhina fruit

Rhus typhina bark

Rhus typhina leaf

Toxicodendron vernix (L.) Kuntze

poison-sumac, swamp-sumac, poison-elderberry, thunderwood

Quick Guide: *Leaves* alternate, pinnately compound, leaflets from 7 to 13 and elliptical, margin entire, rachis and petiole red; *Fruit* a whitish drupe in open drooping clusters; *Bark* gray and smooth, often with dark patches; *Habitat* stream edges and swamps.

Leaves: Pinnately compound, alternate, deciduous, up to 40 cm long. Leaflets from 7 to 13, elliptical to oblong, 5 to 10 cm long, margin entire; rachis and petiole red; autumn color bright red.

Twigs: Stout, brown-gray, glabrous with lenticels; leaf scar large, shield or heart shaped, with prominent bundle scars.

Buds: Terminal bud 1.5 cm long, conical, with purple-brown scales.

Bark: Mottled, brown-gray and smooth with lenticels or warts, often showing black patches from exudation of sap.

Flowers: In arching panicles from leaf axils on current-year shoots after the leaves, petals yellow-green, anthers orange; plants are dioecious.

Fruit: Drupe, gray-white, round, juicy, 6 mm wide, in open dangling clusters persisting over winter.

Form: Shrub or small tree up to 10 m (33 ft) in height with an open crown.

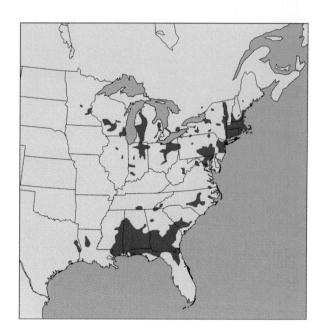

Habitat and Ecology: Found in swamps, bogs, pineland depressions, and on stream edges.

Uses: Caution! The oleoresin in all plant parts may cause a severe rash in humans when touched. Fruit eaten by birds.

Botanical Name: *Toxicodendron* means "poison tree"; *vernix* means "varnish" and refers to the clear sap, which turns black when exposed to air.

Toxicodendron vernix leaf

Toxicodendron vernix twig

Toxicodendron vernix bark

Toxicodendron vernix fruit

Asimina triloba (L.) Dunal

pawpaw, common pawpaw, wild banana tree

Quick Guide: *Leaves* alternate, simple, large, obovate, entire, with a green pepper smell when crushed; *Velvety maroon pubescence* on the leaf underside, bud, and twig; *Flower* petals in threes and purple-maroon; *Fruit* bananalike; *Habitat* moist, fertile soils.

Leaves: Simple, alternate, deciduous, obovate to oblong, drooping, 10 to 30 cm long, apex and base acute, margin entire, with a green pepper smell when crushed; autumn color yellow.

Twigs: Slender, light brown, with soft maroon pubescence; leaf scar raised, cresent shaped, with numerous bundle scars.

Buds: Naked, covered with velvety maroon pubescence; terminal bud conical, 1 cm long; flower buds plump.

Bark: Brown-gray, mottled, smooth and with warts.

Flowers: Perfect, with two layers of dark, purple-maroon petals in groups of three, petals with prominent venation, flower up to 5 cm wide when fully open, appearing before or with the leaves.

Fruit: Berry, large, thick, often curved, up to 13 cm long, with a custardlike taste when mature, maturing in late summer.

Form: Shrub or small understory tree up to 12 m (4 ft) in height, can form thickets.

Habitat and Ecology: Shade tolerant; usually found on moist, fertile soils near streams and in hardwood forests but occasionally on drier sites.

Uses: Inner bark was used for string and fishnets, and fruit was used in custards. Fruit eaten by raccoon, opossum, gray fox, gray squirrel, birds, and small rodents. Larvae of the zebra swallowtail butterfly feed on the foliage. An interesting landscape tree due to flowers and fruit, but will root sprout.

Botanical Name: *Asimina* is believed to be a Native American word meaning "sleeve-shaped

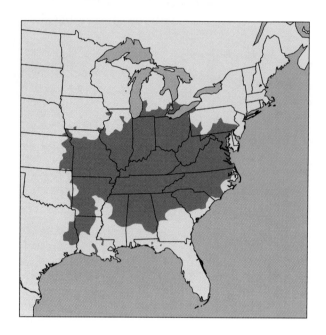

fruit"; *triloba* refers to the petals in groups of three.

Similar Species: Dwarf pawpaw [*Asimina parviflora* (Michx.) Dunal] is a smaller plant and has smaller leaves (up to 20 cm long), smaller flowers (2 cm wide when open), smaller fruit (up to 8 cm long), and an acuminate leaf apex.

Asimina parviflora leaves

Asimina triloba fruit

Asimina triloba fruit

Asimina triloba twig

Asimina triloba bark

Asimina triloba leaves

Annonaceae

Aquifoliaceae Guide

1. Leaves leathery, evergreen, with large spiny teeth on the margin; fruit a red drupe. *Ilex opaca*
2. Leaves leathery, evergreen, apex with a rigid spine, margin with an occasional small sharp tooth; fruit an orange-red or red drupe. *Ilex cassine*
3. Leaves leathery, evergreen, margin with bristle-tipped teeth above the middle; fruit a black drupe. *Ilex coriacea*
4. Similar to *Ilex coriacea* but leaves entire or serrate above the middle and shrub form. *Ilex glabra*
5. Leaves leathery, evergreen, up to 3 cm long, apex with a rigid bristle tip, margin entire; fruit a red drupe. *Ilex myrtifolia*
6. Leaves evergreen, up to 4 cm long, apex lacking a rigid spine, margin with small round teeth; fruit a red drupe. *Ilex vomitoria*
7. Leaves deciduous, apex lacking a rigid spine, margin serrate to crenate; fruit a red drupe. *Ilex decidua*
8. Leaves deciduous, apex acuminate and without a rigid spine, venation prominent, margin serrate except at the base; fruit a red drupe. *Ilex montana*
9. Similar to *Ilex montana* but with more leathery leaves and a smaller fruit. *Ilex verticillata*
10. Similar to *Ilex montana* but leaves with a mostly entire margin or minutely crenate-serrate above the middle, and found in the southern Coastal Plain. *Ilex ambigua*
11. Leaves deciduous, apex without a spine, margin serrate above the middle, underside with dense white pubescence; fruit a red drupe. *Ilex amelanchier*

Ilex glabra pistillate flowers

Ilex opaca staminate flowers

Ilex opaca pistillate flowers

Aquifoliaceae

Ilex amelanchier Curtis ex Chapm.
sarvis holly

Quick Guide: *Leaves* alternate, simple, deciduous, obovate, margin serrate above the middle, underside densely pubescent; *Fruit* a red drupe; *Bark* gray with warty lenticels; *Habitat* swamps and stream edges.

Leaves: Simple, alternate, deciduous, 5 to 9 cm long, obovate to oblanceolate, thin, apex acute to abruptly acuminate, base cuneate to rounded, margin serrate above the middle, underside with dense white pubescence, petiole pubescent; autumn color yellow.

Twigs: Gray-green, lightly pubescent; leaf scar semicircular with one bundle scar.

Buds: Small, rounded, with green or red scales.

Bark: Brown-gray, smooth, with warty lenticels.

Flowers: Small, green-white, blooming in late spring; species is dioecious.

Fruit: Drupe, round, red, up to 1 cm wide, containing several ribbed stones; maturing in autumn.

Form: A shrub or small tree up to 6 m (20 ft) in height.

Habitat and Ecology: Rare, found in swamps, floodplains, and near streams in the Coastal Plain.

Uses: Fruit eaten by birds and small mammals.

Botanical Name: *Ilex* refers to *Quercus ilex* or holly oak, an evergreen Mediterranean tree with toothed leaves; *amelanchier* possibly refers to the similarity of the leaves to the genus *Amelanchier*.

Ilex amelanchier fruit

Ilex amelanchier bark

Ilex amelanchier leaves

Aquifoliaceae

Ilex cassine L.
dahoon

Quick Guide: *Leaves* alternate, simple, evergreen, apex with a rigid spine, margin with an occasional small tooth; *Twig* downy pubescent; *Fruit* an orange-red drupe in fall; *Bark* smooth and gray; *Habitat* swamps and stream edges.

Leaves: Simple, alternate, evergreen, shiny, leathery, elliptical to oblong or oblanceolate but can be variable in shape, 4 to 10 cm long, apex with a rigid spine, margin entire or with an occasional small sharp tooth near the apex, midrib downy pubescent.

Twigs: Mottled green-brown, downy pubescent; leaf scar semicircular with one bundle scar.

Buds: Terminal bud small, ovoid; scales overlapping, green or brown, with pale pubescence.

Bark: Light gray, mottled, mostly smooth with warts.

Flowers: Small, green-white, blooming in late spring; species is dioecious.

Fruit: Drupe, round, usually orange-red to bright red, about 7 mm wide, containing several ribbed stones, maturing in autumn.

Form: Shrub or small tree up to 10 m (33 ft) in height.

Habitat and Ecology: Found on edges of ponds and streams, and in swamps, flatwoods, bogs, and

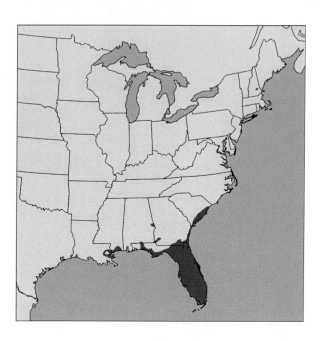

pine barrens with *Gordonia lasianthus, Myrica heterophylla, Nyssa biflora, Persea palustris, Pinus elliottii, Pinus serotina,* and *Taxodium distichum* var. *distichum* and var. *imbricarium*.

Uses: Fruits eaten by birds and small mammals. Planted as an ornamental for the evergreen foliage and attractive fruit.

Botanical Name: *Ilex* refers to *Quercus ilex* or holly oak, an evergreen Mediterranean tree with toothed leaves.

Ilex cassine leaves

Ilex cassine immature fruit

Ilex cassine bark

Ilex cassine flowers

Aquifoliaceae

Ilex coriacea (Pursh) Chapm.
large gallberry, sweet gallberry

Quick Guide: *Leaves* alternate, simple, evergreen, margin thick with an occasional bristle-tipped tooth above the middle; *Fruit* a black drupe; *Bark* gray with warty lenticels; *Habitat* sandy, wet soils.

Leaves: Simple, alternate, evergreen, shiny dark green, 4 to 9 cm long, elliptical to oval or obovate, apex bristle-tipped, margin thickened and entire or with an occasional bristle-tipped tooth mostly above the middle, underside glabrous or pubescent.

Twigs: Gray-green to dark brown, pubescent and sometimes sticky; leaf scar semicircular with one bundle scar.

Buds: Small; scales green or red-brown, overlapping, pubescent.

Bark: Brown-gray to green, smooth, with warty lenticels.

Flowers: Small, green-white, blooming in late spring; species is dioecious.

Fruit: Drupe, round, shiny black, 7 mm wide, sweet, containing several ribbed stones; maturing in autumn but usually not persisting over winter.

Form: A shrub or small tree to up to 6 m (20 ft) in height.

Habitat and Ecology: Found on sandy, wet soils of swamps, bogs, bays, and bottoms.

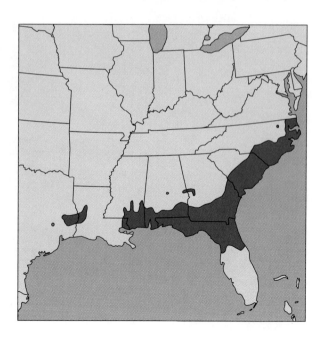

Uses: Fruit eaten by birds and small mammals; leaves browsed by deer.

Botanical Name: *Ilex* refers to *Quercus ilex* or holly oak, an evergreen Mediterranean tree with toothed leaves; *coriacea* means "leathery," referring to the leaves.

Similar Species: Inkberry (*Ilex glabra* (L.) Gray) is a shrub found on similar sites and is distinguished by leaves with an entire margin sometimes serrate rather than spiny above the middle and by a black, dry, bitter drupe that persists over winter.

Ilex glabra leaves and immature fruit

Ilex coriacea leaves

Ilex coriacea bark

Ilex coriacea fruit

Aquifoliaceae

Ilex decidua Walt.

possumhaw, deciduous holly, swamp holly, meadow holly

Quick Guide: *Leaves* alternate, deciduous, simple, margin serrate to crenate, clustered on spur branches; *Fruit* a bright red, juicy drupe prominent after leaf fall; *Bark* smooth and gray with warts; *Habitat* along streams and in floodplains.

Leaves: Simple, alternate, deciduous, elliptical, obovate to oblanceolate but variable, 4 to 8 cm long, base cuneate to rounded, apex acute to rounded, margin serrate to crenate, midrib pubescent, clustered on spur branches; autumn color yellow.

Twigs: Green-brown to gray-white, glabrous, with lenticels; leaf scar semicircular with one bundle scar.

Buds: Terminal bud small; scales overlapping, green, brown, or red-brown, pubescent or glabrous.

Bark: Thin, green-gray to gray-brown, mottled, mostly smooth with warts but may be rough on large trees.

Flowers: Small, green-white, appearing in late spring; species is dioecious.

Fruit: Drupe, orange-red to red, round, 6 to 10 mm wide, juicy, containing several ribbed stones, maturing in autumn and persisting until the following spring.

Form: Shrub or small tree usually only up to 6 m (20 ft) in height but sometimes taller.

Habitat and Ecology: Shade tolerant; found in swamps and floodplains, on moist upland soils, and

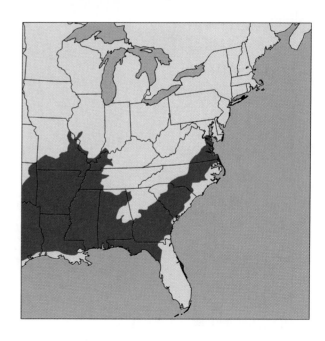

occasionally on drier sites in association with a variety of species.

Uses: A moderately used browse of white-tailed deer; fruit eaten by various song and game birds including northern bobwhite, wild turkey, and ruffed grouse as well as white-tailed deer and gray squirrel. Landscape value limited to naturalizing open areas because of sprouting and messy form. Cultivars are available and selected for fruit and form.

Botanical Name: *Ilex* refers to *Quercus ilex* or holly oak, an evergreen Mediterranean tree with toothed leaves; *decidua* refers to the deciduous leaves.

Ilex decidua fruit

Ilex decidua bark

Ilex decidua leaves and flowers

Ilex montana Torr. & Gray ex Gray.

mountain winterberry, mountain holly

Quick Guide: *Leaves* alternate, simple, thin, venation prominent, margin serrate except at the base; *Fruit* a bright red drupe; *Bark* brown-gray with lenticels; *Habitat* mesic forests.

Leaves: Simple, alternate, deciduous, light green, thin, 5 to 14 cm long, ovate or elliptical, apex acute to acuminate, base acute to rounded, margin sharply serrate except at the base, venation prominent, pubescent below, often clustered on spur branches; autumn color yellow.

Twigs: Red-brown to green-gray, with lenticels; leaf scar semicircular with one bundle scar.

Buds: Terminal bud small; scales overlapping, brown, pubescent.

Bark: Brown-gray with warty lenticels.

Flowers: Small, green-white, flowering in late spring; species is dioecious.

Fruit: Drupe, round, bright red, 5 to 13 mm wide, containing several ribbed stones; maturing in autumn and persisting after leaf fall.

Form: A shrub or small tree to up to 9 m (30 ft) in height.

Habitat and Ecology: Shade tolerant; found on moist soils of upland forests and mountains.

Uses: Fruit eaten by songbirds, waterfowl, and small mammals; foliage browsed by deer, moose, and rabbit.

Botanical Name: *Ilex* refers to *Quercus ilex* or holly oak, an evergreen Mediterranean tree with toothed leaves; *montana* means "of the mountains."

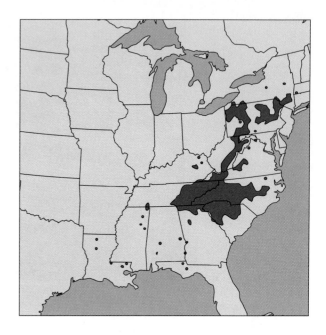

Similar Species: Common winterberry or black-alder (*Ilex verticillata* (L.) Gray) is usually a shrub found on moist, upland sites and in swamps, bogs, and ditches. It is distinguished by thicker often darker leaves with more prominent venation, flowers with 6 to 8 petals versus 4 to 5 in *Ilex montana,* and a smaller (6 mm wide) orange-red drupe. Many cultivars are available for landscaping. Carolina holly or sand holly (*Ilex ambigua* (Michx. Torr.) is found on well-drained sandy soils of the southern Coastal Plain and is identified by leaves less than 10 cm long with an acute apex, a leaf margin entire or minutely serrate/crenate above the middle, and shiny red drupes 5–10 mm wide.

Ilex montana leaves

Ilex verticillata bark

Ilex montana bark

Ilex verticillata leaves and immature fruit

Aquifoliaceae 115

Ilex myrtifolia Walt.
myrtle-leaved holly, myrtle dahoon

Quick Guide: *Leaves* alternate, simple, evergreen, small, margin mostly entire; *Fruit* a red drupe; *Bark* gray with corky warts; *Habitat* sandy, wet soils.

Leaves: Simple, alternate, evergreen, lanceolate, oblong, or oblanceolate, 0.5 to 3 cm long, shiny dark green, apex bristle-tipped, margin entire or with an occasional bristle-tipped tooth, lateral veins obscure, mostly glabrous below, the petiole is very short.

Twigs: Rigid, crooked, gray-brown, lightly pubescent; leaf scar semicircular with one bundle scar.

Buds: Terminal bud small; scales green or red, overlapping, pubescent.

Bark: Brown-gray, smooth, with corky warts.

Flowers: Small, green-white, flowering in late spring; species is dioecious.

Fruit: Drupe, round, red, 5 to 8 mm wide, containing several ribbed stones; maturing in autumn and persisting over winter.

Form: A shrub or small tree up to 6 m (20 ft) in height.

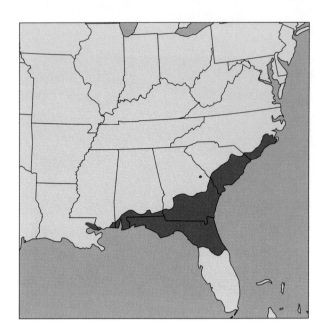

Habitat and Ecology: Found on sandy soils of swampy woods, cypress ponds, flatwoods, and bays in the Coastal Plain.

Uses: Fruit eaten by songbirds and small mammals.

Botanical Name: *Ilex* refers to *Quercus ilex* or holly oak, an evergreen Mediterranean tree with toothed leaves; *myrtifolia* means "myrtle-like leaves."

Ilex myrtifolia leaves

Ilex myrtifolia bark

Ilex myrtifolia bark

Aquifoliaceae

Ilex opaca Ait.

American holly, white holly, Christmas holly, boxwood, evergreen holly

Quick Guide: *Leaves* alternate, simple, evergreen, margin spiny toothed; *Fruit* a bright red drupe in autumn; *Bark* smooth, thin, and gray mottled; *Habitat* a variety of moist sites.

Leaves: Simple, alternate, evergreen (persisting 3 years), dark green, leathery, thick, elliptical to obovate, 5 to 12 cm long, margin with sharp spiny teeth and occasionally entire.

Twigs: Gray-white to brown, mottled, pubescent when young, with lenticels; leaf scar semicircular with one bundle scar.

Buds: Terminal bud small; scales overlapping, green or red, pubescent.

Bark: Light gray, mottled, smooth, with warts.

Flowers: Small, green-white, appearing in late spring; species is dioecious (see page 105).

Fruit: Drupe, round, red or orange-red, 4 to 10 mm wide, containing several ribbed stones, maturing in autumn and persisting over winter.

Form: Up to 15 m (50 ft) in height.

Habitat and Ecology: Shade tolerant; found on a variety of sites including swamps, bottomlands, hammocks, stream edges, well-drained uplands, and occasionally on drier protected sites. Forest associates include *Acer rubrum, Fagus grandifolia, Fraxinus americana, Liquidambar styraciflua, Liriodendron tulipifera, Nyssa sylvatica, Quercus alba,* and *Quercus nigra.*

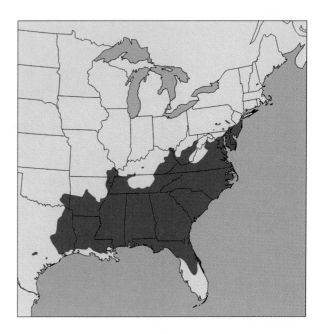

Uses: Wood ivory white, hard, heavy, close grained; used for veneer, turnery, specialty items (including piano keys and scientific instruments), furniture, and cabinets. Fruit eaten by northern bobwhite, wild turkey, numerous songbirds, and small to midsize mammals. Though slow growing, a popular ornamental for large spaces for the evergreen foliage and red fruit, and up to one thousand cultivars have been developed. Branches with fruit are used in Christmas decorations.

Botanical Name: *Ilex* refers to *Quercus ilex* or holly oak, an evergreen Mediterranean tree with toothed leaves; *opaca* means "dark or shaded," referring to the dark green, thick leaves.

Ilex opaca leaves and fruit

Ilex opaca bark

Ilex opaca bark

Ilex opaca leaves

Aquifoliaceae

Ilex vomitoria Ait.

yaupon, Christmas holly, evergreen holly

Quick Guide: *Leaves* alternate, simple, small, oblong to oval, evergreen, margin with rounded teeth; *Fruit* a bright orange-red drupe in fall; *Bark* smooth and white-gray; *Habitat* a variety of sites from flatwoods to upland forests.

Leaves: Simple, alternate, evergreen, shiny, oblong to oval, 1 to 4 cm long, margin crenate.

Twigs: Red-brown, pubescent; older twigs gray-white, stiff, thornlike, often at 90-degree angles; leaf scar semicircular with one bundle scar.

Buds: Terminal bud small; scales overlapping, green or red, pubescent.

Bark: Thin, white-gray to gray-brown, mottled, smooth.

Flowers: Small, green-white, appearing in late spring; species is dioecious.

Fruit: Drupe, round, orange-red, 4 to 8 mm wide, containing several ribbed stones, maturing in autumn.

Form: Shrub or small tree up to 8 m (26 ft) in height, often forming thickets.

Habitat and Ecology: Found on well-drained sites including coastal forests.

Uses: Planted as an ornamental and for hedges for the evergreen foliage and bright fruit, but does root

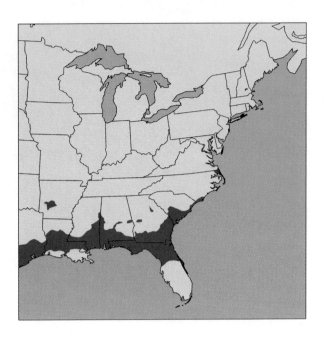

sprout. Fruit eaten by various songbirds and small mammals; important browse for white-tailed deer.

Botanical Name: *Ilex* refers to *Quercus ilex* or holly oak, an evergreen Mediterranean tree with toothed leaves; *vomitoria* refers to the reported ability of the leaves to induce regurgitation when consumed; some Native American tribes used the drink in purification ceremonies.

Ilex vomitoria leaves and fruit

Ilex vomitoria bark

Ilex vomitoria leaves and flowers

Aralia spinosa L.

devil's walkingstick, Hercules-club, prickly-ash, toothache tree, angelica-tree

Quick Guide: *Leaves* alternate, bi- or tripinnately compound and over 1 m long; *Prickles* on the stem; *Flowers* in large white clusters at the stem terminal in summer; *Fruit* a purple juicy drupe in heavy terminal clusters in late summer; *Habitat* adapted to a variety of sites.

Leaves: Bi- or tripinnately compound, alternate, deciduous, up to 1.6 m long, forming a large diamond-shaped silhouette, petiole stout and purple-red. Leaflets mostly ovate, 5 to 10 cm long, margin serrate, apex acute to acuminate, base rounded or inequilateral, midrib prickly; rachis stout, with prickles, and purple-red; autumn color purple-red.

Twigs: Stout, gray-brown, smooth, with lenticels and prickles; leaf scar large, U-shaped with numerous sharp prickles.

Buds: Terminal bud large (up to 2.0 cm long), conical, and scales overlapping, brown-gray, and rough-edged; lateral buds smaller and triangular.

Bark: Gray-brown, with warts and prickles, becoming ridged or scaly on large stems.

Flowers: With green-white petals, clustered in very large compound panicles at the stem terminal in summer.

Fruit: Drupe, round, purple-black, juicy, 5 to 8 mm wide, born on purple-red stalks in drooping terminal clusters, maturing in late summer and early fall.

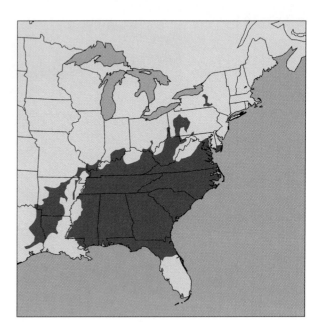

Form: Single-stemmed shrub or small tree up to 10 m (33 ft) in height, often forming thickets.

Habitat and Ecology: Shade intolerant; found on a variety of sites including upland forests and edges of swamps and streams.

Uses: Fruit eaten by numerous songbirds and small to midsize mammals such as eastern chipmunk and striped skunk; foliage moderately browsed by white-tailed deer. The flowers and fruit make this an interesting natural border plant, but it can form thickets.

Botanical Name: *Aralia* is from a Native American name; *spinosa* means "bearing spines."

Aralia spinosa twig

Aralia spinosa twig

Aralia spinosa flowers

Aralia spinosa bark

Aralia spinosa fruit

Aralia spinosa leaf

Araliaceae

Betulaceae Guide

1. Leaf margin finely serrate and wavy, leaves obovate, underside with maroon hairs; fruit a nutlet in persistent woody cone. *Alnus serrulata*
2. Leaf margin doubly serrate, leaves oval, underside with white-gray pubescence, lateral veins sunken; fruit a nutlet in a persistent woody cone. *Alnus rugosa*
3. Leaf margin doubly serrate, leaves diamond shaped or ovate, base wedge shaped; fruit a nutlet in a papery cone; young bark orange-brown to salmon-pink and peeling. *Betula nigra*
4. Leaf margin doubly serrate, leaves ovate, base cordate to inequilateral; twigs with a weak wintergreen odor when cut; fruit a nutlet in a papery cone; bark shiny bronze or gray and peeling horizontally on young trees, on large trees flaky and darker. *Betula alleghaniensis*
5. Leaf margin serrate or doubly serrate, leaves ovate, base cordate to inequilateral; twigs with a strong wintergreen odor when cut; fruit a nutlet in a papery cone; bark maroon-black with horizontal lenticels on young trees, large trees with black scaly plates. *Betula lenta*
6. Leaf margin serrate or doubly serrate, leaves ovate, base rounded to acute; twigs with white lenticels; fruit a nutlet in a papery cone; bark chalky white and peeling. *Betula papyrifera*
7. Leaf margin doubly serrate, leaves triangular, apex very long pointed; twigs with resin glands and white warty lenticels; fruit a nutlet in a papery cone; bark gray-white and not peeling. *Betula populifolia*
8. Leaf margin doubly serrate, leaves ovate to elliptical, base cordate and inequilateral, lateral veins dividing near the leaf margin; buds maroon or green-white striped; fruit a nutlet in a hoplike sac; bark shreddy. *Ostrya virginiana*
9. Leaf margin doubly serrate, leaves ovate to elliptical, base cordate and inequilateral; buds brown-white striped; fruit a nutlet in a three-lobed leafy bract; bark smooth and sinewy. *Carpinus caroliniana*

Carpinus caroliniana pistillate flowers

Carpinus caroliniana staminate flowers

Ostrya virginiana staminate and pistillate flowers

Ostrya virginiana staminate and pistillate flowers

Betulaceae

Alnus rugosa (Du Roi) Spreng.
speckled alder, tag alder, hoary alder, gray alder

Quick Guide: *Leaves* alternate, simple, deeply veined, with velvety white-gray pubescence on the underside; *Buds* stalked with maroon, valvate scales; *Fruit* a nutlet in a woody cone; *Bark* smooth with horizontal lenticels; *Habitat* water edges.

Leaves: Simple, alternate, deciduous, ovate to oval, 5 to 10 cm long, thick, upper surface wrinkled, base rounded to obtuse, apex acute, margin doubly serrate, lateral veins deeply sunken and ending at a margin tooth, underside with white-gray pubescence and prominent veins; autumn color yellow.

Twigs: Slender, red-brown to gray, pubescent, speckled with lenticels; leaf scar cordate to round with three bundle scars.

Buds: True terminal bud lacking; lateral buds 8 mm long, stalked, and with two or three maroon, valvate scales.

Bark: Red-brown to gray, smooth, speckled with horizontal pale or orange lenticels.

Flowers: Imperfect, in separate catkins; mature staminate catkins up to 8 cm long, dangling; pistillate catkins much smaller, erect, short, red-brown when open in spring.

Fruit: Nutlet, enclosed in a woody cone about 1 cm long; maturing in fall and persisting over winter.

Form: A shrub or small tree up to 9 m (30 ft) in height, often forming thickets.

Habitat and Ecology: Shade intolerant; found in ditches and on banks of streams, ponds, and swamps with *Abies balsamea, Larix laricina, Picea glauca, Picea mariana,* and *Populus tremuloides.*

Uses: Red-brown wood was used for bobbins, wooden heels, and to produce charcoal for gunpowder; now used in specialty items and also used in wetland restoration. See *Alnus serrulata* for wildlife uses.

Botanical Name: *Alnus* is Latin for "alder"; *rugosa* means "wrinkled," referring to the leaves. Also named *Alnus incana* (L.) Moench spp. *rugosa* (Du Roi) Clausen.

Alnus rugosa fruit

Alnus rugosa bark

Alnus rugosa bark

Alnus rugosa leaves

Alnus serrulata (Ait.) Willd.

hazel alder, common alder, tag alder, smooth alder

Quick Guide: *Leaves* alternate, simple, finely toothed, obovate, underside with maroon pubescence; *Buds* stalked, scales maroon and valvate; *Fruit* a nutlet in a woody cone; *Bark* brown-gray, sinewy, and smooth; *Habitat* water edges.

Leaves: Simple, alternate, deciduous, obovate to elliptical, 5 to 13 cm long, base rounded to shortly cuneate, apex obtuse, margin finely serrate and somewhat wavy, lateral veins sunken and ending in a margin tooth, underside with maroon or pale pubescence; autumn color yellow or brown.

Twigs: Slender, brown to gray, pubescent; leaf scar half-round with three bundle scars.

Buds: True terminal bud lacking; lateral buds 8 mm long, stalked, with two or three red-maroon valvate scales.

Bark: Brown-gray, mottled, smooth, fluted, sinewy.

Flowers: Imperfect, in separate catkins; mature staminate catkins red-yellow, 8 cm long, dangling; pistillate catkins erect, short, red when open in spring.

Fruit: Nutlet, laterally winged, enclosed in a woody cone 1 cm long, maturing in fall and persisting over winter.

Form: Shrub or small tree up to 6 m (20 ft) in height, forming thickets.

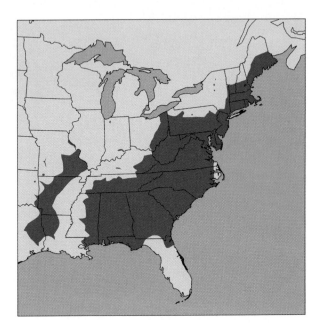

Habitat and Ecology: Shade intolerant; found on stream and pond edges, and in drainage ditches, swamps, and wet woods with a variety of species.

Uses: A nitrogen fixer used in waste site reclamation. Can easily take over pond and stream edges, so landscape value is limited. A browse for moose, beaver, rabbits, muskrat, and white-tailed deer; seed eaten by songbirds; beavers use stems to build dams. Dense thickets valuable as wildlife cover.

Botanical Name: *Alnus* is Latin for "alder"; *serrulata* refers to the finely serrate leaf margin.

Alnus serrulata twig

Alnus serrulata staminate and pistillate flowers

Alnus serrulata bark

Alnus serrulata leaves and fruit

Betula alleghaniensis Britt.

yellow birch, swamp birch, silver birch, gray birch

Quick Guide: *Leaves* alternate, simple, 2 ranked, ovate, base cordate, margin doubly serrate; *Bark* yellow-bronze, shiny, peeling horizontally in thin strips on small trees; *Habitat* moist, cool forests.

Leaves: Simple, alternate, deciduous, 2 ranked, ovate, 8 to 15 cm long, base cordate to inequilateral, apex acute to acuminate, margin sharply doubly serrate, underside and petiole densely to sparsely pubescent; autumn color bright yellow.

Twigs: Slender, flexible, zig-zag, brown-gray to maroon, pubescent or glabrous, with lenticels, weak wintergreen odor and taste when cut. Older branches with spur shoots. Leaf scar widely crescent shaped with three bundle scars.

Buds: True terminal bud lacking; lateral buds acute, up to 1 cm long, more appressed than *Betula lenta*; scales overlapping, green-brown to chestnut brown, shiny, glabrous or pubescent.

Bark: On small trees golden to bronze, gray or maroon and shiny, with horizontal lenticels, peeling horizontally in thin strips; large trees rougher, darker, and scaly.

Flowers: Imperfect, in separate catkins; staminate catkins long and drooping; pistillate catkins shorter, upright, thick; maturing in spring.

Fruit: Laterally winged nutlet in an upright papery cone up to 4 cm long; fruit scale with three-lobed pubescent bracts; maturing in early fall.

Form: Up to 30 m (100 ft) in height and 61 cm (2 ft) in diameter.

Habitat and Ecology: Intermediate shade tolerance; found on cool, moist sites such as mountain slopes, ravines, and coves. It often grows

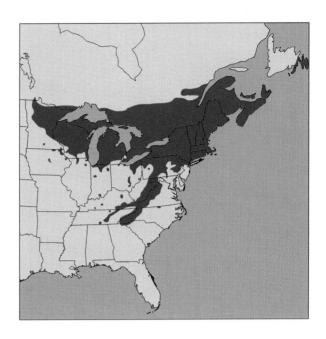

on rotting logs and stumps and will form long, stilted roots. Forest associates include *Abies balsamea, Abies fraseri, Acer pensylvanicum, Acer rubrum, Acer saccharum, Acer spicatum, Aesculus flava, Betula lenta, Betula papyrifera, Fagus grandifolia, Picea rubens, Pinus strobus, Prunus pensylvanica, Prunus serotina, Quercus rubra, Tilia americana,* and *Tsuga canadensis.*

Uses: Wood blond to red-brown, straight grained, hard; used for veneer, plywood, furniture, paneling, turnery, cabinets, and flooring. All the birches used in distillation of wood alcohol. Can be planted as an ornamental in cool climates. A browse for white-tailed deer, elk, and moose; beaver, rabbits, hares, and porcupine eat the bark and spruce grouse eat the catkins; a variety of songbirds and small mammals eat the seeds. The bark is highly flammable.

Botanical Name: *Betula* is Latin for "birch"; *alleghaniensis* refers to the Allegheny Mountains.

Betula alleghaniensis fruit

Betula alleghaniensis leaves

Betula alleghaniensis twig

Betula alleghaniensis bark

Betula alleghaniensis bark

Betula lenta L.

black birch, sweet birch, cherry birch

Quick Guide: *Leaves* alternate, simple, ovate, 2 ranked, base cordate, margin serrate or doubly serrate; *Twigs* glabrous with a strong wintergreen smell when cut; *Bark* black and scaly on large trees; *Habitat* moist, cool forests.

Leaves: Simple, alternate, deciduous, 2 ranked, ovate, 5 to 13 cm long, base cordate or inequilateral, apex acute, margin serrate or doubly serrate, pubescence on veins below, petiole pubescent; autumn color yellow.

Twigs: Slender, zig-zag, flexible, maroon, glabrous, shiny, with lenticels, strong wintergreen odor and taste when cut. Older branches with spur shoots. Leaf scar crescent shaped with three bundle scars.

Buds: Lacking a true terminal bud; lateral buds acute, about 6 mm long, divergent; scales green-brown to yellow-brown, mostly glabrous, shiny, overlapping.

Bark: Maroon to black and smooth with horizontal lenticels on young trees; large trees with gray to black scaly plates.

Flowers: Imperfect, in separate catkins; staminate catkins, long, drooping; pistillate catkins shorter, upright, thick; maturing in spring.

Fruit: Laterally winged nutlet in an upright papery cone up to 4 cm long; fruit scale with three-lobed mostly glabrous bracts; maturing in early fall.

Form: Up to 18 m (60 ft) in height and 61 cm (2 ft) in diameter.

Habitat and Ecology: Shade intolerant; found on moist, cool sites with *Aesculus flava, Acer*

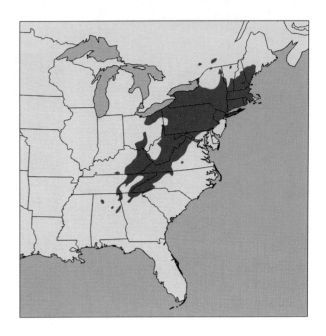

pensylvanicum, Acer rubrum, Acer saccharum, Betula alleghaniensis, Liriodendron tulipifera, Picea rubens, Pinus strobus, Prunus pensylvanica, Prunus serotina, Quercus alba, Quercus rubra, Tsuga canadensis, and *Ulmus americana.*

Uses: Wood blond to red-brown, straight grained, heavy, hard; used for pulpwood, veneer, furniture, paneling, turnery, cabinets, and specialty items. A source of wintergreen oil and birch beer. Planted as an ornamental on moist cool sites and more heat tolerant than *Betula alleghaniensis*. Same wildlife uses as for *Betula alleghaniensis*.

Botanical Name: *Betula* is Latin for "birch"; *lenta* means "pliant," referring to the flexible twigs.

Betula lenta leaves and fruit

Betula lenta twig

Betula lenta bark

Betula lenta bark

Betulaceae

Betula nigra L.
river birch, red birch, water birch

Quick Guide: *Leaves* alternate, simple, 2 ranked, triangular, margin doubly serrate, base truncate; *Bark* with silver, red-brown or black scales, may peel revealing pink or orange inner bark; *Habitat* stream and river margins.

Leaves: Simple, alternate, deciduous, 2 ranked, triangular or ovate, 4 to 10 cm long, base wedge shaped, apex acuminate, margin doubly serrate, underside and petiole pubescent or glabrous; autumn color yellow.

Twigs: Slender, zig-zag, maroon, with lenticels, pubescent or glabrous; leaf scar triangular with three bundle scars.

Buds: True terminal bud lacking; lateral buds conical to triangular, 6 mm long, appressed, with yellow-maroon pubescent or glabrous scales.

Bark: Scaly with silvery, red-brown or gray-brown irregular scales; may peel in wide strips exposing pale pink or orange inner bark on young stems or upper bole.

Flowers: Imperfect, in separate catkins; staminate catkins yellow-green, long and drooping; pistillate catkins upright, red, thick; maturing in spring.

Fruit: Laterally winged nutlet in an upright papery cone up to 4 cm long; fruit scale with three-lobed pubescent bracts; maturing in late spring or early summer.

Form: Up to 24 m (80 ft) in height and 1 m (3 ft) in diameter.

Habitat and Ecology: Shade intolerant; found on edges of rivers and streams, often leaning out

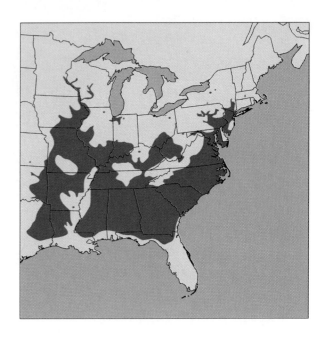

over water. Forest associates include *Acer negundo, Acer rubrum, Acer saccharinum, Celtis laevigata, Celtis occidentalis, Fraxinus* spp., *Platanus occidentalis, Populus deltoides, Quercus lyrata, Quercus michauxii, Quercus nigra, Quercus phellos, Salix nigra,* and *Ulmus americana.*

Uses: Wood blond to red-brown, straight grained, moderately heavy, moderately hard; used for pulpwood and inexpensive furniture. Same wildlife uses as for *Betula alleghaniensis* but not used as extensively. Planted for control of erosion and as an ornamental for the attractive bark and form. Cultivars have been developed to enhance bark colors and disease resistance.

Botanical Name: *Betula* is Latin for "birch"; *nigra* means "dark," referring to the bark.

Betula nigra leaves and fruit

Betula nigra flowers

Betula nigra twig

Betula nigra bark

Betula nigra bark

Betula papyrifera Marsh.

paper birch, canoe birch, white birch

Quick Guide: *Leaves* alternate, simple, ovate, 2 ranked, margin serrate or doubly serrate; *Twigs* with white lenticels; *Bark* chalky white and peeling in horizontal strips; *Habitat* moist sites.

Leaves: Simple, alternate, deciduous, 2 ranked, ovate, 5 to 12 cm long, base rounded to acute, apex acuminate, margin serrate or doubly serrate, glabrous or pubescent on veins below, petiole sometimes pubescent; autumn color yellow.

Twigs: Slender, maroon, with white lenticels, older branches with spur shoots; leaf scar crescent shaped with three bundle scars.

Buds: Lacking a true terminal bud; lateral buds divergent, ovoid-acute, up to 1 cm long; scales green-brown to yellow-brown, resinous, overlapping.

Bark: Maroon-red on very young trees; becoming chalk white and peeling in paperlike strips, large trees with rough dark bark at the base.

Flowers: Imperfect, in separate catkins; staminate catkins, drooping; pistillate catkins shorter, upright, slender; maturing in spring.

Fruit: Laterally winged nutlet in a pendent cone about 4 cm long; fruit scale with three-lobed minutely pubescent bracts; maturing in fall.

Form: Up to 21 m (70 ft) in height and 60 cm (2 ft) in diameter.

Habitat and Ecology: Shade intolerant; found on a wide variety of sites but commonly on moist sites with *Abies balsamea, Acer saccharum, Betula*

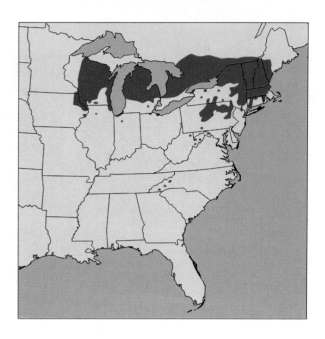

alleghaniensis, Fraxinus nigra, Picea mariana, Picea rubens, Pinus resinosa, Pinus strobus, and *Populus grandidentata.*

Uses: Wood blond to red-brown, straight grained, heavy, hard; used for pulpwood, veneer, furniture, paneling, turnery, cabinets, toothpicks, and specialty items. The waterproof bark was used by Native Americans to sheath canoes. A popular ornamental in cool climates for the attractive bark but susceptible to the bronze birch borer, gypsy moth, and birch leaf miner. Same wildlife uses as for *Betula alleghaniensis* and also a favorite browse of deer and moose.

Botanical Name: *Betula* is Latin for "birch"; *papyrifera* refers to the bark.

Betula papyrifera bark

Betula papyrifera flowers

Betula papyrifera fruit

Betula papyrifera bark

Betula papyrifera leaf

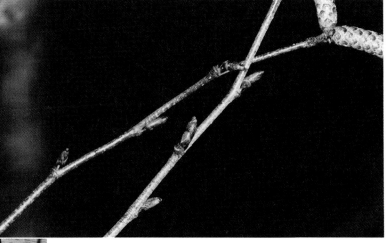
Betula papyrifera twigs

Betulaceae

Betula populifolia Marsh.

gray birch, wire birch, white birch, aspen-leaved birch, old-field birch

Quick Guide: *Leaves* alternate, simple, triangular, apex tapering to a long point, margin doubly serrate; *Bark* chalky white and not peeling; *Habitat* common on disturbed areas and in old fields.

Leaves: Simple, alternate, deciduous, 2 ranked, drooping, triangular, 5 to 10 cm long, base truncate, apex long-acuminate and sometimes curved, margin doubly serrate, glandular above, mostly glabrous below, petiole long and often twisted; autumn color yellow.

Twigs: Slender, maroon-brown to gray, with resin glands and white warty lenticels; leaf scar crescent-shaped with three bundle scars.

Buds: Lacking a true terminal bud; lateral buds ovoid-acute, up to 8 mm long, divergent; scales brown, resinous, overlapping.

Bark: Maroon to black with lenticels and cracked on young trees; becoming chalky white or grayish with horizontal black lenticels, usually not peeling; large trees with rough, dark patches below branches.

Flowers: Imperfect, in separate catkins; staminate catkins long, drooping; pistillate catkins shorter, upright, slender; maturing in spring.

Fruit: Laterally winged nutlet in a pendent cone about 2 cm long; fruit scale with three-lobed, minutely pubescent bracts; maturing in fall.

Form: A small, usually multistemmed tree with poor form up to 12 m (40 ft) in height.

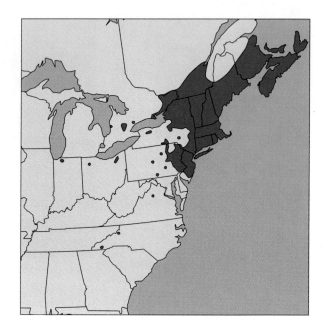

Habitat and Ecology: Shade intolerant; found on a variety of sites but common on disturbed areas, poor soils, and water edges. Forest associates on more mesic sites include *Acer nigrum*, *Acer rubrum*, *Betula lenta*, *Betula papyrifera*, and *Picea mariana*.

Uses: Wood blond to red-brown, light, soft; used for turnery and specialty items. Planted for site reclamation and as an ornamental for the interesting bark but susceptible to leaf miner damage. Wildlife uses as for *Betula alleghaniensis*.

Botanical Name: *Betula* is Latin for "birch"; *populifolia* means "poplar-like leaves."

Betula populifolia bark

Betula populifolia fruit

Betula populifolia twig

Betula populifolia bark

Betula populifolia leaves

Carpinus caroliniana Walt.

hornbeam, blue beech, musclewood, ironwood, American hornbeam

Quick Guide: *Leaves* alternate, simple, 2 ranked, thin, margin doubly serrate; *Fruit* a nutlet attached to a three-lobed leafy bract; *Buds* maroon-white striped; *Bark* smooth, gray, sinewy, and fluted; *Habitat* moist to wet sites.

Leaves: Simple, alternate, deciduous, 2 ranked, thin, ovate to elliptical, 5 to 10 cm long, base cordate, apex acute to acuminate, margin doubly serrate, petiole and underside pubescent; autumn color dull orange-red to brown.

Twigs: Slender, zig-zag, maroon, pubescent or glabrous, with white lenticels; leaf scar crescent shaped to round with three bundle scars.

Buds: True terminal bud lacking; lateral buds 4 mm long, ovoid; scales maroon-white striped, overlapping.

Bark: Thin, gray or blue-gray, smooth, fluted, sinewy.

Flowers: Imperfect, in separate catkins with the leaves; staminate catkins up to 4 cm long and drooping; pistillate flowers with forked red stigmas, obscure, at tips of new shoots (see page 125).

Fruit: Nutlet in a three-lobed leafy bract, hanging in pairs, maturing in late summer.

Form: Understory tree reaching 9 m (30 ft) in height.

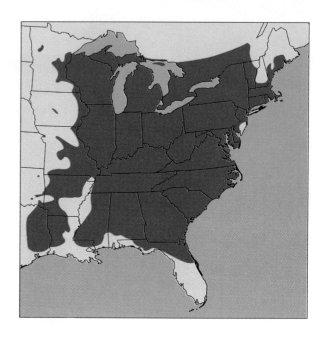

Habitat and Ecology: Shade tolerant; found near streams, in floodplains and swamps, and on cool slopes in many forest cover types.

Uses: Wood whitish, hard, dense; used for specialty items such as tool handles and bowls. Browse for white-tailed deer, rabbit, and beaver, and seed eaten by songbirds, wild turkey, squirrels, and rodents. Can be planted as an ornamental in the shade or full sun.

Botanical Name: *Carpinus* is Latin for "hornbeam"; *caroliniana* refers to the geographic range.

Carpinus caroliniana leaves

Carpinus caroliniana fruit

Carpinus caroliniana twig

Carpinus caroliniana fruit

Carpinus caroliniana bark

Betulaceae

Ostrya virginiana (Mill.) K. Koch
hophornbeam, eastern hophornbeam, American hophornbeam, ironwood

Quick Guide: *Leaves* alternate, simple, 2 ranked, thin, margin doubly serrate, lateral veins dividing at the leaf margin; *Fruit* hoplike; *Buds* maroon or green-brown striped; *Bark* red-brown and shreddy; *Habitat* a variety of sites.

Leaves: Simple, alternate, deciduous, 2 ranked, thin, ovate to lanceolate, 5 to 13 cm long, base cordate, apex acute or acuminate, margin doubly serrate, lateral veins dividing near the leaf margin, underside and petiole pubescent; autumn color yellow.

Twigs: Red-brown, slender, zig-zag, pubescent or glabrous; leaf scar crescent shaped to round with three bundle scars.

Buds: True terminal bud lacking; lateral buds divergent, ovoid-acute, 6 mm long; scales green-brown striped or red-brown, overlapping.

Bark: Gray-brown to red-brown and scaly or shreddy.

Flowers: Imperfect, in separate catkins; staminate catkins up to 7 cm long and drooping; pistillate catkins with forked red stigmas, obscure, at tips of new shoots in spring (see page 125).

Fruit: Nutlet enclosed in a hoplike inflated sac 2.0 cm long, maturing in late summer.

Form: Understory tree usually only up to 12 m (40 ft) in height.

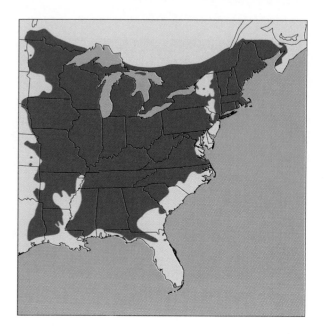

Habitat and Ecology: Shade tolerant; an understory tree found on a wide variety of sites and soils.

Uses: Wood whitish, hard; used for mallets, tool handles, specialty items, and fenceposts. Seed eaten by wild turkey, ruffed grouse, songbirds, and small to midsize mammals; occasionally browsed by white-tailed deer; buds and catkins eaten in winter by many birds, squirrels, and rabbits. Can be planted as an ornamental in the shade or full sun.

Botanical Name: *Ostrya* is Latin for "a tree with hard wood"; *virginiana* refers to the geographic range.

Ostrya virginiana fruit

Ostrya virginiana twig

Ostrya virginiana bark

Ostrya virginiana leaves

Betulaceae

Catalpa bignonioides Walt.

southern catalpa, cigar-tree, Indian-bean, common catalpa, fish-bait tree

Quick Guide: *Leaves* whorled, simple, large, heart shaped, margin entire, underside pubescent; *Flowers* white with maroon and yellow spots; *Fruit* a long beanlike capsule; *Twigs* stout with moon crater leaf scars; *Bark* brown-gray and scaly; *Habitat* moist, rich soils.

Leaves: Simple, whorled or opposite, deciduous, heart shaped, blade 13 to 20 cm long, base cordate or rounded, apex acute or abruptly acuminate, margin entire or sometimes lobed at the base petiole long (13 cm), underside pubescent; autumn color yellow.

Twigs: Stout, light brown, glabrous or lightly pubescent, with lenticels. Leaf scar round, raised, resembling moon craters, varying in size at a given node, bundle scars in a circular pattern.

Buds: True terminal bud lacking; lateral buds moundlike, embedded, with light brown scales.

Bark: Gray-brown and scaly.

Flowers: Perfect, trumpetlike, petals white with maroon and yellow spots, in large erect panicles in early summer.

Fruit: Capsule, green, 2 valved, beanlike, usually up to 30 cm long but sometimes longer, containing seeds with pointed fringed wings (see page 146), maturing in early fall and persisting over winter.

Form: Up to 15 m (50 ft) in height with a low open crown.

Habitat and Ecology: Shade intolerant; in its natural range found on rich soils of riverbanks,

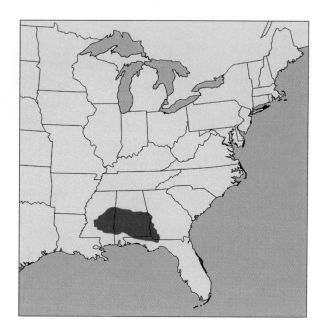

floodplains, and swamp edges. Forest associates include *Acer negundo, Betula nigra, Carya aquatica, Platanus occidentalis, Quercus nigra, Quercus phellos,* and *Ulmus americana.* Its range has spread throughout the eastern United States.

Uses: Wood grayish, soft, durable, coarse grained; used for railroad ties, fenceposts, and specialty items such as picture frames, and was used for telegraph poles. The catalpa sphinx larva that feed on the leaves are used for fish bait (catawba worms). Cultivars are planted as ornamentals for the attractive leaves and flowers and tolerance of stress, but fruit can be messy.

Botanical Name: *Catalpa* is from a Cherokee Indian name; *bignonioides* refers to the flower, which is similar to the trumpet-flower in the genus *Bignonia.*

Catalpa bignonioides fruit

Catalpa bignonioides leaf

Catalpa bignonioides twig

Catalpa bignonioides bark

Catalpa bignonioides flowers

Bignoniaceae

Catalpa speciosa Warder ex Engelm.

northern catalpa, common catalpa, cigar-tree, Indian-bean

Quick Guide: Similar to *Catalpa bignonioides* but a larger tree and *Leaves* lacking an unpleasant odor when crushed; *Bark* ridged; *Flowers* blooming a few weeks earlier and with fewer purple spots; and *Fruit* longer.

Leaves: Simple, whorled or opposite, deciduous, heart shaped, up to 35 cm long, apex acuminate, base cordate, underside pubescent, margin entire or occasionally with a large dentate tooth; autumn color yellow.

Twigs: Similar to *Catalpa bignonioides*.

Buds: Similar to *Catalpa bignonioides*.

Bark: Red-brown to brown-gray and furrowed with flat sometimes scaly ridges.

Flowers: Similar to *Catalpa bignonioides*.

Fruit: Similar to *Catalpa bignonioides* but up to 46 cm long.

Form: Up to 24 m tall (80 ft) with a low crown.

Habitat and Ecology: Shade intolerant; original range was very small but now found throughout the

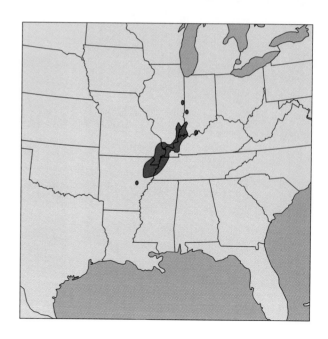

eastern United States on rich, moist soils and around old home sites.

Uses: Similar to *Catalpa bignonioides*.

Botanical Name: *Catalpa* is from a Cherokee Indian name; *speciosa* is Latin for "showy," referring to the flowers.

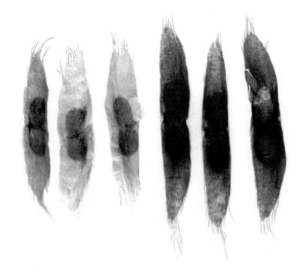

Seeds of *Catalpa speciosa* (left) and *Catalpa bignonioides* (right)

Catalpa speciosa leaf

Catalpa speciosa fruit

Catalpa speciosa bark

Catalpa speciosa flowers

Bignoniaceae

Caesalpiniaceae Guide

1. Leaves simple, heart shaped, margin entire, petiole swollen at both ends; twigs dark and zig-zag; flowers in pink showy clusters in early spring; bark red-brown to black, shallowly ridged or scaly. *Cercis canadensis*
2. Leaves pinnately or bipinnately compound, leaflets 12 to 20 per leaf; twigs with branched thorns; fruit kidney shaped; bark with clumps of thorns; swamps. *Gleditsia aquatica*
3. Leaves pinnately or bipinnately compound, leaflets 14 to 30 per leaf; twigs with branched thorns; fruit long and often twisted; bark with clumps of thorns; moist soils. *Gleditsia triacanthos*
4. Leaves bipinnately compound, leaflets up to 70 per leaf and alternate on the rachis; branches stout, crooked, and flaky; fruit large and leathery; bark with raised scaly ridges. *Gymnocladus dioicus*

Gleditsia triacanthos flowers

Gleditsia triacanthos leaves

Gleditsia triacanthos fruit

Cercis canadensis flowers

Gymnocladus dioicus flowers

Caesalpiniaceae

Cercis canadensis L.

eastern redbud, Judas-tree, redbud

Quick Guide: *Leaves* alternate, simple, heart-shaped, glabrous, margin entire, petiole swollen at both ends; *Flowers* in pink showy clusters in early spring; *Fruit* podlike; *Bark* red-brown to black, shallowly ridged or scaly; *Habitat* a variety of sites from dry to moist.

Leaves: Simple, alternate, deciduous, kidney or heart shaped, blade 8 to 12 cm long, palmately veined, apex abruptly acute, base cordate, margin entire, petiole up to 9 cm long and swollen at both ends, underside glabrous; autumn color yellow.

Twigs: Zig-zag, slender, red-brown to black, glabrous, with lenticels; leaf scar heart shaped, raised, with three bundle scars.

Buds: True terminal bud lacking; lateral buds maroon to black, 2 mm long, triangular, appressed; flower buds plump and above the leaf buds.

Bark: Red-brown to black and smooth on small trees; shallowly ridged or scaly on large trees.

Flowers: Perfect, similar to sweet pea, about 1 cm long, petals rose pink to purplish, clustered along the branch before the leaves (see page 149).

Fruit: Legume, flat, red-brown to black, up to 10 cm long, maturing in late summer.

Form: Understory tree up to 12 m (40 ft) in height with a low open crown.

Habitat and Ecology: Shade tolerant; commonly found on rich moist soils especially near

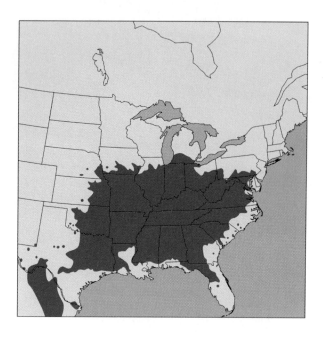

streams but also on drier upland sites with a wide variety of species.

Uses: Seed eaten by birds (including northern bobwhite), white-tailed deer, and small mammals; flowers used in salads. Popular ornamental because of the early spring flowers that bloom even on small trees and tolerance of a range of conditions. Cultivars offer more deeply colored flowers, white flowers, and purple foliage.

Botanical Name: *Cercis* is the classic Greek name for the Judas-tree of southern Europe and Asia; *canadensis* refers to the New World. Also grouped in Fabaceae.

Caesalpiniaceae

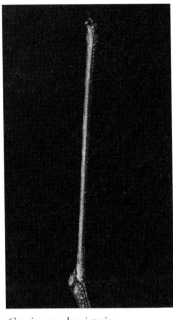
Cercis canadensis twig

Cercis canadensis flowers

Cercis canadensis leaf

Cercis canadensis bark

Cercis canadensis bark

Cercis canadensis fruit

Gleditsia aquatica Marsh.
waterlocust

Quick Guide: Similar to *Gleditsia triacanthos* but *Fruit* kidney shaped; *Habitat* swamps and floodplains.

Leaves: Similar to *Gleditsia aquatica* but leaves mostly glabrous and with 12 to 20 leaflets.

Twigs: Similar to *Gleditsia aquatica* but thorns less frequently branched.

Buds: Similar to *Gleditsia aquatica*.

Bark: Mottled gray-brown, mostly smooth with warty lenticels and clumps of thorns.

Flowers: Similar to *Gleditsia aquatica*.

Fruit: Legume, kidney shaped, flat, up to 5 cm long with one to three seeds, lacking pulp around the seeds.

Form: Up to 20 m (66 ft) tall.

Habitat and Ecology: Found in swamps, hammocks, and floodplains in association with *Fraxinus caroliniana*, *Fraxinus profunda*, *Nyssa aquatica*, *Planera aquatica*, and *Taxodium distichum* var. *distichum*.

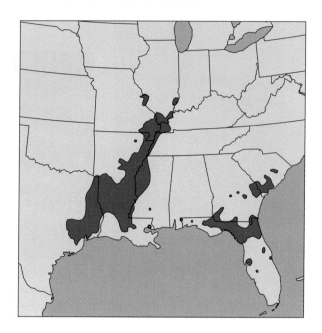

Uses: The durable wood is sometimes used for specialty items and furniture.

Botanical Name: *Gleditsia* is in honor of the botanist Johann Gleditsch; *aquatica* refers to the habitat. Also grouped in Fabaceae.

Gleditsia aquatica fruit

Gleditsia aquatica leaves

Gleditsia aquatica bark

Gleditsia aquatica bark

Gleditsia aquatica leaves and twig

Caesalpiniaceae

Gleditsia triacanthos L.

honeylocust, thorny-locust, three-thorned acacia, sweet-locust

Quick Guide: *Leaves* alternate, pinnately or bipinnately compound, leaflets up to 30 in number; *Twigs* with branched thorns; *Fruit* podlike, long, and often twisted; *Bark* often with clumps of thorns, gray-brown, smooth or with large scaly plates; *Habitat* moist soils.

Leaves: Pinnately or bipinnately compound, alternate, deciduous, up to 20 cm long. Leaflets dark green, shiny, 14 to 30 in number, elliptical, about 2 cm long, subopposite to alternate on the rachis, margin entire to lightly toothed, sessile, underside pubescent; rachis pubescent; autumn color yellow (see page 148).

Twigs: Stout, zig-zag, green to red-brown, glabrous, with lenticels and branched thorns; leaf scar heart shaped with three bundle scars.

Buds: True terminal bud lacking; lateral buds almost completely embedded, sunken.

Bark: Gray-brown and smooth on small trees; large trees with large loose plates. Small and large trees often with dense clusters of thorns protruding from the bark.

Flowers: Perfect and imperfect, with five yellow-green petals and greatly exserted stamens, in dangling racemes in late spring after the leaves (see page 149).

Fruit: Legume, red-brown, flat, up to 45 cm long, often twisted or curved (see page 149); pulp around the kidney-shaped flattened seeds is sweet in late summer when the pod is yellow-brown.

Form: Up to 24 m (80 ft) in height and 1 m (3 ft) in diameter.

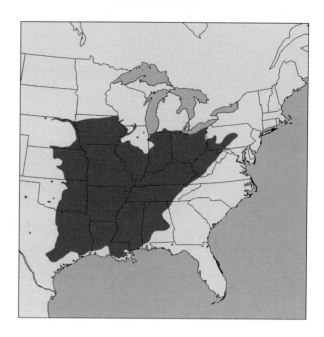

Habitat and Ecology: Shade intolerant; naturalized throughout eastern areas and found on a variety of sites but commonly near streams and in bottomlands.

Uses: Wood red-brown to yellow, very heavy, very hard, durable; used for fenceposts, furniture and specialty items, and was used for wheel hubs. Fruit eaten by birds (including northern bobwhite), small mammals, and cattle; entire pods relished by white-tailed deer; flowers popular with bees. Thornless cultivars with enhanced disease resistance are commonly planted.

Botanical Name: *Gleditsia* is in honor of the botanist Johann Gleditsch; *triacanthos* refers to the three-branched thorns. Also grouped in Fabaceae.

Gleditsia triacanthos fruit

Gleditsia triacanthos flowers

Gleditsia triacanthos twig

Gleditsia triacanthos bark

Gleditsia triacanthos bark

Caesalpiniaceae

Gymnocladus dioicus (L.) K. Koch
Kentucky coffeetree

Quick Guide: *Leaves* alternate, mostly bipinnately compound, leaflets up to 70 in number and alternate on the rachis; *Twigs* stout, crooked, and mottled with two to three superposed lateral buds; *Fruit* large, leathery, and podlike, persisting over winter; *Bark* gray-brown with scaly, ridges; *Habitat* moist, fertile soils.

Leaves: Bipinnately compound, alternate, deciduous, on average 50 cm long but up to 1 m long. Leaflets up to 70, ovate, 3 to 8 cm long, alternate on the rachis, apex acute, base rounded to inequilateral, margin entire, underside glabrous; autumn color yellow.

Twigs: Stout, stiff, crooked, mottled, with white patches, and with lenticels; leaf scar large, heart shaped, with up to five bundle scars.

Buds: True terminal bud lacking; lateral buds round, 3 mm long, with silky chestnut hairs, often widely superposed, almost completely embedded.

Bark: Gray-brown with raised, scaly ridges.

Flowers: Species is dioecious, petals green-white and widely splayed, in erect terminal clusters with the leaves (see page 149).

Fruit: Legume, purple to red-brown, flat, wide, curved, leathery, up to 25 cm long and 5 cm wide; seeds large (up to 3 cm long), dark, hard, rounded and somewhat flattened; maturing in fall and persistent over winter.

Form: Up to 30 m (100 ft) in height with a narrow crown.

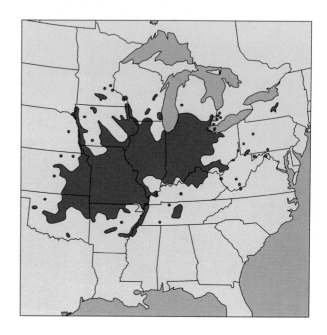

Habitat and Ecology: Found on moist fertile soils, with *Acer saccharum*, *Carya ovata*, *Cladrastis kentukea*, *Fagus grandifolia*, *Fraxinus quadrangulata*, and *Juglans nigra*.

Uses: Wood red-brown to yellow, strong, heavy, hard; used for fencing, furniture, and cabinets. Seeds were roasted and used as a coffee substitute, but raw seeds are poisonous. Foliage and seeds poisonous to livestock. Planted as an ornamental for the bark, form, and stress tolerance.

Botanical Name: *Gymnocladus* is from Greek for "naked branch"; *dioicus* means "dioecious." Also grouped in Fabaceae.

Gymnocladus dioicus fruit

Gymnocladus dioicus leaf

Gymnocladus dioicus twig

Gymnocladus dioicus bark

Gymnocladus dioicus bark

Caesalpiniaceae

Caprifoliaceae Guide

1. Leaves pinnately compound, leaflets usually 5 to 7, rachis grooved and pubescent. *Sambucus canadensis*
2. Similar to *Sambucus canadensis* but leaflets with a downy underside and fruit bright red. *Sambucus pubens*
3. Leaves simple, obovate to oval, margin finely serrate, underside with rusty pubescence. *Viburnum rufidulum*
4. Leaves simple, elliptical to oval, margin finely serrate, underside mostly glabrous. *Viburnum prunifolium*
5. Leaves simple, obovate, margin mostly entire, underside glandular or pubescent, nearly sessile. *Viburnum obovatum*
6. Leaves simple, elliptical, margin entire, petiole and midrib rusty. *Viburnum nudum*
7. Leaves simple, elliptical, margin irregularly serrate, underside lacking rusty pubescence. *Viburnum cassinoides*
8. Leaves simple, nearly round, margin serrate, underside with pubescence. *Viburnum alnifolium*
9. Leaves simple, ovate, margin sharply serrate, apex often acuminate, underside with dark dots, petiole winged. *Viburnum lentago*
10. Leaves simple, nearly round, margin dentate. *Viburnum dentatum*
11. Leaves simple, palmately three lobed, margin coarsely toothed. *Viburnum acerifolium*
12. Leaves simple, palmately three lobed, margin coarsely toothed except at the base and in the sinuses, petiole with glands. *Viburnum trilobum*

Viburnum obovatum leaves and fruit

Viburnum alnifolium leaf

Sambucus pubens fruit

Viburnum cassinoides leaves

Sambucus canadensis L.

elderberry, American elder, common elder, white elder

Quick Guide: *Leaves* opposite, pinnately compound, leaflets 5 to 7 but up to 11, rachis grooved and pubescent; *Twigs* with warty lenticels; *Flowers* in white flat-topped clusters in early summer; *Fruit* a juicy purple drupe in drooping clusters in late summer; *Habitat* water edges and other wet sites.

Leaves: Pinnately compound, opposite, deciduous, up to 30 cm long, rachis grooved and pubescent. Leaflets mostly 5 to 7 but up to 11, lowest leaflets sometimes lobed or bipinnately compound, elliptical to lanceolate or ovate, 4 to 12 cm long, margin serrate; autumn color yellow.

Twigs: Stout, gray, with warty lenticels, glabrous, pith very large, emitting an unpleasant odor when bruised. Leaf scar crescent shaped with five bundle scars, opposite leaf scars connected.

Buds: True terminal bud lacking; lateral buds ovoid, 2 mm long, with red-brown scales.

Bark: Gray-brown and smooth with warty lenticels, becoming rough on large stems.

Flowers: Perfect, white, in flat-topped cymes in early summer.

Fruit: Drupe, juicy, dark purple, 6 mm wide, appearing in drooping clusters in late summer.

Form: Small tree up to 6 m (20 ft) in height, more commonly a sprawling shrub.

Habitat and Ecology: Shade intolerant; found on a variety of wet sites including edges of streams, ponds and swamps, and in drainage ditches and bottoms.

Uses: Fruit eaten by numerous songbirds and game birds including northern bobwhite, wild turkey, ruffed grouse, ring-necked pheasant, and mourning dove; small mammals also eat the fruit and white-tailed deer occasionally browse the foliage; flowers popular with bees and butterflies; provides nesting cover for birds; fruit used in preserves and wine, and flowers used in teas. Can be used to naturalize wet areas.

Botanical Name: *Sambucus* is derived from the Greek word *Sambuke,* referring to the hollow stems used as musical wind instruments; *canadensis* refers to the New World. Also named *Sambucus nigra* spp. *canadensis* L. (L.) R. Bolli.

Similar Species: Scarlet elder (*Sambucus pubens* Michx.) is a shrub found on moist, cool sites in the northeast and as far south as northern Georgia and has leaves with a downy underside when young and bright red drupes (see page 159).

Sambucus canadensis flowers

Sambucus canadensis fruit

Sambucus canadensis bark

Sambucus canadensis leaves

Viburnum lentago L.

nannyberry, sheepberry, nannyplum, sweet viburnum

Quick Guide: *Leaves* simple, opposite, ovate, sharply serrate, underside with dark spots, petiole winged and with small teeth; *Flower buds* up to 3 cm long, purplish and swollen at the base; *Fruit* a blue-black drupe in red-stalked clusters; *Habitat* moist, fertile soils.

Leaves: Simple, opposite, deciduous, ovate to elliptical, 5 to 16 cm long, shiny, apex acute to acuminate, base rounded, margin with sharp teeth, underside with small dark dots, petiole with wavy wings and small teeth; autumn color orange to red.

Twigs: Slender, green-brown to gray, glabrous or scurfy-pubescent, with lenticels; leaf scar crescent-shaped with three bundle scars.

Buds: Terminal bud up to 3 cm long; flower buds the largest, very long-pointed and bulbous at the base; scales valvate, pale brown to purplish, scurfy-pubescent.

Bark: Red-brown to dark gray and scaly with orange lenticels; on large trees becoming gray-brown with blocky or scaly plates.

Flowers: White, fragrant, in large rounded clusters (cymes) at branch tips in spring.

Fruit: Drupe, nearly round, blue-black, about 1 cm wide, sweet, in red-stalked clusters in early autumn, persisting over winter.

Form: A shrub or small crooked tree up to 10 m (30 ft) in height.

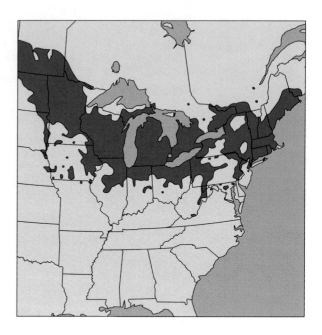

Habitat and Ecology: Shade tolerant; found on rich, moist soils of riverbanks and lakeshores and tolerant of low pH. Sometimes found on drier sites.

Uses: Wood brown-yellow, heavy, and hard. Planted as an ornamental but forms root sprouts and susceptible to mildew. Fruit eaten by many species of birds and small and large mammals; leaves and twigs browsed by deer, moose, and beaver; provides shelter for wildlife.

Botanical Name: *Viburnum* is the Latin name of the wayfaringtree of Eurasia; *lentago* means "flexible," referring to the arching branches.

Viburnum lentago immature fruit

Viburnum lentago flower bud

Viburnum lentago flowers

Viburnum lentago bark

Viburnum lentago leaves

Caprifoliaceae

Viburnum nudum L.
possumhaw, swamphaw

Quick Guide: *Leaves* opposite, simple, elliptical, margin mostly entire, midrib and petiole rusty scurfy; *Twigs* rusty scurfy; *Fruit* a blue-black drupe; *Bark* smooth; *Habitat* moist sites.

Leaves: Simple, opposite, deciduous, elliptical to obovate, shiny green, 6 to 15 cm long, apex and base acute to rounded, margin mostly entire or wavy, midrib rusty scurfy, petiole rusty scurfy; autumn color purple-red.

Twigs: Slender, brown, rusty scurfy when young, with lenticels; leaf scar V-shaped with three bundle scars.

Buds: Terminal bud long and narrow, about 1.0 cm long; scales valvate and pale rusty scurfy.

Bark: Gray-brown and smooth except on very large stems.

Flowers: Perfect, white, in flat-topped clusters in spring.

Fruit: Drupe, nearly round, blue-black when mature, about 1 cm wide; maturing in fall.

Form: A shrub or small tree to 6 m (20 ft) in height.

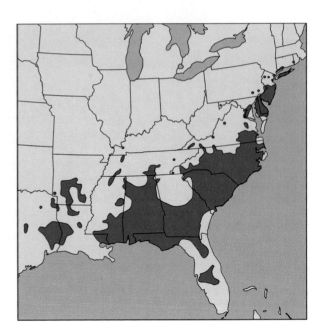

Habitat and Ecology: Found on moist soils of swamp and stream margins and cool forest slopes.

Uses: Similar to *Viburnum rufidulum*.

Botanical Name: *Viburnum* is the Latin name of the wayfaringtree of Eurasia; *nudum* means "bare, naked."

Viburnum nudum twig

Viburnum nudum leaves

Viburnum nudum bark

Viburnum nudum fruit

Caprifoliaceae

Viburnum prunifolium L.
blackhaw

Quick Guide: *Leaves* opposite, simple, elliptical to oval, margin finely serrate, underside mostly glabrous; *Twigs* glaucous; *Fruit* a blue-black drupe; *Bark* with scaly ridges; *Habitat* well-drained soils.

Leaves: Simple, opposite, deciduous, elliptical to oval, ovate or obovate, shiny green, 5 to 8 cm long, apex and base acute to obtuse, margin finely serrate, underside mostly glabrous, petiole pubescent; autumn color red or purple.

Twigs: Slender, stiff, green-brown to purplish, glaucous, with lenticels; leaf scar crescent-shaped with three bundle scars.

Buds: Terminal bud conical, flower bud up to 1.5 cm long, purplish, with two valvate and scurfy scales.

Bark: Gray-brown to black with blocky, scaly ridges.

Flowers: Perfect, white, in flat-topped clusters in spring.

Fruit: Drupe, elliptical, blue-black, about 1 cm long, in drooping clusters in early summer.

Form: A shrub or small tree to 8 m (26 ft) in height.

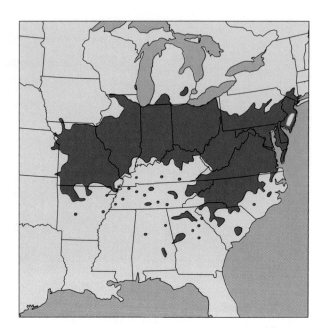

Habitat and Ecology: Found on dry soils on a variety of sites including margins of woods and along fences.

Uses: Similar to *Viburnum rufidulum*.

Botanical Name: *Viburnum* is the Latin name of the wayfaringtree of Eurasia; *prunifolium* refers to the plumlike leaves.

Viburnum prunifolium immature fruit

Viburnum prunifolium twig

Viburnum prunifolium leaves

Viburnum prunifolium bark

Caprifoliaceae

Viburnum rufidulum Raf.

rusty blackhaw, bluehaw, rusty nannyberry, rusty blackhaw viburnum

Quick Guide: *Leaves* opposite, simple, oval to obovate, margin finely serrate, underside rusty pubescent; *Twigs* with rusty pubesence; *Fruit* a blue drupe in red-stalked clusters; *Bark* blocky; *Habitat* a variety of sites.

Leaves: Simple, opposite, deciduous, shiny green, oval to obovate, 5 to 10 cm long, base cuneate to rounded, apex rounded or notched, margin finely serrate, underside with rusty pubescence; autumn color purple-red. Petiole grooved and red with rusty pubescence, often winged.

Twigs: Slender, green-brown, with rusty pubescence, becoming red-brown to gray and glabrous, with lenticels; leaf scar U-shaped with three bundle scars.

Buds: Terminal bud 1 cm long, with two valvate scales covered with rusty scurfy pubescence.

Bark: Dark brown on small trees; blocky on large trees.

Flowers: Perfect, white, in flat-topped clusters in early spring.

Fruit: Drupe, nearly round, turning from red to dark blue, 1 cm long, red stalked, appearing in drooping clusters in early summer.

Form: Shrub or small tree up to 6 m (20 ft) in height.

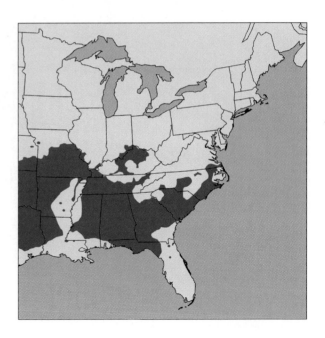

Habitat and Ecology: Found on moist to dry sites, mostly in well-drained woodlands.

Uses: Moderately browsed by white-tailed deer, moose, and beaver; fruit eaten by numerous songbirds and game birds including northern bobwhite, pheasant, and wild turkey; fruit also eaten by coyote, white-tailed deer, beaver, skunk, and small mammals. The viburnums are attractive landscape trees and shrubs, and many cultivars are available.

Botanical Name: *Viburnum* is the Latin name of wayfaringtree of Eurasia; *rufidulum* (rufus) refers to the red-brown pubescence on leaves, twigs, and buds.

Viburnum rufidulum flowers

Viburnum rufidulum twig

Viburnum rufidulum leaves and flowers

Viburnum rufidulum bark

Caprifoliaceae

Other Viburnums

Viburnum acerifolium L.

Maple-leaved viburnum is found on well-drained soils in the eastern United States as far south as Georgia and has palmately three-lobed leaves 5 to 12 cm long that are coarsely toothed and maplelike.

Viburnum alnifolium Marsh.

Hobblebush or witch hobble is found on moist, cool sites in the eastern United States as far south as the southern Appalachians. It is distinguished by cordate leaves 5 to 18 cm long with a serrate or doubly serrate margin and scurfy-brown pubescence on the underside, and buds that are pale, hairy, and curved (see page 159). Also named *Viburnum lantanoides* Michx.

Viburnum cassinoides L.

Witherod or wild raisin is found from Maine to Alabama on moist to wet sites and has elliptical leaves 5 to 15 cm long with an irregularly serrate or dentate-crenate margin and flattened, orange scurfy buds (see page 159). Also named *Viburnum nudum* var. *cassinoides* (L.) Torr. & Gray.

Viburnum dentatum L.

Arrowwood is found throughout the eastern United States on a variety of sites and has nearly round leaves 5 to 10 cm long with a dentate or serrate teeth on the margin.

Viburnum obovatum Walt.

Small viburnum or small-leaf viburnum is found on wet sites in the Coastal Plain from Virginia to Florida and has obovate leaves up to 5 cm long that are sessile or nearly so with margins entire or dentate-crenate above the middle and a glandular or minutely pubescent underside (see page 159).

Viburnum trilobum Marsh.

Highbush cranberry is found in New England and is similar to *Viburnum acerifolium* but found on wet sites and has leaves lacking margin teeth in the sinuses and at the base and has glands on the grooved petiole near the blade. The fruit is orange-red rather than blue-black. Also named *Viburnum opulus* var. *americanum* L. Ait.

Viburnum dentatum leaves

Viburnum acerifolium flowers

Viburnum acerifolium leaves

Viburnum dentatum flowers

Caprifoliaceae

Euonymus atropurpureus Jacq.

eastern wahoo, wahoo, burningbush, spindletree

Quick Guide: *Leaves* opposite, simple, elliptical, finely toothed; *Twigs* bright green and 4 angled when young; *Fruit* a pink to red capsule with bright red seeds; *Habitat* a variety of sites.

Leaves: Simple, opposite, elliptical, 5 to 12 cm long, apex short acuminate, base acute, margin finely serrate, underside glabrous or lightly pubescent; autumn color pink, bright red, or yellow.

Twigs: Slender, bright green the first year, 4 sided when young, with lenticels; leaf scar pale, shield shaped, with one bundle scar.

Buds: Acute, with green to purple overlapping scales.

Bark: Green to gray with vertical streaks or splits.

Flowers: Perfect, with four dark purple petals, appearing in May or June.

Fruit: Capsule, four lobed, 1 to 2 cm wide, pale pink to red or purplish, with bright red seeds, maturing in fall.

Form: A shrub or small tree to 8 m (25 ft) tall.

Habitat and Ecology: Found on a variety of sites with well-drained soils including open woods, stream edges, and wooded slopes.

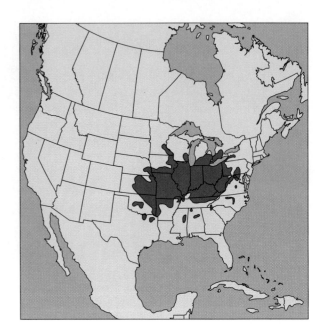

Uses: Fruit eaten by birds such as American robin, cardinal, cedar waxwing, and turkey. Foliage browsed by deer. Used in landscaping for the attractive fruit and colorful fall foliage.

Botanical Name: *Euonymus* means "of good name"; *atropurpureus* means "dark purple," referring to the flowers or fruit. Also named *Euonymus atropurpurea* Jacq.

Euonymus atropurpureus leaves

Euonymus atropurpureus immature fruit

Cornaceae Guide

1. Leaves alternate or appearing opposite or whorled, long petioled, with arcuate venation; flowers in white flat-topped clusters; fruit a purple-blue drupe on bright red stalks; young bark green with white streaks. *Cornus alternifolia*
2. Leaves clearly opposite or whorled, with arcuate venation; flowers with four petal-like white bracts; fruit a shiny red drupe; bark blocky. *Cornus florida*
3. Similar to *Cornus florida*, but flowers in white flat-topped clusters; fruit a purple-blue drupe in long-stalked clusters; pith white; bark ridged rather than blocky. *Cornus stricta*
4. Similar to *Cornus stricta*, but leaf underside and twig with pale or rusty pubescence; pith brown. *Cornus amomum*
5. Similar to *Cornus stricta*, but upper leaf surface and twig with stiff hairs; fruit white. *Cornus drummondii*
6. Similar to *Cornus stricta*, but leaves densely pubescent below; twigs gabrous; fruit white. *Cornus racemosa*
7. Similar to *Cornus stricta*, but twigs blood red and fruit white. *Cornus stolonifera*
8. Similar to *Cornus stricta*, but leaves nearly round with dense pubescence below. *Cornus rugosa*
9. Leaves alternate, most leaves greater than 12 cm long; fruit a long-stalked, large, juicy drupe with a longitudinal ridged stone. *Nyssa aquatica*
10. Leaves alternate, most leaves greater than 12 cm long, apex more rounded than in *Nyssa aquatica*; fruit a short-stalked, larger, juicy, sour drupe with a papery winged stone. *Nyssa ogeche*
11. Leaves alternate, most leaves equal to or less than 12 cm long, obovate to oval; drupes in groups of three to five, stone shallowly ribbed. *Nyssa sylvatica*
12. Leaves alternate, most leaves equal to or less than 12 cm long, elliptical or lanceolate; drupes in pairs, stone ribbed. *Nyssa biflora*

Nyssa biflora twig

Nyssa sylvatica twig

Nyssa ogeche twig

Nyssa ogeche leaves

Cornaceae 175

Cornus alternifolia L. f.

alternate-leaf dogwood, pagoda dogwood

Quick Guide: *Leaves* alternate but may appear opposite; *Flowers* in white flat-topped clusters; *Fruit* a purple-black drupe in red-stalked clusters; *Bark* green and white streaked; *Habitat* rich, moist soils.

Leaves: Simple, alternate, clustered at twig ends so may appear opposite or whorled, deciduous, ovate, oval, or elliptical, 5 to 14 cm long, apex acute to acuminate, base rounded to cuneate, margin minutely serrate or entire and wavey, venation arcuate, underside pubescent, petiole long and often red; autumn color yellow, red, or maroon.

Twigs: Slender, bright green to maroon, glabrous; leaf scar V-shaped with three bundle scars.

Buds: Terminal bud 6 mm long, ovoid, with two to three glabrous or finely pubescent scales.

Bark: Green-brown, smooth, and vertically streaked on small trees; gray-brown and ridged on large trees.

Flowers: Perfect, in white, compact, flat-topped clusters after the leaves in late spring.

Fruit: Drupe, round, purple-black, 5 to 10 mm wide, in erect red-stalked clusters, maturing in summer.

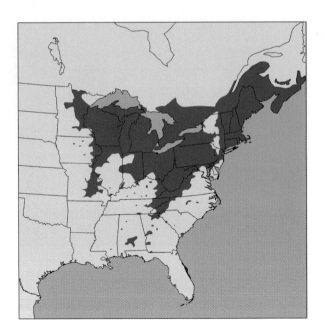

Form: Shrub or small understory tree up to 9 m (30 ft) in height.

Habitat and Ecology: Shade tolerant; found on rich, moist soils of upland forests.

Uses: Browsed by white-tailed deer, and fruit eaten by birds and small mammals. Can be planted as an ornamental on moist acidic soils with shade.

Botanical Name: *Cornus* is Latin for "horn," referring to the hard wood; *alternifolia* refers to the alternate leaves.

Cornus alternifolia twig

Cornus alternifolia leaves

Cornus alternifolia fruit

Cornus alternifolia bark

Cornus alternifolia flowers

Cornus amomum Mill.

silky dogwood, swamp dogwood, pale dogwood, silky cornel

Quick Guide: *Leaves* opposite, simple, venation arcuate, underside with pale or rusty pubescence; *Twigs* densely pubescent when young and pith brown; *Fruit* a blue drupe in red-stalked clusters; *Habitat* wet sites.

Leaves: Simple, opposite, deciduous, ovate, oval or elliptical, 5 to 15 cm long, apex acute on abruptly acuminate, base rounded to acute, margin entire, venation arcuate, underside with pale or rusty pubescence especially along the veins, petiole pubescent; autumn color yellow to maroon.

Twigs: Slender, red, with silky gray or maroon pubescence, pith brown; leaf scar V-shaped with three bundle scars.

Buds: Terminal bud conical, 5 mm long, maroon, with densely pubescent scales.

Bark: Red-brown or green-brown, becoming brown and shallowly ridged on larger trees.

Flowers: Perfect, in white, compact, flat-topped clusters on pubescent stalks after the leaves in late spring.

Fruit: Drupe, round, blue, 5 to 9 mm wide, on pubescent red stalks in late summer.

Form: An understory multistem shrub or small tree up to 5 m (16 ft) in height.

Habitat and Ecology: Found on a variety of wet sites including stream banks, swamp edges, and wet thickets throughout the eastern United States.

Uses: Browsed by white-tailed deer and provides cover for wood ducks and woodcock; fruit eaten by birds and small mammals. Can be planted on wet sites with partial shade.

Botanical Name: *Cornus* is Latin for "horn," referring to the hard wood; *amomum* is Latin for "shrub."

Cornus amomum leaves

Cornus amomum immature fruit

Cornus amomum twig

Cornus amomum bark

Cornus amomum flowers

Cornus drummondii C. A. Mey.

roughleaf dogwood, swamp dogwood

Quick Guide: *Leaves* opposite, simple, ovate, upper surface rough with stiff hairs; *Fruit* a white drupe in red-stalked clusters; *Bark* red-brown to gray; *Habitat* a variety of sites.

Leaves: Simple, opposite, deciduous, ovate to oval or elliptical, 4 to 14 cm long, apex acuminate, base rounded, margin entire, venation arcuate, upper surface rough with stiff hairs, underside densely pubescent with rough hairs, petiole red and pubescent; autumn color yellow to maroon.

Twigs: Slender, green to maroon, covered with stiff hairs when young, pith brown; leaf scar V-shaped with three bundle scars.

Buds: Terminal bud acute, 3 to 5 mm long, with brown pubescent scales.

Bark: Red-brown to gray, becoming shallowly grooved or blocky.

Flowers: Perfect, in white, compact, flat-topped clusters after the leaves in late spring.

Fruit: Drupe, round, white, 7 mm wide, in red-stalked clusters in late summer.

Form: An understory multistem shrub or small tree up to 9 m (30 ft) in height.

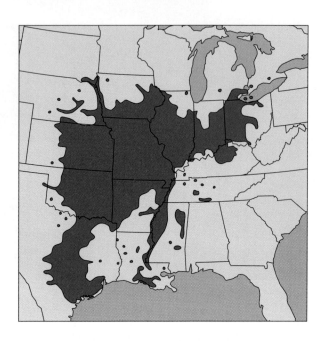

Habitat and Ecology: Found on a variety of sites from stream banks to dry upland forests and the prairie border.

Uses: Browsed by white-tailed deer and good for wildlife cover; fruit eaten by birds and small mammals. The hard wood is used for tool handles and this tree is used in landscaping because of its tolerance of a range of site conditions.

Botanical Name: *Cornus* is Latin for "horn," referring to the hard wood; *drummondii* is for Thomas Drummond, a Scottish botanist.

Cornus drummondii leaves

Cornus drummondii fruit

Cornus drummondii bark

Cornus drummondii flowers

Cornaceae

Cornus florida L.

flowering dogwood, flowering cornel, boxwood

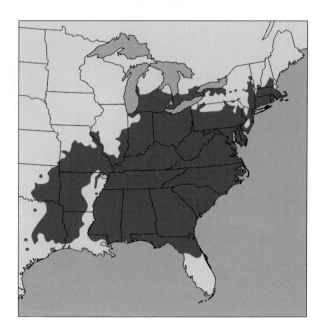

Quick Guide: *Leaves* opposite, simple, ovate, with arcuate venation; *Twigs* bright green; *Flowers* with four white petal-like bracts in early spring; *Fruit* a bright red drupe in fall; *Bark* dark and shallowly blocky; *Habitat* moist soils.

Leaves: Simple, opposite, deciduous, mainly ovate or oval, 7 to 13 cm long, apex acute, base rounded, margin entire, venation arcuate, underside pubescent; autumn color yellow, red, or maroon.

Twigs: Slender, bright green or reddish, glabrous, new growth "telescopic" to previous growth; leaf scar V-shaped with three bundle scars.

Buds: Terminal vegetative bud acute, 6 mm long, with two green valvate and pubescent scales; flower bud larger, pink-green, onion shaped.

Bark: Gray-black, shallowly blocky.

Flowers: Perfect, in yellow-green clusters surrounded by four white, petal-like, notched bracts, appearing before or with the leaves.

Fruit: Drupe, ovoid, shiny red, about 1 cm long, in erect clusters maturing in late summer or early fall.

Form: Usually an understory tree up to 12 m (40 ft) in height and 30 cm (1 ft) in diameter.

Habitat and Ecology: Shade tolerant; found on stream edges and in moist upland forests in association with a wide variety of species.

Uses: Wood reddish brown, hard, heavy; was used for tool handles, mallet heads, spools, golf club heads, and in woodcarving. A red dye and guinine like drug were derived from the bark. Browsed by white-tailed deer; beaver eat the bark; fruit eaten by a large number of birds and mammals including waterfowl, ruffed grouse, northern bobwhite, wild turkey, numerous songbirds, fox and gray squirrels, and black bear. Because of its wide distribution and palatability, a staple food for numerous forest-dwelling wildlife species. A popular ornamental, but suffers from insects and disease if not planted on moist acidic soils in partial shade. Cultivars offer a range in flower colors.

Botanical Name: *Cornus* is Latin for "horn," referring to the hard wood; *florida* means "abounding in flowers."

Cornus florida flower

Cornus florida leaves

Cornus florida fruit

Cornus florida bark

Cornus florida flower buds

Cornus racemosa Lam.

gray dogwood, panicled dogwood, gray-stemmed dogwood

Quick Guide: *Leaves* opposite, simple, ovate, venation arcuate, underside with dense white pubescence; *Flowers* in more rounded clusters relative to other dogwoods; *Fruit* a white drupe in red-stalked clusters; *Habitat* stream edges and upland forests.

Leaves: Simple, opposite, deciduous, ovate or elliptical, 5 to 13 cm long, apex acuminate, base rounded, margin entire, venation arcuate, underside with dense white pubescence, petiole pubescent; autumn color yellow to maroon.

Twigs: Slender, maroon to gray, glabrous, pith light brown; leaf scar V-shaped with three bundle scars.

Buds: Terminal bud acute with dark brown, valvate scales.

Bark: Brown-gray with scaly ridges.

Flowers: Perfect, white, in rounded rather than flat clusters, appearing after the leaves in late spring.

Fruit: Drupe, round, white, about 7 mm wide, in red-stalked clusters in late summer.

Form: An understory, multistem shrub often forming thickets or small tree up to 4 m (12 ft) in height.

Habitat and Ecology: Found on moist soils of stream banks and upland forests.

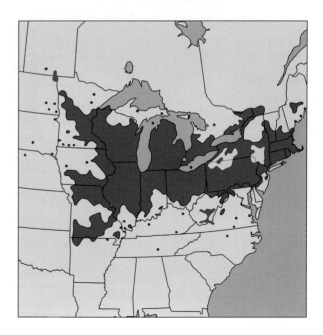

Uses: Browsed by white-tailed deer and provides cover for woodcock and ruffed grouse; fruit eaten by birds and small mammals.

Botanical Name: *Cornus* is Latin for "horn," referring to the hard wood; *racemosa* refers to the flowers.

Similar Species: Other shrublike species of dogwood include red-osier dogwood (*Cornus stolonifera* Michx. or *Cornus sericea* L.), which is found on wet sites in the northeast and has blood red twigs (see facing page) with a white pith and white fruit. Roundleaf dogwood (*Cornus rugosa* Lam.) is found in New England on well-drained soils and has nearly round leaves with dense pubescence below and light blue or white fruit.

Cornus stolonifera twig

Cornus racemosa leaves

Cornus racemosa imature fruit

Cornus racemosa bark

Cornus racemosa flowers

Cornaceae

Cornus stricta Lam.

swamp dogwood, stiff cornel, stiff dogwood

Quick Guide: *Leaves* opposite, simple, venation arcuate, underside glabrous or pubescent; *Pith* white; *Flowers* in white, open, flat-topped clusters; *Fruit* a purple-blue drupe in drooping, long-stalked clusters; *Bark* gray-brown and ridged; *Habitat* bottoms, swamps, and stream edges.

Leaves: Simple, opposite, deciduous, ovate to elliptical, 7 to 13 cm long, apex acute to acuminate, base acute to cuneate, margin entire, venation arcuate, underside glabrous or with pale pubescence; autumn color yellow, red, or maroon.

Twigs: Slender, bright green to maroon and glabrous or lightly pubescent, upright or stiff, pith white; leaf scar V-shaped with three bundle scars.

Buds: Terminal bud 5 mm long, ovoid, with rusty pubescent green or red scales.

Bark: Gray-brown, smooth or ridged.

Flowers: Perfect, in white, open, flat-topped clusters after the leaves in late spring.

Fruit: Drupe, round, purple-blue, 6 mm wide, in long-stalked drooping clusters, maturing in summer.

Form: Shrub or small understory tree up to 8 m (26 ft) in height with stiff branches.

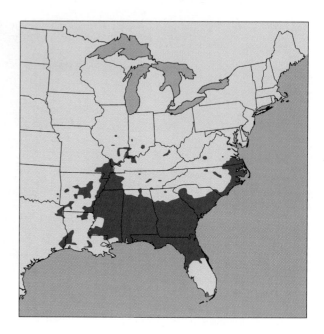

Habitat and Ecology: Shade tolerant; found in bottomlands, wet thickets, swamps, and along streams.

Uses: Browsed by white-tailed deer and fruit eaten by birds and small mammals. Can be planted as an ornamental on moist acidic soils with shade.

Botanical Name: *Cornus* is Latin for "horn," referring to the hard wood; *stricta* means "stiff," and refers to the stiff branches. Also named *Cornus foemina* Mill.

Cornus stricta flowers and leaves

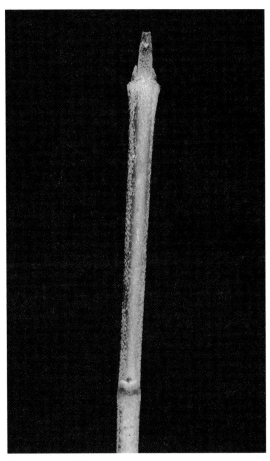

Cornus stricta twig

Cornus stricta bark

Cornus stricta fruit

Cornaceae

Nyssa aquatica L.

water tupelo, cotton-gum, water-gum, swamp tupelo, sourgum

Quick Guide: *Leaves* alternate, simple, large, some obovate, margin entire or with an occasional large dentate tooth, petiole often long and pubescent; *Fruit* a large purple-black drupe with a prominently ridged stone; *Bark* gray-brown with scaly or blocky ridges, trunk swollen at the base; *Habitat* swamps.

Leaves: Simple, alternate, deciduous, elliptical to oblong and ovate or obovate, blade 7 to 30 cm long, apex acute to acuminate, base rounded to cuneate, margin entire or with an occasional large dentate tooth, petiole long (3 to 5 cm) grooved and pubescent, underside pale and pubescent; autumn color yellow. Cotton-gum is for the cottonlike hair on unfolding leaves.

Twigs: Stout, red-brown, glabrous or pubescent, diaphragmed, with lenticels; leaf scar heart shaped with three bundle scars.

Buds: Terminal bud ovoid to round, 3 mm long, with green or red overlapping scales.

Bark: Gray-brown and shallowly grooved with scaly, blocky or flattened ridges.

Flowers: Imperfect or perfect but staminate often infertile, yellow-green, appearing in spring with the leaves; staminate flowers clustered in round heads; pistillate flowers with a prominent style and stigma, born alone on long (4 cm) stalks.

Fruit: Drupe, purple-black, white-speckled, juicy, ovoid, 3 cm long, on long drooping stalks, stone with about 10 longitudinal ridges, maturing in late summer to early fall.

Form: Up to 30 m (100 ft) in height and 1 m (3 ft) in diameter; trunk swollen at the base.

Habitat and Ecology: Shade intolerant; found in deep swamps, wet flats, and floodplain forests in pure stands or with *Acer rubrum, Carya aquatica, Fraxinus profunda, Gleditsia aquatica, Liquidambar styraciflua, Nyssa biflora, Populus heterophylla, Quercus lyrata, Salix nigra, Taxodium distichum* var. *distichum* and var. *imbricarium*.

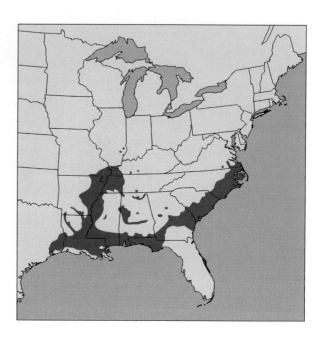

Uses: Wood white to brown-gray, cross grained, moderately heavy, moderately hard; used for pulpwood, pallets, crates, boxes, and furniture. A bee flower and source of tupelo honey. See *Nyssa sylvatica* for wildlife uses. Can be planted as an ornamental on wet sites.

Botanical Name: *Nyssa* refers to a water-loving nymph in classical mythology; *aquatica* refers to the wet habitat. Also grouped in Nyssaceae.

Nyssa aquatica fruit

Nyssa aquatica leaves

Nyssa aquatica bark

Nyssa aquatica bark

Nyssa aquatica leaves

Nyssa biflora Walt.

swamp tupelo, swamp blackgum, two-flowered tupelo, blackgum

Quick Guide: *Leaves* alternate, simple, many elliptical, margin entire, smaller than *Nyssa aquatica* and *Nyssa ogeche*; *Fruit* a purple-black drupe with a ridged stone, often in pairs; *Bark* gray-brown and shallowly grooved; *Habitat* bottoms and swamps.

Leaves: Simple, alternate, deciduous, elliptical to lanceolate or oblanceolate, 4 to 15 cm long, apex acute, base mostly cuneate, margin entire, underside glabrous or with fine pubescence; autumn color bright red.

Twigs: Slender, red-brown, glabrous, diaphragmed, with lenticels; leaf scar almost round with three bundle scars.

Buds: Terminal bud acute, 4 mm long, with green or reddish overlapping scales (see page 175).

Bark: Gray-brown, shallowly grooved with somewhat blocky or flattened ridges.

Flowers: Imperfect or perfect but staminate often infertile, yellow-green, appearing in spring after the leaves; staminate flowers clustered in round heads; pistillate flowers with a prominent style and stigma, usually in pairs on long stalks.

Fruit: Drupe, blue-black, juicy, nearly round, 1 to 1.5 cm long, usually in pairs on long stalks, stone prominently ribbed, maturing in late summer to early fall.

Form: Up to 37 m (120 ft) in height and 1 m (3 ft) in diameter; base of the trunk usually swollen.

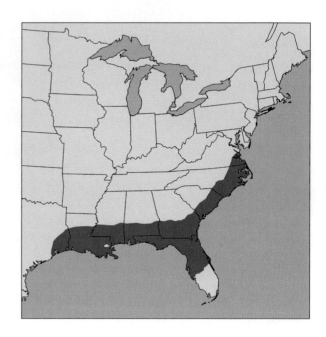

Habitat and Ecology: Shade intolerant; found in frequently inundated bottomlands and swamps with *Acer rubrum, Chamaecyparis thyoides, Gordonia lasianthus, Magnolia virginiana, Nyssa aquatica, Pinus serotina, Quercus laurifolia, Taxodium distichum* var. *distichum* and var. *imbricarium*.

Uses: Wood white to brown-gray, cross grained, moderately heavy, moderately hard; used for pulpwood, pallets, crates, boxes, and furniture. A bee flower and source of tupelo honey. See *Nyssa sylvatica* for wildlife uses.

Botanical Name: *Nyssa* refers to a water-loving nymph in classical mythology; *biflora* refers to the flowers born in pairs. Also grouped in Nyssaceae.

Nyssa biflora immature fruit

Nyssa biflora bark

Nyssa biflora bark

Nyssa biflora leaves

Cornaceae

Nyssa ogeche Bartr. ex Marsh.
Ogeechee tupelo, sour tupelo, Ogeechee-lime, white tupelo, bee-tupelo

Quick Guide: *Leaves* alternate, simple, large, elliptical to obovate, margin entire or with an occasional large dentate tooth, petiole shorter and leaf apex more rounded compared with *Nyssa aquatica*; *Fruit* a large, purple-red, sour drupe with a winged stone; *Bark* gray-brown and shallowly grooved, trunk swollen at the base; *Habitat* swamps and floodplains.

Leaves: Simple, alternate, deciduous, elliptical to obovate or oblong, 7 to 18 cm long, apex acute to rounded possibly with a mucronate tip, base cuneate to blunt, margin entire but may show an occasional large dentate tooth, petiole usually shorter than for *Nyssa aquatica* and pubescent, underside pale and pubescent; autumn color yellow (see page 175).

Twigs: Stout, mottled green-red-brown, glaucous or downy pubescent, diaphragmed, with lenticels; leaf scar heart shaped with three bundle scars (see page 175).

Buds: Terminal bud round, 3 mm long, and scales overlapping, red-brown, and pubescent; lateral buds smaller.

Bark: Gray-brown, shallowly grooved with flattened ridges, can be somewhat blocky or scaly.

Flowers: Imperfect or perfect but staminate often infertile, yellow-green, appearing in spring with the leaves; staminate flowers in round heads on long (4 cm) stalks; pistillate flowers with a prominent style and stigma, solitary, born on shorter (2 cm) stalks.

Fruit: Drupe, purple-red, juicy, sour, ovoid, up to 4 cm long, on short stiff stalks, stone attached to the skin with papery longitudinal wings, maturing in late summer or early fall.

Form: Crooked, up to 15 m (50 ft) in height and 61 cm (2 ft) in diameter; base of trunk swollen.

Habitat and Ecology: Shade intolerant; found in swamps and floodplain forests periodically or permanently inundated. Forest associates include *Acer rubrum, Chamaecyparis thyoides, Cyrilla racemiflora, Cliftonia monophylla, Magnolia virginiana, Nyssa aquatica, Pinus serotina, Quercus laurifolia, Quercus virginiana, Taxodium distichum* var. *distichum* and var. *imbricarium*.

Uses: Wood white to brown-gray, cross grained, moderately heavy, moderately hard; used for pulpwood, pallets, crates, boxes, and furniture. A favorite bee flower and important source of tupelo honey. See *Nyssa sylvatica* for wildlife uses. The juice of the fruit has been used as a lime substitute (Ogeechee-lime).

Botanical Name: *Nyssa* refers to a water-loving nymph in classical mythology; *ogeche* refers to the Ogeechee River Valley where this tree is found. Also grouped in Nyssaceae.

Nyssa ogeche leaves and fruit

Nyssa ogeche pistillate flowers

Nyssa ogeche staminate flowers

Nyssa ogeche bark

Nyssa ogeche and *Taxodium distichum* var. *imbricarium*

Nyssa sylvatica Marsh.

blackgum, black tupelo, sourgum, tupelo, tupelo-gum, pepperidge

Quick Guide: *Leaves* alternate, simple, many obovate, margin entire; *Fruit* a blue-black drupe with a shallowly ridged stone; *Leaf scar* with three bundle scars; *Bark* variable, commonly gray-brown and thickly ridged or blocky; *Habitat* dry uplands to swamp edges.

Leaves: Simple, alternate, deciduous, elliptical to obovate or oval, 5 to 16 cm long, apex acute or abruptly acuminate to rounded, base cuneate to blunt, margin entire or with an occasional large dentate tooth on saplings, underside glabrous or with fine pubescence; autumn color bright red.

Twigs: Slender, red-brown, glabrous, diaphragmed, with lenticels; leaf scar almost round with three bundle scars.

Buds: Terminal bud ovoid, 6 mm long; scales overlapping, red-brown to green-brown, glabrous or with pale or golden pubescence (see page 175).

Bark: Gray-brown, variable, from deeply grooved and blocky to shallowly grooved with flattened or scaly ridges.

Flowers: Imperfect or perfect but staminate often infertile, yellow-green, appearing in spring after the leaves; staminate flowers clustered in round heads; pistillate flowers with a prominent style and stigma, in groups of up to five clustered at the ends of long stalks.

Fruit: Drupe, ovoid, blue-black, juicy, about 1.3 cm long, in groups of three to five on long stalks, stone shallowly ribbed, maturing in late summer to early fall.

Form: Up to 37 m (120 ft) in height and 1 m (3 ft) in diameter; large branches often at a 90-degree angle to the trunk.

Habitat and Ecology: Shade tolerant; found on a variety of sites ranging from dry uplands to edges of streams and swamps and a component of many forest cover types.

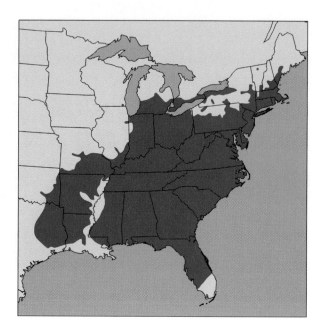

Uses: Wood white to gray-brown, moderately hard, moderately heavy; used for pulpwood, veneer, containers, pallets, railroad ties, woodenware, and gunstocks. A white-tailed deer browse and a den tree; fruit eaten by a wide variety of songbirds and game birds, including northern bobwhite, wild turkey, ruffed grouse, red-cockaded woodpecker, and wood ducks; fruit also eaten by a variety of mammals including black bear, fox and gray squirrels, gray fox, opossum, and raccoon; bark eaten by beaver; also a bee flower and source of tupelo honey. Of all the *Nyssa* spp., *Nyssa sylvatica* is the most important to wildlife because of its widespread distribution. An attractive ornamental because of good form, brilliant fall color, and tolerance of a range of conditions.

Botanical Name: *Nyssa* refers to a water-loving nymph in classical mythology; *sylvatica* means "of the forest," referring to the ubiquitous range of this species. Also grouped in Nyssaceae.

Nyssa sylvatica staminate and pistillate flowers

Nyssa sylvatica leaves

Nyssa sylvatica bark

Nyssa sylvatica bark

Nyssa sylvatica fruit

Cliftonia monophylla (Lam.) Britt. ex Sarg.

buckwheat-tree, black titi, buckwheat-bush

Quick Guide: *Leaves* alternate, simple, evergreen, leathery, margin entire, lateral veins obscure; *Flowers* in fragrant, white racemes; *Fruit* buckwheatlike; *Habitat* swamps and bogs.

Leaves: Simple, alternate, evergreen, leathery, shiny, oblanceolate to elliptical, 2.5 to 10 cm long, margin entire, lateral veins obscure, nearly sessile, underside pale.

Twigs: Slender, red-brown; leaf scar shield shaped with one bundle scar.

Buds: Terminal bud ovoid, lateral buds appressed.

Bark: Gray-brown, mottled, and smooth on small trees; large trees more ridged or scaly.

Flowers: Perfect, very fragrant, white to pink, bell shaped, in racemes at twig ends in early spring.

Fruit: Drupe, green to yellow or brown, two to five winged, similar to the triangular fruits of buckwheat, each 6 mm long, appearing in late summer and clusters of dry fruits persistent over winter.

Form: Shrub or small tree up to 9 m (30 ft) in height often forming thickets.

Habitat and Ecology: Found in floodplains, flatwoods, swamp edges, and acidic bogs with *Acer*

rubrum, Chamaecyparis thyoides, Cyrilla racemiflora, Magnolia virginiana, Nyssa biflora, Nyssa ogeche, Pinus glabra, Pinus serotina, Quercus laurifolia, Taxodium distichum var. *distichum* and var. *imbricarium*.

Uses: A preferred browse by white-tailed deer and livestock, and a bee flower.

Botanical Name: *Cliftonia* is for William Clifton, who collected this species in 1764; *monophylla* means "simple leaf."

Cliftonia monophylla flowers

Cliftonia monophylla flowers

Cliftonia monophylla bark

Cliftonia monophylla bark

Cliftonia monophylla fruit and leaf

Cyrillaceae

Cyrillaceae 197

Cyrilla racemiflora L.

swamp cyrilla, white titi, black titi

Quick Guide: *Leaves* alternate, simple, leathery, margin entire, with arcuate venation; *Twigs* three-angled; *Flowers* white in curved racemes; *Fruit* a drupe in stiff clusters persisting over winter; *Habitat* water edges and bottoms.

Leaves: Simple, alternate, tardily deciduous, leathery, elliptical to oblanceolate or obovate, 5 to 10 cm long, margin entire, venation arcuate, glabrous below; autumn color orange or red.

Twigs: Slender, yellow-brown to gray, three-angled; leaf scar raised, shield shaped, with one bundle scar.

Buds: Terminal bud ovoid, 5 mm long, with brown overlapping scales.

Bark: Gray-brown, mottled, smooth, becoming scaly on large trees.

Flowers: Perfect, white or yellowish, in stiff curved racemes up to 15 cm long, in whorls, blooming in early summer.

Fruit: Drupe, unwinged, ovoid, dry, brown, 3 mm long, in stiff clusters up to 15 cm long, maturing in summer and persistent over winter.

Form: Shrub or small tree up to 9 m (30 ft) in height often forming thickets.

Habitat and Ecology: Found along stream edges and in floodplains, flatwoods, swamps, and acid bogs with *Acer rubrum, Chamaecyparis*

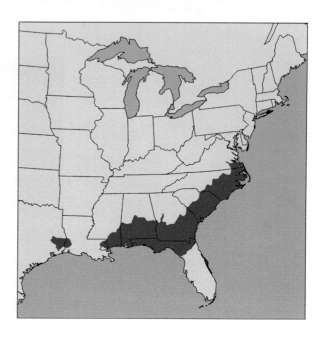

thyoides, Cliftonia monophylla, Magnolia virginiana, Nyssa biflora, Nyssa ogeche, Pinus glabra, Pinus serotina, Quercus laurifolia, Taxodium distichum var. *distichum* and var. *imbricarium*.

Uses: A preferred white-tailed deer browse. Dense thickets form important cover for midsize and large mammals such as white-tailed deer and black bear. A bee flower.

Botanical Name: *Cyrilla* is for the Italian botanist Domenico Cirillo; *racemiflora* refers to the flowers born in racemes.

Cyrilla racemiflora leaves

Cyrilla racemiflora flowers

Cyrilla racemiflora flowers

Cyrilla racemiflora bark

Cyrilla racemiflora fruit

Cyrillaceae

Diospyros virginiana L.

common persimmon, possumwood, simmon

Quick Guide: *Leaves* alternate, simple, lanceolate, margin entire; *Buds* black and triangular; *Leaf scar* with one bundle scar; *Fruit* a large, yellow-orange berry; *Bark* gray-black, blocky, alligatorlike; *Habitat* moist to dry forests.

Leaves: Simple, alternate, deciduous, lanceolate to elliptical or oblong, 5 to 15 cm long, apex acute or abruptly acuminate, base rounded, margin entire, with black spots in fall, underside mostly glabrous; autumn color yellow.

Twigs: Zig-zag, brown, glabrous or pubescent, with lenticels; leaf scars elevated, crescent shaped, with one slitlike bundle scar; old bud scales often remaining at branch junctions.

Buds: True terminal bud lacking; lateral buds black, triangular ("snake head"), 3 mm long, with two overlapping scales.

Bark: Shallowly grooved with orange in the grooves on small trees; large trees gray-black, thick, very blocky.

Flowers: Species is dioecious, flowers yellow-green, appearing after the leaves in late spring; pistillate flowers bell shaped, solitary from leaf axils; staminate flowers urn shaped, in clusters of two to four.

Fruit: Berry, pulpy, round, plumlike, yellow to orange, 4 cm wide, edible and tasting like apricot when mature otherwise with an astrigent taste, maturing in fall.

Form: Up to 24 m (80 ft) in height and 61 cm (2 ft) in diameter.

Habitat and Ecology: Shade tolerant; found on a variety of sites ranging from bottomlands to dry sandy soils. Forest associates are numerous and include *Liriodendron tulipifera, Nyssa sylvatica, Pinus echinata, Pinus taeda, Quercus coccinea,*

Quercus falcata, Quercus marilandica, and *Sassafras albidum* on upland sites; scrub oaks on very sandy sites; and on damp soils *Acer rubrum, Fraxinus pennsylvanica, Nyssa biflora, Quercus lyrata, Quercus michauxii, Quercus nigra, Ulmus crassifolia,* and *Ulmus americana.*

Uses: Wood white to light brown, heavy, hard; used for furniture, veneer, and turnery. The dark heartwood was used for golf clubs and billiard cues. Fruit relished by raccoon, red and gray fox, coyote, white-tailed deer, opossum, coyote, skunk, and the domestic dog; seeds of fruit eaten by wild turkey, northern bobwhite, and small rodents; flowers popular with bees. Fruit used in jams, wine, and baking. Roasted seeds were used as a coffee substitute. Limited value in landscaping due to leaf spot. Cultivars with seedless fruit and brilliant fall foliage have been developed.

Botanical Name: *Diospyros* means "divine fruit"; *virginiana* refers to the geographic range.

Diospyros virginiana leaves and pistillate flowers

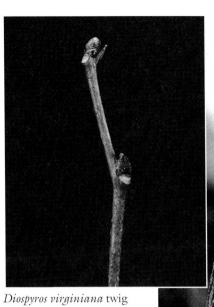

Diospyros virginiana twig

Diospyros virginiana bark

Diospyros virginiana bark

Diospyros virginiana fruit

Ericaceae Guide

1. Leaves elliptical, thin, deciduous, margin with fine teeth, midrib with hairs. *Oxydendrum arboreum*
2. Leaves oval, leathery, evergreen or tardily deciduous, shiny, margin entire or with small teeth, 2 to 7 cm long. *Vaccinium arboreum*
3. Leaves elliptical, leathery, evergreen, margin entire, sometimes appearing opposite or whorled, 5–12 cm long, glabrous below. *Kalmia latifolia*
4. Leaves leathery, evergreen, margin entire and revolute, greater than 10 cm long, underside pale with maroon hairs. *Rhododendron maximum*.
5. Leaves similar to *Rhododendron maximum* but smaller and underside lacking maroon hairs; flowers pink or purple rather than white-pink. *Rhododendron catawbiense*.

Oxydendrum arboreum bark

Vaccinium arboreum bark

Oxydendrum arboreum twig

Rhododendron maximum flower bud

Oxydendrum arboreum twig

Ericaceae

Kalmia latifolia L.
mountain laurel, mountain ivy

Quick Guide: *Leaves* alternate but may appear opposite or whorled, simple, elliptical, evergreen, and glabrous; *Flowers* showy white-pink in spring; *Fruit* a dry capsule; *Leaf scar* with one bundle scar; *Habitat* a variety of sites.

Leaves: Simple, usually alternate but may appear opposite or whorled, evergreen, leathery, waxy, elliptical to oblanceolate, 5 to 12 cm long, margin entire, apex sometimes bristle-tipped, underside glabrous when mature.

Twigs: Slender, red-brown or red-green; leaf scar small, shield shaped, with one bundle scar.

Buds: Flower buds large with glandular hairs; leaf buds smaller, naked, beneath the flower buds.

Bark: Red-brown to gray, loosely ridged or shreddy.

Flowers: Perfect, showy, saucer shaped, each up to 2.5 cm wide, white-pink petals with a red-pink star-shaped pattern, filaments in petal pouches, in large clusters blooming in late spring.

Fruit: Capsule, dry, 6 mm wide, maturing in early fall.

Form: Small, crooked tree up to 9 m (30 ft) tall but usually a shrub, often forming dense thickets.

Habitat and Ecology: Found on a variety of sites such as mountain slopes in the southern Appalachians, stream edges and in ravines in the Piedmont and Coastal Plain, and bogs in the northeastern United States.

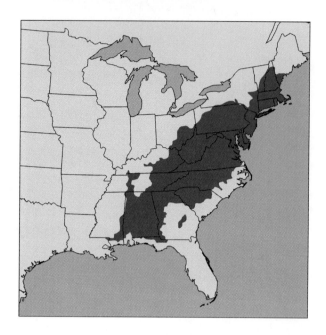

Uses: Wood used for rustic furniture and wooden utensils. Although reported to be poisonous to domestic animals, leaves and twigs browsed by white-tailed deer; leaves and buds eaten by ruffed grouse. Thickets form excellent cover for a variety of wildlife. A popular ornamental due to outstanding spring flowers, but requires well-drained acidic soils and partial shade.

Botanical Name: *Kalmia* is for the Swedish botanist Pehr Kalm; *latifolia* means with "broad leaves."

Kalmia latifolia flowers

Kalmia latifolia bark

Kalmia latifolia leaves and fruit

Oxydendrum arboreum (L.) DC.

sourwood, sorrel tree, lily-of-the-valley-tree

Quick Guide: *Leaves* alternate, simple, elliptical, sour-tasting, margin with fine teeth, midrib with hairs; *Fruit* a stiff drooping cluster of capsules persistent over winter; *Bark* dark and deeply grooved with orange-red within the grooves; *Habitat* slopes and ridges.

Leaves: Simple, alternate, deciduous, mostly elliptical, 10 to 18 cm long, margin with fine teeth, midrib with hairs, sour tasting; autumn color bright red.

Twigs: Zig-zag, yellow-green to red, mottled, with black lenticels, sour tasting; leaf scar elevated, shield shaped, with one bundle scar (see page 203).

Buds: True terminal bud lacking; lateral buds brown, small, 1 mm long, round, embedded.

Bark: Gray-black, deeply grooved, with orange-red within the grooves (see page 203); on large trees more blocky.

Flowers: Perfect, white, urn shaped, in drooping curved racemes up to 20 cm long in midsummer.

Fruit: Capsule, woody, 3 to 8 mm long, in stiff clusters, maturing in early fall and persisting over winter.

Form: Up to 24 m (80 ft) in height and 61 cm (2 ft) in diameter, trunk often curved.

Habitat and Ecology: Intermediate shade tolerance; found on a variety of well-drained sites but more commonly on dry, rocky soils with *Acer*

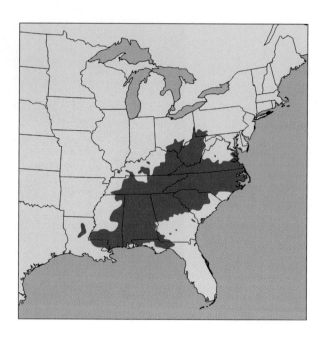

rubrum, *Carya* spp., *Pinus echinata*, *Pinus rigida*, *Pinus taeda*, *Pinus virginiana*, *Quercus coccinea*, *Quercus falcata*, *Quercus marilandica*, *Quercus prinus*, *Quercus stellata*, *Quercus velutina*, and *Sassafras albidum*.

Uses: Wood light yellow-brown to reddish brown, heavy, hard; used for pulpwood and specialty items such as tool handles. An attractive landscape tree for the flowers, fall color, and stress tolerance. A browse for white-tailed deer, and flowers an important source of nectar for bees (sourwood honey). Leaves used in teas.

Botanical Name: *Oxydendrum* means "sour tree"; *arboreum* means "treelike."

Oxydendrum arboreum flowers

Oxydendrum arboreum leaves

Oxydendrum arboreum fruit

Oxydendrum arboreum bark

Oxydendrum arboreum form

Ericaceae

Rhododendron maximum L.

rosebay rhododendron, great rhododendron, rosebay, great laurel

Quick Guide: *Leaves* alternate, simple, evergreen, leathery, margin entire and revolute, maroon pubescence below; *Flowers* in showy pink-white clusters in summer; *Fruit* a capsule with sticky scales; *Habitat* a variety of sites.

Leaves: Simple, alternate, evergreen, oblong to elliptical and oblanceolate, 10 to 25 cm long, dark green, leathery, stiff, apex acute, base acute or cuneate, margin entire and revolute, petiole with maroon pubescence, underside pale with maroon hairs, appearing wilted in winter.

Twigs: Stout, green to red-brown with maroon hair; leaf scar variable.

Buds: Flower buds large, 4 cm long, with sticky pubescent scales (see page 203).

Bark: Red-brown to gray, becoming scaly or shreddy.

Flowers: Perfect, white or pinkish, 4 cm wide, with five petals, in showy terminal clusters in summer.

Fruit: Capsule, sticky, woody, 1.3 cm long, maturing in fall.

Form: Understory shrub or small twisted tree up to 9 m (30 ft) tall, forming dense thickets.

Habitat and Ecology: Found on a variety of cool, moist sites including stream edges, mountain slopes, swamps, and coves. Forming dense thickets in the southern Appalachians.

Uses: Can be used as an ornamental on moist acid soils with good drainage. Foliage browsed by white-

tailed deer; ruffed grouse eat the leaves, buds, and seeds. Dense thickets ("green hells") form excellent cover for a variety of birds and mammals including ruffed grouse, songbirds, white-tailed deer, black bear, snowshoe hare, and cottontail rabbit.

Botanical Name: *Rhododendron* means "rose tree"; *maximum* refers to its size.

Similar Species: Purple rhododendron or mountain rosebay (*Rhododendron catawbiense* Michx.) is found at high elevations in the southern Appalachians and is distinguished by smaller light green leaves with a rounded base and apex, a pale underside lacking maroon hair, and purple-pink to lilac flowers. A better species for landscaping because of more colorful flowers and greater stress tolerance.

Rhododendron maximum leaves and flowers

Rhododendron maximum thicket

Rhododendron maximum bark

Rhododendron catawbiense leaves

Vaccinium arboreum Marsh.

sparkleberry, farkleberry, tree sparkleberry, tree huckleberry

Quick Guide: *Leaves* alternate, simple, small, glossy, margin entire or with small teeth; *Fruit* a juicy blue-black berry; *Bark* red-brown and shreddy; *Habitat* well-drained to xeric soils.

Leaves: Simple, alternate, tardily deciduous, oval to elliptical, 2 to 7 cm long, dark green and shiny, apex sometimes mucronate, margin entire or with small glandular teeth, nearly sessile; autumn color red.

Twigs: Slender, red-brown; leaf scar shield shaped with one bundle scar.

Buds: True terminal bud lacking; lateral buds round, small, 1 mm long, embedded.

Bark: Red-brown to gray, shreddy, revealing red inner bark on large stems.

Flowers: Perfect, white, bell shaped, in attractive racemes in spring after the leaves.

Fruit: Berry, red-blue to blue-black, juicy, round, 6 mm wide, maturing in summer or early fall.

Form: Shrub or small twisted tree up to 9 m (30 ft) tall.

Habitat and Ecology: Found in a variety of sites such as sand dunes, dry slopes, forest edges, and stream margins.

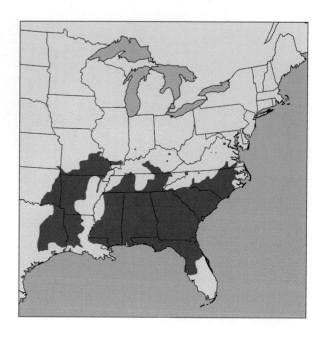

Uses: Wood close grained, hard; used for woodenware, pipes, and walking canes. Fruit eaten by a variety of birds and mammals but of lower preference compared with other *Vaccinium* spp. because the fruit tends to be gritty.

Botanical Name: *Vaccinium* is Latin for "blueberry"; *arboreum* means "treelike."

Vaccinium arboreum immature fruit

Vaccinium arboreum leaves

Vaccinium arboreum bark

Vaccinium arboreum flowers

Ericaceae

Sapium sebiferum (L.) Roxb.

Chinese tallowtree, popcorn tree, tallowtree, Florida aspen

Quick Guide: *Leaves* alternate, simple, rhombic, apex long tapered, petiole with glands at the blade; *Flowers* in drooping bright yellow clusters in spring; *Fruit* a popcornlike capsule in fall; *Bark* gray-brown with orange-red in the grooves; *Habitat* a variety of sites.

Leaves: Simple, alternate, deciduous, rhombic or deltoid, blade about 7 cm long, apex tapering to a very long point, margin entire, petiole long with a pair of glands at the blade; autumn color yellow, orange, or red.

Twigs: Green-brown, glaucous, with lenticels; leaf scar shield shaped with three bundle scars; sap milky, poisonous.

Buds: Terminal bud triangular, naked.

Bark: Gray-brown, grooved, with interlacing ridges and orange-red between the ridges.

Flowers: Imperfect, in bright yellow drooping spikes up to 20 cm long in spring after the leaves; spike mostly with staminate flowers and a few pistillate flowers at the base.

Fruit: Capsule, three-lobed, containing three white waxy seeds about 1 cm long that resemble popcorn kernels, maturing in fall.

Form: Up to 15 m (50 ft) in height.

Habitat and Ecology: A native of eastern Asia and planted as an ornamental. Aggressively forming thickets on a variety of sites. This tree is listed as legally noxious and considered a harmful invasive in many states, particularly in the Southeast.

Uses: In China, the wax from the seed coat was used in making candles. A direct competitor of other valuable wildlife plants.

Botanical Name: *Sapium* is Latin for "the lower part of a fir tree," perhaps referring to similarities in bark between fir and Chinese tallowtree; *sebiferum* means "tallowly or waxy," referring to the fruit. Also named *Triadica sebifera* (L.) Small.

Sapium sebiferum fruit

Sapium sebiferum leaves and flowers

Sapium sebiferum bark

Sapium sebiferum flowers and leaves

Euphorbiaceae

Cladrastis kentukea (Dum.-Cours.) Rudd

yellowwood, Virgilia, gopherwood, American yellowwood

Quick Guide: *Leaves* alternate, pinnately compound, leaflets seven to eleven and subopposite to widely alternate on the rachis; *Buds* hidden by the swollen petiole; *Flowers* fragrant and white in pendent clusters in spring; *Fruit* podlike; *Bark* smooth and gray; *Habitat* rich, moist soils.

Leaves: Pinnately compound, alternate, deciduous, 20 to 30 cm long. Leaflets seven to eleven, terminal leaflet obovate, 7 to 10 cm long, can be widely alternate on the rachis, apex acute or abruptly acuminate, base obtuse to rounded, margin entire; petiole swollen at the base and covering the bud; autumn color yellow.

Twigs: Zig-zag, red-brown, brittle, glabrous; leaf scar narrow, nearly encircling the bud.

Buds: True terminal bud lacking; lateral bud conelike and naked with golden silky pubescence.

Bark: Gray-brown, smooth, "beechlike."

Flowers: Perfect, white, pealike, fragrant, in pendent clusters up to 40 cm long, similar to *Robinia pseudoacacia*, blooming in spring, usually does not bloom in successive years.

Fruit: Legume, brown, flat, often constricted, up to 10 cm long, maturing in fall.

Form: Up to 15 m (50 ft) in height with low branches.

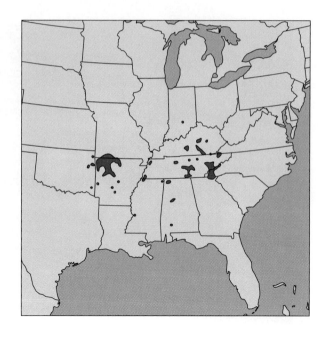

Habitat and Ecology: Found on stream banks and in coves with *Acer saccharum, Carya ovata, Fagus grandifolia, Gymnocladus dioicus,* and *Juglans nigra*.

Uses: Sapwood white; heartwood yellow, hard, heavy; was used in gunstocks and sap was used to make a yellow dye. Can be planted as an ornamental on moist cool sites and a pink cultivar has been developed.

Botanical Name: *Cladrastis* is from Greek for "brittle branch"; *kentukea* refers to the geographic range.

Cladrastis kentukea fruit

Cladrastis kentukea flowers

Cladrastis kentukea twig

Cladrastis kentukea bark

Cladrastis kentukea leaf

Fabaceae

Robinia pseudoacacia L.

black locust, yellow locust, yellow ash, false acacia

Quick Guide: *Leaves* alternate, pinnately compound, leaflets 7 to 19, margin entire; *Twigs* with a stout pair of spines at each node; *Flowers* white, fragrant, and pealike in pendent clusters; *Fruit* a legume; *Bark* light brown with thick interlacing ridges: *Habitat* a variety of sites.

Leaves: Pinnately compound, alternate, deciduous, 20 to 36 cm long. Leaflets 7 to 19, oblong or oval, 2 to 5 cm long, apex rounded to emarginate, base obtuse to rounded, margin entire; autumn color yellow.

Twigs: Zig-zag, moderately stout, angular, red-brown, with a pair of stout spines at the node; leaf scar triangular to three lobed, with three bundle scars.

Buds: True terminal bud lacking; lateral buds minute, naked, embedded, hidden under the leaf scar.

Bark: Light brown to gray and furrowed with thick, interlacing, corky or scaly ridges.

Flowers: Perfect, white, pealike, wonderfully fragrant, in pendent racemes in spring after the leaves.

Fruit: Legume, brown, flat, up to 10 cm long, maturing in fall.

Form: Up to 21 m (70 ft) in height and 61 cm (2 ft) in diameter, can form thickets.

Habitat and Ecology: Shade intolerant; grows best on moist limestone soils but is found on a variety of sites ranging from moist slopes and stream banks to dry rocky soils and old fields, often coming in after disturbance. Forest associates include *Acer rubrum, Carya glabra, Carya tomentosa, Juniperus*

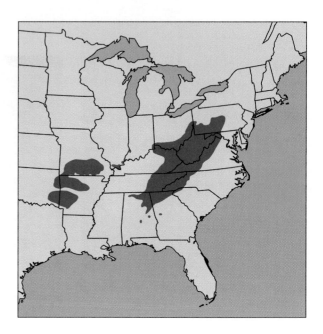

virginiana, Liriodendron tulipifera, Pinus echinata, Pinus pungens, Pinus virginiana, Quercus alba, Quercus coccinea, Quercus rubra, and *Quercus stellata.*

Uses: Wood yellow-green to red-brown, very hard, very heavy; once used for wooden nails and ship-building, now used for pulpwood, boxes, woodenware, fuel, and railroad ties. Used in waste site reclamation because of ability to fix nitrogen. Cultivars have been developed to improve form and leaf color. Browsed by white-tailed deer and fruit eaten by northern bobwhite, mourning dove, fox, opossum, and white-tailed deer; a bee flower.

Botanical Name: *Robinia* is for Jean and Vespasion Robin who first cultivated locust in Europe; *pseudoacacia* refers to the leaves and spines, which are similar to the genus *Acacia.*

Robinia pseudoacacia flowers and fruit

Robinia pseudoacacia bark

Robinia pseudoacacia bark

Robinia pseudoacacia leaf

Robinia pseudoacacia twig

Fabaceae

Fagaceae Guide

I. Fruit a nut in a spiny bur; branch terminal with a single bud. The Chestnuts and American Beech

1. Leaf margin with curved bristle-tipped teeth; leaf underside, twigs, and buds glabrous; two to three nuts enclosed in a spiny bur. *Castanea dentata*
2. Leaf leathery and margin coarsely serrate, leaf underside with tomentose hair; two to three nuts enclosed in a large spiny bur. *Castanea mollissima*
3. Leaf margin with coarse bristle-tipped teeth; leaf underside and twigs pubescent to tomentose; one nut enclosed in a spiny bur. *Castanea pumila*
4. Leaf margin serrate, parallel lateral veins ending at a tooth; two small triangular nuts in a weakly spiny bur; buds long and sharp; bark smooth even on large trees. *Fagus grandifolia*

II. Fruit an acorn; branch terminal with a cluster of buds. The Oaks

A. The Lobed Red Oaks: Lobes with bristle tips; acorn maturing in two seasons

1. Leaves shallowly three lobed at the apex, fan shaped, leathery, underside with dense orangish hair; habitat dry sites. *Quercus marilandica*
2. Leaves shallowly three lobed at the apex, fan shaped, thin, underside glabrous; habitat dry sites. *Quercus arkansas*
3. Leaves irregularly three to five lobed, sinuses shallow; habitat uplands. *Quercus georgiana*
4. Leaves three to seven lobed, sinuses deep, base bell shaped, leaves drooping; habitat dry sites. *Quercus falcata*
5. Leaves three to seven lobed, sinuses deep, leaves often vertical, base acute; habitat sandhills. *Quercus laevis*
6. Leaves five to seven lobed, lobing irregular on the same leaf, sinuses deep, terminal lobe elongated; habitat bottomlands. *Quercus texana*
7. Leaves five to seven lobed, lobing and sinus depth variable from lower to upper canopy leaves; habitat uplands. *Quercus velutina*
8. Leaves five to seven lobed, basal lobes wide, sun leaves like *Quercus falcata* except at the base, densely pubescent below; habitat bottomlands. *Quercus pagoda*
9. Leaves five to nine lobed, sinuses shallow, underside with gray hair; twigs with gray pubescence; habitat dry soils. *Quercus ilicifolia*
10. Leaves five to nine lobed, sinuses deep, terminal lobe not elongated; habitat bottomlands. *Quercus palustris*
11. Leaves five to nine lobed, lobes with many bristle tips, sinuses deep, underside glabrous or with tufts or hair in vein axils; habitat bottomlands. *Quercus shumardii*
12. Leaves five to nine lobed, sinuses deep; acorn cap covering a third of the nut; habitat uplands. *Quercus ellipsoidalis*
13. Leaves seven to nine lobed, sinuses deep, underside glabrous or with tufts of hair in vein axils; acorn nut may have concentric grooves at the tip; habitat dry uplands. *Quercus coccinea*
14. Leaves seven to eleven lobed, sinuses less than halfway to the midrib, underside glabrous or with tufts of hair in vein axils; habitat moist uplands. *Quercus rubra*

B. The Unlobed Red Oaks: Apex with a bristle tip; acorn maturing in two seasons

1. Leaves oblong with dense blue-gray pubescence on the underside, margin mostly entire; habitat sandhills. *Quercus incana*
2. Leaves obovate but variable, leathery, margin entire and revolute, underside mostly glabrous; habitat scrub forests. *Quercus myrtifolia*
3. Leaves oblanceolate to elliptical, shiny, leathery, lacking pubescence on the midrib, margin mostly entire; habitat uplands. *Quercus hemisphaerica*
4. Leaves spatulate, obovate, oblanceolate and subrhombic, mostly glabrous below, margin mostly entire; habitat bottomlands and swamps. *Quercus laurifolia*
5. Leaves linear to lanceolate, thin, often with yellow pubescence on the midrib, margin entire; habitat bottomlands. *Quercus phellos*
6. Leaves spatulate to obovate and occasionally lobed, margin mostly entire, mostly glabrous below; habitat uplands to bottomlands. *Quercus nigra*
7. Leaves oblong, leathery, underside pubescent, margin entire; habitat moist sites. *Quercus imbricaria*
8. Leaf margin with coarse bristle-tipped teeth; acorn cap with long curved scales; habitat planted. *Quercus acutissima*

C. The Lobed White Oaks: Lobes and apex lacking bristle tips; acorn maturing in one season

1. Lobes three to seven and irregular, underside glabrous; habitat river bluffs. *Quercus austrina*
2. Lobes five and variable, some central lobes cruciform, underside pubescent; habitat dry soils. *Quercus stellata*
3. Lobes three to five and variable, usually pointing toward the apex, central lobes only slightly cruciform, underside pubescent; habitat dry soils. *Quercus margaretta*
4. Lobes five to nine, lobes and sinuses very irregular, underside pubescent; acorn cap almost completely covering the nut; habitat bottomlands. *Quercus lyrata*
5. Lobes five to nine, apex fan shaped, middle sinus often the deepest, underside pubescent, acorn large with a mossy cap; habitat a variety of sites. *Quercus macrocarpa*
6. Lobes seven to nine, sinuses extending halfway or more to the midrib, lobing comparatively regular, underside pale and glabrous; acorn with a knobby cap; habitat well-drained soils. *Quercus alba*

D. The Unlobed White Oaks: Leaves of most species lacking bristles at the apex and margin; acorn maturing in one season

1. Leaf margin entire or with an occasional bristle, leaves oblong to oblanceolate, very leathery, apex with or without a rigid spine, underside pubescent; acorn cap long-stalked; habitat sandy soils. *Quercus virginiana*
2. Leaf margin entire and often rolled under, leaves elliptical and very leathery, underside with gray pubescence; acorn short-stalked and in pairs; habitat sandy soils. *Quercus geminata*
3. Leaf margin sinuate, leaves obovate to spatulate, underside with dense pubescence, petiole yellow; habitat limestone soils and bottomlands. *Quercus durandii*

4. Leaf margin with rounded callous-tipped teeth, obovate throughout the canopy, underside pubescent; habitat bottomlands. *Quercus michauxii*
5. Leaf margin with curved callous-tipped teeth, elliptical in the upper canopy and obovate in shade leaves, underside pubescent; habitat alkaline soils. *Quercus muehlenbergii*
6. Leaf margin with rounded smooth teeth, obovate to elliptical; bark deeply grooved (# 3.–5. with scaly bark); habitat uplands. *Quercus prinus*
7. Leaves obovate to broadly spatulate, leathery, sometimes shallowy lobed at the apex, margin wavy, underside pubescent; habitat scrub oak forests. *Quercus chapmanii*
8. Leaves elliptical to oblanceolate, margin entire or wavy, underside with yellow pubescence; habitat bottomlands. *Quercus oglethorpensis*
9. Leaves obovate, margin with rounded or blunt irregular teeth, underside densely white pubescent; acorn long stalked; habitat bottomlands. *Quercus bicolor*

Fagus grandifolia fruit

Quercus muehlenbergii pistillate flowers

Quercus virginiana staminate flowers

Quercus phellos staminate flowers

Quercus nigra pistillate flower

Quercus alba staminate flowers

Castanea dentata (Marsh.) Borkh.
American chestnut

Quick Guide: *Leaves* alternate, simple, elliptical, margin with bristle-tipped teeth; *Twigs* and buds glabrous; *Fruit* a triangular nut, two to three enclosed in a large bur; *Bark* of small trees smooth and brown-gray to black; *Habitat* a variety of sites.

Leaves: Simple, alternate, deciduous, oblong to elliptical or lanceolate, 13 to 27 cm long, apex acuminate, base cuneate to rounded, margin with coarse bristle-tipped teeth, underside mostly glabrous, petiole glabrous; autumn color yellow.

Twigs: Slender, red-brown to brown-gray, glabrous; leaf scar crescent shaped to oval with numerous bundle scars.

Buds: True terminal bud lacking; lateral buds chestnut brown to orange-brown, ovoid, 6 mm long, glabrous, with two to three overlapping scales.

Bark: Red-brown to gray-brown or black and smooth, becoming ridged on larger sprouts. On large trees the bark was gray-brown to dark brown with broad, interlacing ridges.

Flowers: Imperfect, staminate catkins long and stiff; pistillate flowers at the base of staminate catkins or from leaf axils; appearing after the leaves.

Fruit: Nut, shiny brown, nearly round, flattened, up to 3 cm wide, two to three nuts enclosed in a spiny bur, maturing in fall.

Form: Once a large tree up to 30 m (100 ft) in height but, because of chestnut blight, now only sprouts or small trees up to about 9 m (30 ft) tall are usually found.

Habitat and Ecology: Large trees (> 3 m in diameter) existed in the hardwood forests of the

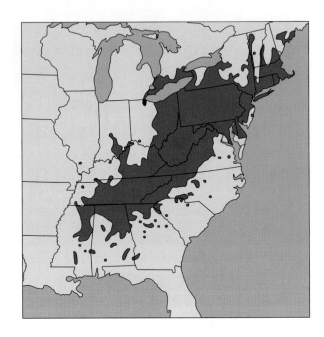

Appalachians and their logs can still be seen on the forest floor. The blight fungus survives on oaks and infects chestnut trees when the bark becomes furrowed. Sprouts are found on moist to well-drained soils.

Uses: Commercially important before the blight, the hard and durable red-brown wood was used for a range of products such as construction, flooring, paneling, furniture, caskets, and trim work; was an important shade tree. Nuts were an important food source for wild turkey, fox, white-tailed deer, small and large mammals, songbirds, and people. Leaves were used in medicinal teas and salves.

Botanical Name: *Castanea* is Latin for "chestnut tree"; *dentata* means "tooth," referring to the leaves.

Castanea dentata twig

Castanea dentata leaves

Castanea dentata bark of a small tree

Castanea dentata leaves and fruit

Fagaceae

Castanea pumila (L.) P. Mill.
Allegheny chinkapin, dwarf chestnut, chinquapin

Quick Guide: *Leaves* alternate, simple, elliptical, margin with coarse bristle-tipped teeth, underside pubescent to tomentose; *Twigs* tomentose; *Fruit* a round nut, only one enclosed in a spiny bur; *Bark* red-brown to gray, becoming fissured; *Habitat* dry, upland forests.

Leaves: Simple, alternate, deciduous, oblong to elliptical or oblanceolate, 8 to 15 cm long, apex rounded to acute, base cuneate to rounded, margin with coarse bristle-tipped teeth, underside and petiole pubescent to tomentose; autumn color yellow-brown.

Twigs: Slender, red-brown to gray, pubescent to tomentose, leaf scar crescent shaped to oval with numerous bundle scars.

Buds: Buds ovoid to round, 4 mm long, with two to three red-brown, hairy scales.

Bark: Red-brown to gray, smooth, becoming shallowly fissured.

Flowers: Imperfect, staminate catkins long and stiff; pistillate flowers at the base of staminate catkins or from leaf axils; appearing after the leaves.

Fruit: Nut, shiny brown, ovoid, up to 2 cm long, only one nut enclosed in a spiny bur, maturing in fall.

Form: Shrub or small tree up to 6 m (20 ft) in height.

Habitat and Ecology: Shade intolerant; found on sandy ridges and dry upland forests with *Carya glabra, Carya tomentosa, Juniperus virginiana,*

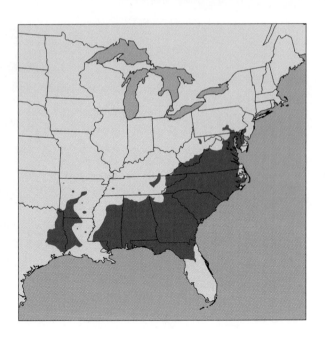

Pinus echinata, Quercus coccinea, Quercus falcata, Quercus prinus, Quercus velutina, and *Ulmus alata*.

Uses: Wood used for fences and railroad ties. Nuts eaten by wild turkey, fox, white-tailed deer, small and midsize mammals, songbirds, and people.

Botanical Name: *Castanea* is Latin for "chestnut tree"; *pumila* means "dwarf," relative to *Castanea dentata*.

Similar Species: Chinese chestnut (*Castanea mollissima* Blume) has been introduced and identified by leathery leaves with white tomentose hair on the underside and buds with white hairs. *Castanea mollissima* is planted as an ornamental for its edible nuts, but the flowers are foul smelling and the burs are very spiny.

Castanea mollissima leaf

Castanea mollissima flowers

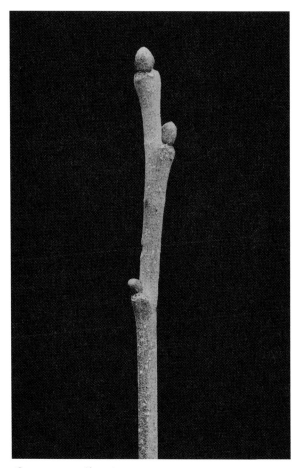
Castanea pumila twig

Castanea pumila leaves

Fagaceae

Fagus grandifolia Ehrh.
American beech, beechnut tree, beech

Quick Guide: *Leaves* alternate, simple, elliptical, each lateral vein ending at a margin tooth; *Buds* large, long pointed, cigarlike; *Fruit* a triangular nut, two to three enclosed in a weakly spiny bur; *Bark* smooth and blue-gray; *Habitat* fertile, moist soils.

Leaves: Simple, alternate, deciduous, elliptical to ovate, 8 to 13 cm long, at first very thin and pubescent but becoming more leathery, shiny and glabrous throughout the summer, apex acuminate to rounded, base obtuse, margin sharply serrate, lateral veins parallel with each vein ending at a tooth; autumn color orange-yellow, brown leaves persisting over winter.

Twigs: Slender, zig-zag, yellow-brown to red-brown, glabrous, with lenticels; leaf scar crescent shaped to half round with three or more bundle scars.

Buds: Terminal bud cigarlike, up to 2.5 cm long, acute, sharp-pointed; scales overlapping, shiny, yellow-brown to red-brown.

Bark: Blue-gray, thin, smooth.

Flowers: Imperfect, in spring with the young leaves; staminate flowers in long-stalked, drooping, ball-like clusters; pistillate spikes much smaller, born in leaf axils at the shoot tip.

Fruit: Nut, about 1.5 cm long, triangular, sweet, usually two nuts enclosed in a woody and weakly spiny bur (see page 220), maturing in fall.

Form: Up to 30 m (100 ft) in height and 1.3 m (4 ft) in diameter with a low wide crown.

Habitat and Ecology: Shade tolerant; found on fertile, moist, loamy soils of bottomlands, coves, stream edges, and north facing slopes. Forest associates include *Acer rubrum*, *Acer saccharum*, *Betula alleghaniensis*, *Liquidambar styraciflua*,

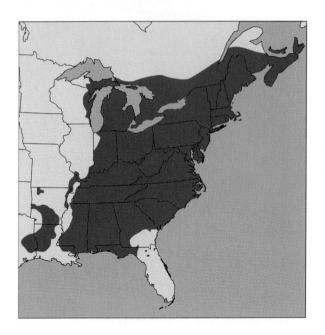

Liriodendron tulipifera, *Magnolia grandiflora*, *Magnolia virginiana*, *Picea rubens*, *Pinus strobus*, *Prunus serotina*, *Quercus* spp., *Tilia americana*, and *Ulmus americana*.

Uses: Wood yellow-brown, heavy, hard, heartwood with tyloses, growth rings distinct; used for furniture (in particular chairs), veneer, flooring, tool handles, boxes, and turnery. Porcupine and beaver gnaw the bark; nuts are relished by wild turkey, gray and red fox, ruffed grouse, wood duck, numerous songbirds, black bear, all squirrel species, and various small mammals. Would be even more valuable to wildlife but for frequent mast failures. Planted as an ornamental and can tolerate a wide range of conditions.

Botanical Name: *Fagus* is Latin for "beech tree," which comes from a Greek word for "to eat"; *grandifolia* means "large leaved."

Fagus grandifolia staminate and pistillate flowers

Fagus grandifolia bark

Fagus grandifolia twig

Fagus grandifolia leaves and fruit

Fagaceae

Quercus acutissima Carr.
sawtooth oak

Quick Guide: Leaves alternate, simple, oblong, margin with bristle-tipped teeth; *Acorn* cap with long curved scales; *Bark* dark and grooved with orange in the grooves; *Habitat* from Asia.

Leaves: Simple, alternate, deciduous, oblong to lanceolate, up to 18 cm long, margin with bristle-tipped coarse teeth, base rounded to acute, apex acute to acuminate; autumn color yellow.

Twigs: Moderately stout with lenticels; leaf scar crescent shaped to oval with numerous bundle scars.

Buds: Terminal buds 3 to 9 mm long and acute; scales red-brown with white pubescence on the margins, appearing striped.

Bark: Dark gray to black and grooved with orange in the grooves; becoming deeply grooved and corky on large trees.

Flowers: Imperfect, in spring with the leaves; staminate flowers in drooping catkins; pistillate spikes in leaf axils.

Fruit: Acorn, 3 cm long; cap with long, curved, stiff scales covering half of the nut; maturing in two seasons.

Form: Up to 21 m (70 ft) in height.

Habitat and Ecology: Imported from Asia.

Uses: Planted for white-tailed deer, wild turkey, and other acorn eaters because of mast production as early as 5 years under ideal conditions. Planted as an ornamental for the attractive form and tolerance of stress. Several cultivars are available.

Botanical Name: *Quercus* is Latin for "oak tree"; *acutissima* means "sharp point," referring to the leaves.

Quercus acutissima fruit

Quercus acutissima twig

Quercus acutissima leaves

Quercus acutissima bark

Quercus acutissima bark

Quercus alba L.
white oak, stave oak

Quick Guide: *Leaves* alternate, simple, with seven to nine round and bristleless lobes; *Acorn* cap knobby; *Bark* gray-white, loosely plated or scaly; *Habitat* a wide range of sites.

Leaves: Simple, alternate, deciduous, 10 to 18 cm long, lobes seven to nine and rounded and bristleless, sinuses extending halfway or almost to the midrib, base cuneate, apex rounded, underside pale and glabrous; autumn color dull red or orange-yellow.

Twigs: Moderately stout, red-brown, glabrous, becoming scaly on large branches; leaf scar crescent shaped to oval with numerous bundle scars.

Buds: Terminal buds 4 mm long, ovoid to round; scales overlapping, red-brown, mostly glabrous.

Bark: Gray-white to gray-brown, shallowly grooved, with small rectangular scales or large loose plates; becoming deeply grooved at the base of large trees.

Flowers: Imperfect, in spring with the leaves; staminate flowers in drooping catkins; pistillate spikes in leaf axils.

Fruit: Acorn, 2 to 3 cm long; nut shiny, yellow to light brown; cap bowl shaped with knobby glabrous scales, covering one-fourth to one-third of the nut; maturing in one season.

Form: Commonly to 30 m (100 ft) in height and 1.2 m (4 ft) in diameter.

Habitat and Ecology: Intermediate shade tolerance; found on a variety of sites including ridges, coves, sandy plains, dry slopes, and bottomlands. Forest associates are numerous and include *Acer saccharum, Fagus grandifolia, Fraxinus americana, Liquidambar styraciflua, Liriodendron tulipifera, Nyssa sylvatica, Pinus echinata, Pinus strobus, Pinus taeda, Quercus* spp., and *Tsuga canadensis*.

Uses: Wood yellow-brown, heavy, hard, heartwood with tyloses; an important commercial wood, used for flooring, furniture, trim work, ships, and staves for barrels. Bark was used in tanneries. A nice landscape tree in large areas due to bark, foliage, and form. Acorns of all the oaks rank at the top of wildlife food plants because of their wide distribution, availability, and palatability. They are considered as the staff of life for many wildlife species. Acorns of the white oak group are generally preferred over the red oak group because of lower tannic acid levels. Smaller birds and mammals prefer the species producing the smaller acorns, whereas larger birds and mammals will eat all sizes including the largest. A partial list of acorn eaters would include mallard, black, pintail, and wood ducks; ruffed and prairie grouse; northern bobwhite; wild turkey; flying fox, and gray squirrel; all species of deer; black bear; peccary; raccoon; numerous songbirds; and numerous small mammals.

Botanical Name: *Quercus* is Latin for "oak tree"; *alba* means "white," referring to the bark.

Quercus alba twig

Quercus alba leaf

Quercus alba bark

Quercus alba bark

Quercus alba fruit

Fagaceae

Quercus austrina Small.
bluff oak, bastard white oak

Quick Guide: *Leaves* alternate, simple, with three to seven rounded and bristleless lobes, sinuses and lobing variable; *Acorn* cap with thin scales; *Bark* gray-white and scaly; *Habitat* river bluffs.

Leaves: Simple, alternate, deciduous, shiny, 5 to 15 cm long, lobes three to seven and bristleless, some leaves may be unlobed, sinuses and lobes variable, base cuneate, apex rounded to acute, underside glabrous; autumn color dull red or orange-yellow.

Twigs: Red-brown, pubescent or glabrous; leaf scar crescent shaped to oval with numerous bundle scars.

Buds: Terminal buds, 4 mm long, ovoid; scales overlapping, red-brown to gray, pubescent.

Bark: Gray-white to gray-brown, loosely ridged; becoming scaly on large trees.

Flowers: Imperfect, in spring with the leaves; staminate flowers in drooping catkins; pistillate spikes in leaf axils.

Fruit: Acorn, 2.0 cm long; cap with thin pubescent scales, covering one-fourth to one-third of the nut; maturing in one season.

Form: Up to 23 m (75 ft) in height.

Habitat and Ecology: An occasional tree found on rich soils of stream edges and river bluffs in the Coastal Plain. Forest associates include *Kalmia latifolia*, *Osmanthus americanus*, *Ostrya virginiana*, *Persea borbonia*, *Prunus caroliniana*, *Quercus hemisphaerica*, and *Vaccinium arboreum*.

Uses: White oak lumber. See *Quercus alba* for wildlife uses.

Botanical Name: *Quercus* is Latin for "oak tree"; *austrina* means "south."

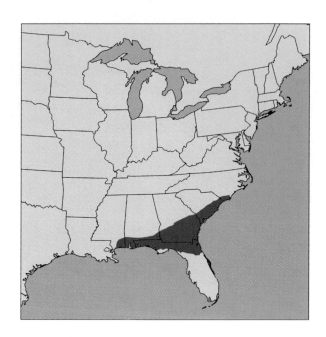

Quercus austrina fruit

Quercus austrina leaves

Quercus austrina bark

Quercus austrina twig

Quercus austrina leaves

Fagaceae

Quercus bicolor Willd.
swamp white oak

Quick Guide: *Leaves* alternate, simple, obovate, with bristleless irregular teeth, underside with white pubescence; *Acorn* born on a long stalk; *Bark* gray-brown with scaly ridges; *Habitat* bottomlands and swamps.

Leaves: Simple, alternate, deciduous, 10 to 23 cm long, obovate, with rounded or blunt bristleless teeth, base cuneate, apex abruptly acute, underside densely white or silvery pubescent; autumn color yellow.

Twigs: Stout, brown to red-brown, glabrous or pubescent; leaf scar crescent shaped to oval with numerous bundle scars. Branches are scaly and peeling.

Buds: Terminal buds 3 mm long, round; scales overlapping, brown, with gray pubescence; stipules often evident.

Bark: Gray-brown with scaly ridges on small trees; large trees more grooved with thick ridges.

Flowers: Imperfect, in spring with the leaves; staminate flowers in drooping catkins; pistillate spikes in leaf axils.

Fruit: Acorn, 2 to 3 cm long, long stalked; cap bowl shaped and fringed at the margin, covering one-third to one-half of the nut; maturing in one season.

Form: Up to 23 m (75 ft) in height and 1 m (3 ft) in diameter.

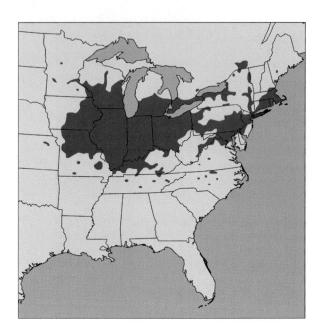

Habitat and Ecology: Intermediate shade tolerance; found in bottomlands, swamps, and along streams with *Acer rubrum, Acer saccharinum, Carya ovata, Fraxinus* spp., *Liquidambar styraciflua, Platanus occidentalis, Populus deltoides, Quercus palustris, Quercus texana, Salix nigra,* and *Ulmus americana.*

Uses: White oak lumber. See *Quercus alba* for wildlife uses. Can be used for naturalizing wet acidic soils.

Botanical Name: *Quercus* is Latin for "oak tree"; *bicolor* means "two colors," referring to the green top of the leaf and pale underside.

Quercus bicolor leaves and fruit

Quercus bicolor bark

Quercus bicolor twig

Quercus bicolor bark

Quercus bicolor fruit

Fagaceae

Quercus chapmanii Sarg.

Chapman oak

Quick Guide: *Leaves* alternate, simple, leathery, mostly unlobed but some leaves with "shoulders," apex and margin lacking bristle tips, margin wavy; *Acorn* cap covering half of the nut; *Bark* gray-brown and scaly; *Habitat* scrub oak forests.

Leaves: Simple, alternate, tardily deciduous or evergreen, leathery, dark green and shiny, oblong to obovate or spatulate, 4 to 11 cm long, usually unlobed but some leaves may show a few shallow lobes near the apex, margin often wavy, base cuneate to rounded, apex rounded without a bristle tip, underside densely or sparsely pubescent.

Twigs: Brown-gray, with pubescence, leaf scar crescent shaped to oval with numerous bundle scars.

Buds: Terminal buds 5 mm long, acute; scales overlapping, red-brown, pubescent.

Bark: Gray-brown and grooved on small trees; scaly on large trees.

Flowers: Imperfect, in spring with the leaves; staminate flowers in drooping catkins; pistillate spikes in leaf axils.

Fruit: Acorn, 1.5 to 2.5 cm long; cap deep and bowl shaped with pubescent scales, covering half or more of the nut; maturing in one season.

Form: Shrub or small tree up to 9 m (30 ft) in height.

Habitat and Ecology: An occasional tree found in scrub oak forests with *Pinus clausa, Quercus geminata, Quercus incana, Quercus laevis, Quercus marilandica, Quercus margaretta,* and *Quercus myrtifolia.*

Uses: See *Quercus alba* for wildlife uses.

Botanical Name: *Quercus* is Latin for "oak tree"; *chapmanii* is for Alan Chapman, who first named this species in the 1800s.

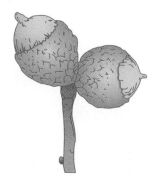

Quercus chapmanii leaves

Quercus chapmanii twig

Quercus chapmanii bark

Quercus chapmanii leaves

Fagaceae

Quercus coccinea Muenchh.

scarlet oak, red oak, Spanish oak

Quick Guide: *Leaves* alternate, simple, lobes seven to nine and bristle tipped, sinuses deep; *Acorn* often with concentric grooves on the nut apex; *Buds* with pubescence on the upper half; *Bark* gray with flat ridges, or dark and rough with white streaks; *Habitat* dry sites.

Leaves: Simple, alternate, deciduous, 10 to 20 cm long, lobes seven to nine and bristle tipped, sinuses extending more than halfway to the midrib and deepest in sun leaves, base truncate to acute, vein axils with tufts of hair; autumn color scarlet.

Twigs: Red-brown to gray-brown, glabrous; leaf scar crescent shaped to oval with numerous bundle scars.

Buds: Terminal buds 6 mm long, acute, slightly angled; scales overlapping, brown, with white or light brown pubescence on the upper half of the bud.

Bark: Gray with flat ridges; or gray-black and rough at the base and with white streaks on the middle and upper trunk; inner bark orange-brown.

Flowers: Imperfect, in spring with the leaves; staminate flowers in drooping catkins; pistillate spikes in leaf axils.

Fruit: Acorn, 1.3 to 2.5 cm long; nut pubescent, may have concentric thin grooves around the apex; cap bowl-like with shiny appressed scales, covering one-third to one-half of the nut; maturing in two seasons.

Form: Up to 30 m (100 ft) in height and 1 m (3 ft) in diameter; branches nearly horizontal, dead branches may persist in the lower canopy.

Habitat and Ecology: Shade intolerant; found on a variety of sites but commonly on dry, lightly sandy, or rocky soils with *Nyssa sylvatica*, *Oxydendrum arboreum*, *Pinus echinata*, *Pinus rigida*, *Pinus strobus*, *Pinus taeda*, *Pinus virginiana*,

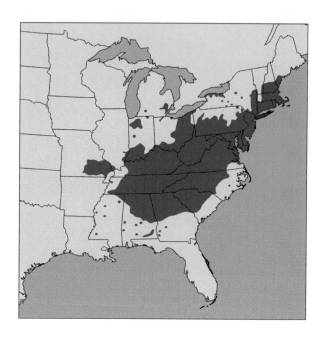

Quercus alba, *Quercus falcata*, *Quercus marilandica*, *Quercus prinus*, *Quercus rubra*, *Quercus stellata*, and *Quercus velutina*.

Uses: Red oak lumber, trim, flooring, and furniture. Planted as an ornamental for the fall color and site adaptability. See *Quercus alba* for wildlife uses.

Botanical Name: *Quercus* is Latin for "oak tree"; *coccinea* means "scarlet," referring to the autumn leaves.

Quercus coccinea fruit

Quercus coccinea shade leaf

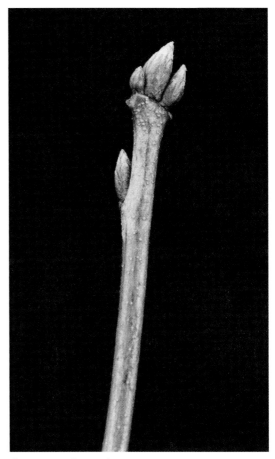

Quercus coccinea twig

Quercus coccinea bark

Quercus coccinea bark

Fagaceae

Quercus durandii Buckl.

Durand oak, Durand white oak

Quick Guide: *Leaves* alternate, simple, spatulate, unlobed or shallowly lobed near the apex, lacking bristle tips, underside densely hairy; *Acorn* cap covering the base of the nut; *Bark* gray-brown and scaly; *Habitat* limestone soils and bottomlands.

Leaves: Simple, alternate, tardily deciduous, dark green and shiny, obovate to spatulate, 12 to 18 cm long, margin sinuate, unlobed or shallowly lobed at the apex, base cuneate, apex rounded without a bristle tip, underside with dense stellate pubescence, midrib and petiole yellow.

Twigs: Gray-brown, glabrous; leaf scar crescent shaped to oval with numerous bundle scars.

Buds: Terminal buds 5 mm long, ovoid; scales overlapping, red-brown, glabrous or slightly pubescent.

Bark: Gray-brown, scaly.

Flowers: Imperfect, in spring with the leaves; staminate flowers in drooping catkins; pistillate spikes in leaf axils.

Fruit: Acorn, 1.3 to 2 cm long; nut brown, flat topped; cap saucer shaped with pubescent scales, covering only the base of the nut; maturing in one season.

Form: Up to 27 m (90 ft) in height.

Habitat and Ecology: An occasional tree found on limestone soils, and in hummocks and bottomlands of the Coastal Plain.

Uses: White oak lumber and as a shade tree; see *Quercus alba* for wildlife uses.

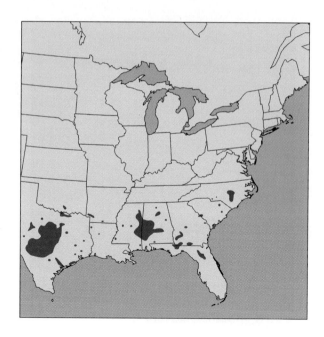

Botanical Name: *Quercus* is Latin for "oak tree"; *durandii* is for the botanist Elias Durand. The taxonamy of this species is confusing. Some consider *Quercus austrina* as *Quercus durandii* or both as *Quercus sinuata* Walt. *Quercus durandii* is also used for Bigelow oak and short lobed oak.

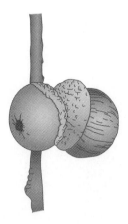

Quercus durandii leaves

Quercus durandii bark

Quercus durandii twig

Fagaceae

Quercus ellipsoidalis E. J. Hill

northern pin oak, jack oak, black oak, Hill's oak

Quick Guide: Separated from *Quercus palustris* by *Acorns* longer and ellipsoidal with the cap covering one-third of the nut; *Habitat* drier sites.

Leaves: Simple, alternate, deciduous, 8 to 18 cm long, lobes five to seven and occasionally nine and bristle tipped, sinuses extending more than halfway to the midrib and C or U shaped, base truncate, vein axils with tufts of hair; autumn color red.

Twigs: Red-brown, pubescent or glabrous; leaf scar crescent shaped to oval with numerous bundle scars.

Buds: Terminal buds 5 mm long, ovoid-acute; scales overlapping, brown, glabrous or lightly pubescent.

Bark: Gray-brown and smooth on small trees, becoming grooved and darker on large trees, inner bark yellow compared to the reddish inner bark of *Quercus palustris*.

Flowers: Imperfect, in spring with the new leaves; staminate flowers in drooping catkins; pistillate spikes in leaf axils.

Fruit: Acorn, 1.2 to 2 cm long, ellipsoidal, the cap covering one-third to one-half of the brown nut; maturing in two seasons.

Form: Up to 21 m (70 ft) in height with some branches forming 90-degree angles and the lower crown often retaining drooping dead branches.

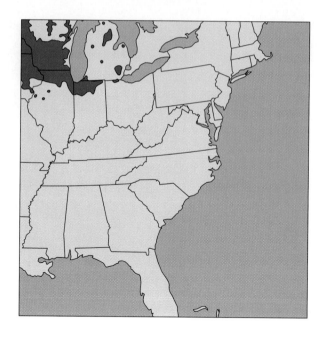

Habitat and Ecology: Shade intolerant; found on sandy and clayey upland soils with *Quercus alba*, *Quercus coccinea*, *Quercus macrocarpa*, *Quercus marilandica*, *Quercus muehlenbergii*, *Quercus prinus*, *Quercus rubra*, *Quercus velutina*, *Pinus banksiana*, *Pinus resinosa*, *Pinus strobus*, *Populus grandidentata*, and *Populus tremuloides*.

Uses: The heavy strong wood is used for shingles, flooring, furniture, and trim. See *Quercus alba* for wildlife uses.

Botanical Name: *Quercus* is Latin for "oak tree"; *ellipsoidalis* refers to the shape of the acorn.

Quercus ellipsoidalis fruit

Quercus ellipsoidalis bark

Quercus ellipsoidalis leaves

Quercus falcata Michx.
southern red oak, Spanish oak

Quick Guide: *Leaves* alternate, simple, drooping, lobes three to seven and bristle tipped, terminal lobe elongated, underside densely hairy, base bell shaped; *Acorn* cap covering up to half of the pubescent nut; *Bark* dark and rough; *Habitat* dry, upland forests.

Leaves: Simple, alternate, deciduous, drooping, 10 to 23 cm long, lobes three to seven and bristle tipped, terminal lobe elongated and sickle shaped, sinuses long and irregular, base bell shaped to obtuse but more obtuse on juvenile trees, underside with dense yellow or rusty pubescence or tomentum; autumn color yellow-brown. Three lobed leaves typical on seedlings.

Twigs: Gray-brown, pubescent, leaf scar crescent shaped to oval with numerous bundle scars.

Buds: Terminal buds up to 6 mm long, ovoid to acute, slightly angled; scales overlapping, red-brown, pubescent.

Bark: Dark gray to black and rough on small trees; deeply grooved and ridged on large trees; inner bark slightly yellow.

Flowers: Imperfect, in spring with the leaves; staminate flowers in drooping catkins; pistillate spikes in leaf axils.

Fruit: Acorn, 1.3 cm long; nut pubescent, sometimes striped; cap bowl-like with appressed pubescent scales, covering one-third to one-half of the nut; maturing in two seasons.

Form: Up to 24 m (80 ft) in height and 1 m (3 ft) in diameter.

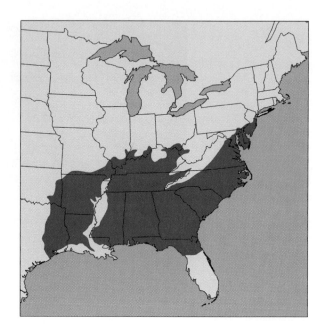

Habitat and Ecology: Intermediate shade tolerance; found mainly in dry upland forests with *Nyssa sylvatica, Pinus echinata, Pinus rigida, Pinus taeda, Pinus virginiana, Quercus alba, Quercus coccinea, Quercus marilandica, Quercus prinus, Quercus stellata,* and *Quercus velutina.*

Uses: Red oak lumber, construction, and furniture. See *Quercus alba* for wildlife uses.

Botanical Name: *Quercus* is Latin for "oak tree"; *falcata* means "scythe or sickle shaped," referring to the leaves.

Quercus falcata leaves

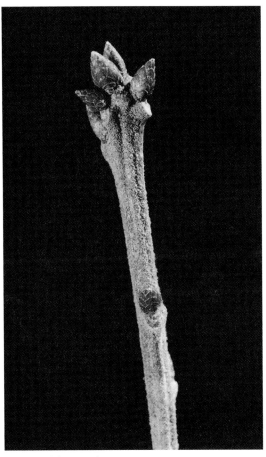
Quercus falcata twig

Quercus falcata bark

Quercus falcata fruit

Fagaceae

Quercus geminata Small
sand live oak

Quick Guide: *Leaves* alternate, simple, very leathery, evergreen, oblong to elliptical, margin entire and revolute, with dense gray pubescence below; *Bark* gray and deeply furrowed; *Acorn* in pairs, the cap covering one-third of the ellipsoidal nut; *Habitat* deep, sandy soils.

Leaves: Simple, alternate, evergreen, very leathery, oblong to elliptical, dark green, glabrous or with brown-gray pubescence on top, 3.5 to 12 cm long, margin entire and revolute with the leaf cupped downward, veins appear sunken, underside with dense gray pubescence, brown pubescence on the midrib.

Twigs: Brown with gray-brown pubescence; leaf scar crescent shaped to oval with numerous bundle scars.

Buds: Small, with red-brown pubescence.

Bark: Brown to dark gray and deeply furrowed.

Flowers: Imperfect, in spring with the new leaves; staminate flowers in drooping catkins; pistillate spikes in leaf axils.

Fruit: Acorn, up to 2 cm long, often in pairs; nut shiny dark brown to black when mature; cap covering one-third of the nut and short-stalked, maturing in one season.

Form: Up to 15 m (50 ft) in height but usually smaller.

Habitat and Ecology: Found in open forests on deep, sandy soils in the Coastal Plain from southeast North Carolina south through Florida and west to Mississippi and Louisiana. Forest associates include *Pinus clausa, Pinus palustris, Quercus laevis, Quercus margaretta,* and *Quercus virginiana.*

Uses: Wood used for charcoal. See *Quercus alba* for wildlife uses.

Botanical Name: *Quercus* is Latin for "oak tree"; *geminata* means "in pairs," referring to the acorns.

Quercus geminata leaves

Quercus geminata bark

Quercus geminata bark

Quercus geminata fruit

Fagaceae

Quercus georgiana Curtis

Georgia oak, Stone Mountain oak

Quick Guide: *Leaves* alternate, simple, lobes three to five with bristle tips and very irregular; *Acorn* cap covering the base of the small nut; *Bark* gray-brown, smooth or shallowly fissured; *Habitat* upland forests.

Leaves: Simple, alternate, deciduous, dark green and shiny, 5 to 10 cm long, lobes three to five with bristle tips and very irregular in size and shape, sinuses variable in length and usually shallow, base cuneate, pubescence in vein axils.

Twigs: Gray-brown, glabrous; leaf scar crescent shaped to oval with numerous bundle scars.

Buds: Terminal buds 5 mm long, acute; scales overlapping, brown, pubescent.

Bark: Gray-brown, smooth or shallowly grooved.

Flowers: Imperfect, in spring with the leaves; staminate flowers in drooping catkins; pistillate spikes in leaf axils.

Fruit: Acorn, 1.3 cm long; nut flat topped; cap saucer shaped covering only the base of the nut; maturing in two seasons.

Form: Shrub or small tree up to 9 m (30 ft) in height.

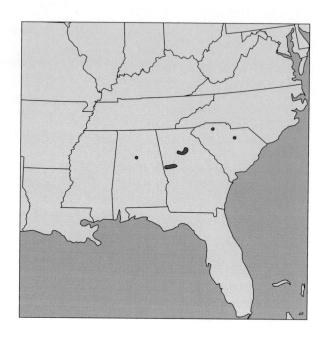

Habitat and Ecology: A rare tree found in oak-pine forests.

Uses: See *Quercus alba* for wildlife uses.

Botanical Name: *Quercus* is Latin for "oak tree"; *georgiana* is for Georgia.

Quercus georgiana fruit

Quercus georgiana twig

Quercus georgiana bark

Quercus georgiana leaves

Fagaceae

Quercus hemisphaerica Bartr. ex Willd.

laurel oak, Darlington oak, water oak, swamp laurel oak, diamond leaf oak

Quick Guide: *Leaves* alternate, simple, unlobed, oblanceolate to elliptical, margin entire or occasionally toothed, apex bristle tipped, vein axils mostly glabrous; *Acorn* cap covering the base of the small nut; *Bark* gray, smooth or shallowly ridged; *Habitat* well-drained soils.

Leaves: Simple, alternate, evergreen or tardily deciduous, leathery, shiny, elliptical to oblanceolate and sometimes obovate, 5 to 12 cm long, apex acute with a bristle tip, base acute to cuneate, margin usually entire but sometimes with a tooth or small lobe especially on saplings, midrib yellow and mostly glabrous, nearly sessile.

Twigs: Slender, gray-brown, glabrous; leaf scar crescent shaped to oval with numerous bundle scars.

Buds: Terminal buds 3 mm long, ovoid; scales overlapping, red-brown, pubescent on the margins.

Bark: Brown-gray to gray-black and smooth or shallowly ridged; very large trees with rougher ridges.

Flowers: Imperfect, in spring with the leaves; staminate flowers in drooping catkins; pistillate spikes in leaf axils.

Fruit: Acorn, 1.3 cm long; nut brown, flat topped, pubescent; cap flat with appressed pubescent scales, covering one-fourth or less of the nut; maturing in two seasons.

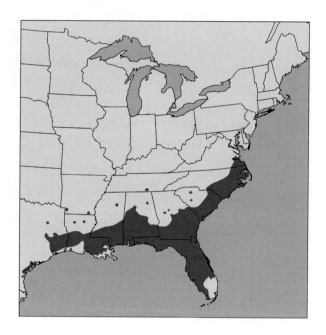

Form: Up to 24 m (80 ft) in height.

Habitat and Ecology: Moderately shade tolerant; found on edges of swamps and rivers in the Coastal Plain and on dry uplands in the Piedmont in association with *Pinus taeda* and *Pinus palustris*.

Uses: Red oak lumber, pulpwood, and fuel. Planted as an ornamental for the persistent shiny leaves. See *Quercus alba* for wildlife uses.

Botanical Name: *Quercus* is Latin for "oak tree"; *hemisphaerica* refers to the shape of the nut.

Quercus hemisphaerica leaves

Quercus hemisphaerica bark

Quercus hemisphaerica bark

Quercus hemisphaerica fruit

Fagaceae 251

Quercus ilicifolia Wangenh.

bear oak, scrub oak, dwarf black oak

Quick Guide: *Leaves* alternate, simple, lobes five to nine with bristle tips, sinuses shallow, underside with downy gray-white hair; *Acorn* cap covering up to half of the nut; *Habitat* acidic, rocky soils and disturbed areas.

Leaves: Simple, alternate, deciduous, 5 to 12 cm long, lobes mostly five but from three to nine and bristle tipped, upper lobes the widest, sinuses at most halfway to the midrib, base cuneate to obtuse, underside with gray-white velvety hair.

Twigs: Slender, brown, with gray pubescence; leaf scar crescent shaped to oval with numerous bundle scars.

Buds: Terminal buds 5 mm long, ovoid; scales overlapping, dark brown, mostly glabrous.

Bark: Gray-brown, mostly smooth.

Flowers: Imperfect, in spring with the leaves; staminate flowers in drooping catkins; pistillate spikes in leaf axils.

Fruit: Acorn, 1.3 cm long, the cap covering one-third to one-half of the nut, found even on small sprouts; maturing in two seasons.

Form: A shrub or small tree often with poor form up to 6 m (20 ft) in height and forming thickets.

Habitat and Ecology: Intermediate shade tolerance; found on acidic, dry, rocky or sandy soils and in disturbed areas. Forest associates include *Acer rubrum, Castanea pumila, Kalmia latifolia,*

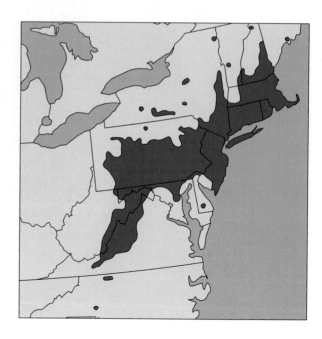

Nyssa sylvatica, Pinus echinata, Pinus pungens, Pinus rigida, Pinus virginiana, Quercus coccinea, Quercus prinus, Quercus velutina, and *Robinia pseudoacacia.*

Uses: Wood not commercially important. Fruit eaten by song- and game birds such as turkey, grouse, and quail, and by deer, bear, and small mammals. Thickets provide cover for wildlife.

Botanical Name: *Quercus* is Latin for "oak tree"; *ilicifolia* means "holly-leaved."

Quercus ilicifolia leaf

Quercus ilicifolia fruit

Quercus ilicifolia twig

Quercus ilicifolia bark

Quercus ilicifolia leaves

Fagaceae

Quercus imbricaria Michx.
shingle oak, laurel oak

Quick Guide: *Leaves* alternate, simple, unlobed, elliptical to oblong, margin entire, apex bristle tipped, underside pubescent; *Acorn* cap covering up to one-half of the nut; *Bark* gray-brown and shallowly fissured; *Habitat* moist, fertile soils.

Leaves: Simple, alternate, deciduous, leathery, shiny, dark green, elliptical to oblong or oblanceolate, 8 to 18 cm long, apex with a bristle tip, base acute to cuneate, margin entire, underside pale pubescent, nearly sessile; autumn color dull red.

Twigs: Slender, red-brown to gray, glabrous; leaf scar crescent shaped to oval with numerous bundle scars.

Buds: Terminal buds 4 mm long, acute, some angled; scales overlapping, red-brown, glabrous or pubescent.

Bark: Brown-gray to almost black, shallowly grooved and ridged on large trees.

Flowers: Imperfect, in spring with the leaves; staminate flowers in drooping catkins; pistillate spikes in leaf axils.

Fruit: Acorn, 2.0 cm long; nut pubescent; cap bowl shaped with appressed pubescent scales, covering one-third to one-half of the nut; maturing in two seasons.

Form: Up to 24 m (80 ft) in height.

Habitat and Ecology: Found on rich, moist soils of bottoms and stream banks with *Quercus lyrata*, *Quercus palustris*, and *Ulmus americana*.

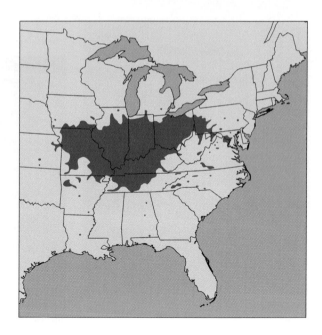

Uses: Red oak lumber, pulpwood, construction, and fuel. Planted as an ornamental for the low, round crown. See *Quercus alba* for wildlife uses.

Botanical Name: *Quercus* is Latin for "oak tree"; *imbricaria* means "overlapping," and refers to the past use of the wood for shingles.

Quercus imbricaria fruit

Quercus imbricaria bark

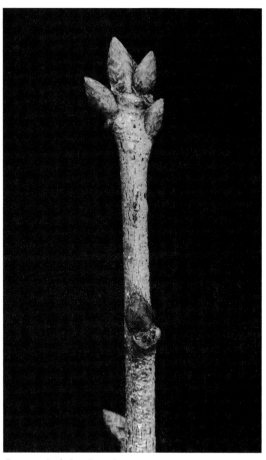
Quercus imbricaria twig

Quercus imbricaria bark

Quercus imbricaria leaves

Fagaceae

Quercus incana Bartr.
bluejack oak, sand jack oak, scrub oak

Quick Guide: *Leaves* alternate, simple, oblong or oblanceolate, margin entire, apex bristle tipped, underside with blue-gray pubescence; *Acorn* cap covering up to one-third of the small nut; *Bark* dark, rough, and blocky; *Habitat* well-drained, sandy soils.

Leaves: Simple, alternate, deciduous, leathery, shiny, oblong to elliptical oblanceolate or lanceolate, 5 to 12 cm long, apex obtuse with a bristle tip, base acute, margin entire or with an occasional tooth or small lobe particularly on seedlings, blue-green above, underside with prominent blue-gray or white pubescence, nearly sessile; autumn color yellow-brown.

Twigs: Slender, gray-brown, with blue-gray pubescence; leaf scar crescent shaped to oval with numerous bundle scars.

Buds: Terminal buds 5 mm long, acute; scales overlapping, red-brown, lightly pubescent.

Bark: Gray-black, rough, blocky.

Flowers: Imperfect, in spring with the leaves; staminate flowers in drooping catkins; pistillate spikes in leaf axils.

Fruit: Acorn, about 1 cm long; nut pubescent; cap bowl shaped with pubescent appressed scales, covering one-fourth to one-third of the nut; maturing in two seasons.

Form: Small tree up to 9 m (30 ft) tall.

Habitat and Ecology: Shade intolerant; found on dry, sandy soils with *Carya pallida, Pinus clausa, Pinus palustris, Quercus laevis, Quercus margaretta,* and *Quercus marilandica*.

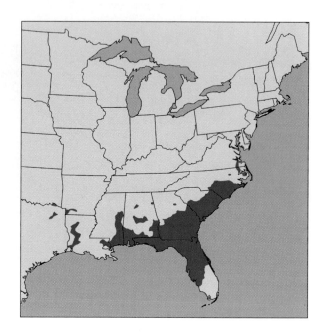

Uses: Red oak lumber, fuel, and railroad ties. See *Quercus alba* for wildlife uses.

Botanical Name: *Quercus* is Latin for "oak tree"; *incana* means "gray," and refers to the grayish pubescence on the leaf underside.

Quercus incana fruit

Quercus incana leaves

Quercus incana bark

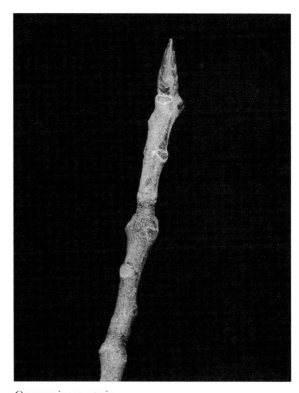

Quercus incana twig

Fagaceae

Quercus laevis Walt.
turkey oak, scrub oak

Quick Guide: *Leaves* alternate, simple, petiole twisted, vertical, lobes three to seven, three-lobed leaves resemble a turkey foot track, base tapered; *Acorn* cap with a rolled lip; *Bark* dark and rough; *Habitat* well-drained, sandy soils.

Leaves: Simple, alternate, deciduous, 8 to 20 cm long, shiny, often hanging vertically, lobes three to seven and bristle tipped, lateral lobes curved, three-lobed leaves resembling a turkey foot track, midrib curved on some leaves, sinuses irregular and may extend more than halfway to the midrib, base acute to cuneate, shiny above, underside pubescent in vein axils; autumn color red.

Twigs: Stout, red-brown, glabrous or pubescent; leaf scar crescent shaped to oval with numerous bundle scars.

Buds: Terminal buds 1.2 cm long, acute; scales overlapping, red-brown, with rusty pubescence.

Bark: Dark gray to black, rough, possibly blocky.

Flowers: Imperfect, in spring with the leaves; staminate flowers in drooping catkins; pistillate spikes in leaf axils.

Fruit: Acorn, 2 cm long; nut pubescent; cap bowl-like with appressed pubescent scales and a rolled edge, covering one-third of the nut; maturing in two seasons.

Form: Up to 15 m (50 ft) tall but usually smaller.

Habitat and Ecology: Shade intolerant; found on well-drained, sandy soils with *Carya pallida*, *Quercus incana*, *Quercus margaretta*, *Quercus marilandica*, *Pinus clausa*, and *Pinus palustris*.

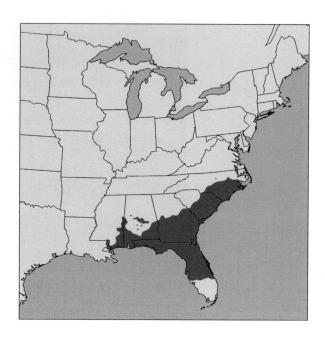

Uses: Red oak lumber and fuel. See *Quercus alba* for wildlife uses.

Botanical Name: *Quercus* is Latin for "oak tree"; *laevis* means "smooth," perhaps referring to the shiny leaves.

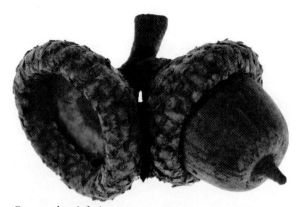

Quercus laevis fruit

Quercus laevis leaf

Quercus laevis twig

Quercus laevis bark

Quercus laevis leaves

Fagaceae

Quercus laurifolia Michx.

swamp laurel oak, Darlington oak, water oak, laurel oak, diamond leaf oak

Quick Guide: *Leaves* alternate, simple, leathery, variable in shape but often spatulate or subrhombic, margin entire; *Bark* gray-black and shallowly ridged; *Habitat* bottomlands and swamps.

Leaves: Simple, alternate, persistent, leathery, variable in shape (spatulate, obovate, subrhombic, oblanceolate), 5 to 14 cm long, midrib yellow, apex with or without a bristle tip, base cuneate, margin usually entire or occasionally with a coarse tooth or shallow lobe, nearly sessile, vein axils with tufts of hair.

Twigs: Moderately stout, gray-brown, glabrous; leaf scar crescent shaped to oval with numerous bundle scars.

Buds: Terminal buds 3 mm long, ovoid; scales overlapping, red-brown, with pubescent margins.

Bark: Gray-brown to black with flattened ridges; blocky on very large trees.

Flowers: Imperfect, in spring with the leaves; staminate flowers in drooping catkins; pistillate spikes in leaf axils.

Fruit: Acorn, 1.5 cm long; nut brown, flat topped, pubescent; cap flat with appressed pubescent scales, covering one-fourth to one-half of the nut; maturing in two seasons.

Form: Up to 30 m (100 ft) in height, often with a swollen buttressed trunk.

Habitat and Ecology: Moderately shade tolerant; found in poorly drained areas, in particular low flats, with *Acer rubrum, Cliftonia monophylla, Cyrilla racemiflora, Magnolia virginiana, Nyssa biflora, Pinus serotina, Quercus lyrata, Quercus michauxii,* and *Quercus virginiana.*

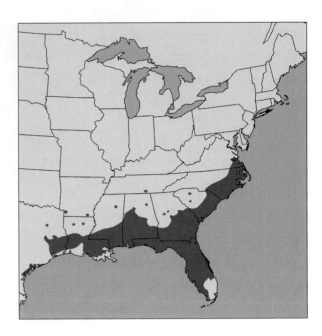

Uses: Red oak lumber, pulpwood, and fuel. See *Quercus alba* for wildlife uses.

Botanical Name: *Quercus* is Latin for "oak tree"; *laurifolia* means "laurel tree leaves."

Quercus laurifolia fruit

Quercus laurifolia leaves

Quercus laurifolia bark

Quercus laurifolia bark

Quercus laurifolia leaves

Quercus lyrata Walt.

overcup oak, swamp post oak, water white oak

Quick Guide: *Leaves* alternate, simple, lobes five to nine without bristles and irregular in shape, sinuses variable; *Acorn* round and almost completely enclosed by a thin cap with fused scales; *Bark* gray-brown with scaly ridges; *Habitat* bottomlands and water edges.

Leaves: Simple, alternate, deciduous, 12 to 20 cm long, lobes five to nine and bristleless and very irregular, sinuses irregular and may extend more than halfway to the midrib, base cuneate, underside pubescent; autumn color yellow-brown.

Twigs: Slender, brown-gray, glabrous; leaf scar crescent shaped to oval with numerous bundle scars.

Buds: Terminal buds 3 mm long, round; scales overlapping, brown, pubescent.

Bark: Gray-brown, shallowly grooved, with scaly ridges and plates.

Flowers: Imperfect, in spring with the leaves; staminate flowers in drooping catkins; pistillate spikes in leaf axils.

Fruit: Acorn, 2 cm long, buoyant, round; nut shiny brown; cap loose and thin with fused scales, sometimes splitting, almost completely covering the nut; maturing in one season.

Form: Up to 30 m (100 ft) in height and 1 m (3 ft) in diameter.

Habitat and Ecology: Intermediate shade tolerance; found in wet bottomlands, floodplains, and shallow swamps, and along edges of deep

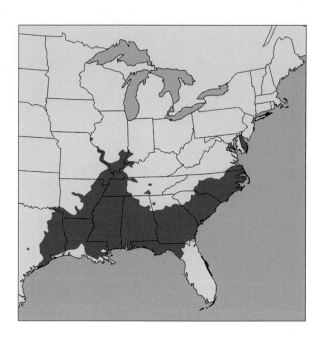

swamps, rivers, sloughs, and sink holes. Forest associates include *Acer rubrum, Carya aquatica, Carya cordiformis, Carya ovata, Celtis laevigata, Diospyros virginiana, Fraxinus* spp., *Liquidambar styraciflua, Nyssa* spp., *Quercus laurifolia, Quercus pagoda, Quercus phellos, Quercus texana, Quercus virginiana,* and *Ulmus americana.*

Uses: White oak lumber. See *Quercus alba* for wildlife uses. Planted as an ornamental but acorns can be messy.

Botanical Name: *Quercus* is Latin for "oak tree"; *lyrata* means "lyrelike."

Quercus lyrata twig

Quercus lyrata leaves

Quercus lyrata bark

Quercus lyrata fruit

Quercus macrocarpa Michx.

bur oak, mossycup oak, mossy-overcup oak

Quick Guide: *Leaves* alternate, simple, leathery, lobes five to nine and without bristles, apex fan shaped, middle sinus the deepest; *Twigs* with corky wings; *Acorn* large with a fringed cap; *Bark* gray-brown with scaly ridges; *Habitat* a variety of sites.

Leaves: Simple, alternate, deciduous, leathery, 10 to 25 cm long, lobes five to nine and bristleless, terminal lobe usually fan shaped, sinuses irregular, on some leaves the middle sinus deepest and possibly elongated, base acute, underside pubescent; autumn color yellow.

Twigs: Stout, gray to red-brown, older twigs with corky ridges; leaf scar crescent shaped to oval with numerous bundle scars.

Buds: Terminal buds 5 mm long, ovoid; scales overlapping, red-brown, with tawny pubescence; stipules often present.

Bark: Gray-brown and shallowly furrowed, ridges may be scaly.

Flowers: Imperfect, in spring with the leaves; staminate flowers in drooping catkins; pistillate spikes in leaf axils.

Fruit: Acorn, large (largest of the native oaks), 2 to 5 cm long; cap prominently "mossy fringed," pubescent, bowl shaped, and covering half to most of the downy nut; maturing in one season.

Form: Up to 30 m (100 ft) in height and 1.2 m (4 ft) in diameter with a broad crown.

Habitat and Ecology: Intermediate shade tolerance; found on a variety of sites including rock bluffs, limestone soils, and bottomlands. It is a component of many forest types including the oak-basswood, upland oak-hickory, sugar maple-basswood, and mixed oak communities and can form bur oak savannas. Best growth is on moist sites with *Acer rubrum*, *Acer saccharum*, *Carya* spp., *Fraxinus nigra*, *Populus deltoides*, *Quercus bicolor*,

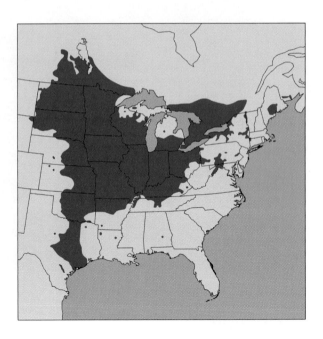

Quercus ellipsoidalis, *Quercus lyrata*, *Quercus palustris*, *Quercus phellos*, *Quercus texana*, *Quercus michauxii*, and *Ulmus americana*.

Uses: White oak lumber. Commonly planted as an ornamental. See *Quercus alba* for wildlife uses.

Botanical Name: *Quercus* is Latin for "oak tree"; *macrocarpa* means "large fruit," referring to the acorn.

Quercus macrocarpa fruit

Quercus macrocarpa leaves

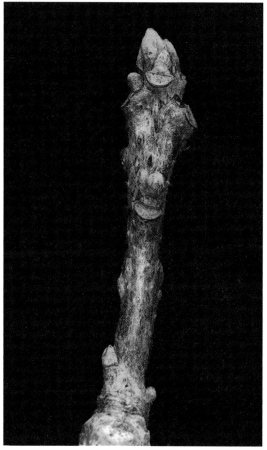

Quercus macrocarpa twig

Quercus macrocarpa bark

Quercus macrocarpa leaves

Fagaceae

Quercus margaretta Ashe

sand post oak, scrubby post oak, dwarf post oak

Quick Guide: *Leaves* alternate, simple, leathery, lobes without bristles and pointing toward the apex, underside pubescent; *Twigs* mostly glabrous; *Bark* gray-brown with scaly ridges; *Habitat* well-drained sandy soils.

Leaves: Simple, alternate, deciduous, leathery, 2 to 15 cm long, lobes three to five and lacking bristles, lobes usually pointing toward the apex, central lobes only occasionally crosslike, some leaves may be unlobed, sinuses shallow and variable, base rounded to cuneate, pubescent below; autumn color yellow-brown or dull red.

Twigs: Slender, red-brown, glabrous or with some pubescence; leaf scar crescent shaped to oval with numerous bundle scars.

Buds: Terminal buds 3 mm long, ovoid, angled; scales overlapping, red-brown, glabrous or with pubescence.

Bark: Gray-brown with scaly ridges.

Flowers: Imperfect, in spring with the leaves; staminate flowers in drooping catkins; pistillate spikes in leaf axils.

Fruit: Acorn, 1.3 to 2 cm long; nut pubescent; cap bowl shaped with pubescent appressed scales, covering one-third to one-half of the nut; maturing in one season.

Form: Up to 18 m (60 ft) in height but usually smaller, can form thickets.

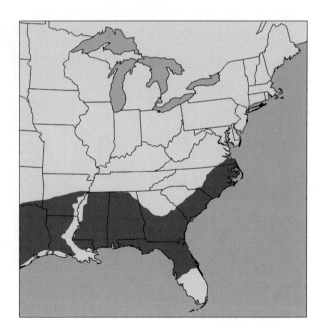

Habitat and Ecology: Shade intolerant; found on well-drained upland sites and on dry, sandy soils with *Carya pallida, Pinus clausa, Pinus palustris, Quercus incana, Quercus laevis,* and *Quercus marilandica.*

Uses: Wood similar to *Quercus stellata* but not commercially important. See *Quercus alba* for wildlife uses.

Botanical Name: *Quercus* is Latin for "oak tree"; *margaretta* is for Margaret Ashe. Also named *Quercus margarettiae* Ashe ex Small.

Quercus margaretta leaves

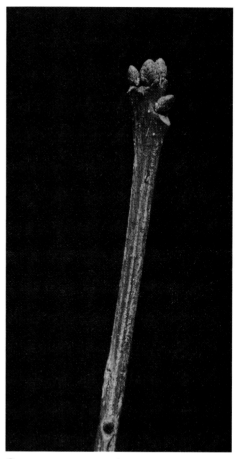

Quercus margaretta twig

Quercus margaretta bark

Quercus margaretta fruit

Quercus marilandica Muenchh.
blackjack oak, black oak, barren oak, jack oak

Quick Guide: *Leaves* alternate, simple, fan shaped, leathery, apex with three broad and bristle-tipped lobes; *Acorn* cap densely hairy and covering up to one-half of the nut; *Bark* dark, rough, and blocky; *Habitat* sandhills and dry upland forests.

Leaves: Simple, alternate, deciduous, T shaped or fan shaped, very leathery, 8 to 20 cm long, apex with three broad and bristle-tipped lobes, base acute to cordate, underside with orangish or yellow pubescence; autumn color yellow-brown.

Twigs: Stout, brown-gray, pubescent; leaf scar crescent shaped to oval with numerous bundle scars.

Buds: Terminal buds 6 mm long, acute, angled; scales overlapping, red-brown, pubescent.

Bark: Gray-black, rough, blocky.

Flowers: Imperfect, in spring after the leaves; staminate flowers in drooping catkins; pistillate spikes in leaf axils.

Fruit: Acorn, 2 cm long; nut pubescent with a rigid point; cap bowl shaped with shaggy pubescent scales, covering one-third to one-half of the nut; maturing in two seasons.

Form: Up to 15 m (50 ft) in height.

Habitat and Ecology: Shade intolerant; found on dry, sandy soils with *Carya pallida, Pinus clausa, Pinus palustris, Quercus incana, Quercus laevis,* and *Quercus margaretta*. Also found on thin, rocky, upland soils with *Oxydendrum arboreum, Quercus falcata, Quercus prinus,* and *Pinus echinata*.

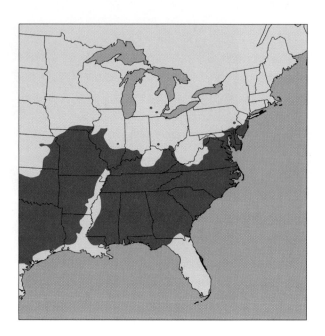

Uses: Red oak lumber, fuel, fences, and railroad ties. See *Quercus alba* for wildlife uses.

Botanical Name: *Quercus* is Latin for "oak tree"; *marilandica* refers to the geographic range.

Similar Species: Arkansas oak (*Quercus arkansana* Sarg.) is an uncommon tree up to 20 m tall found on well-drained, sandy soils in central Alabama, southern Georgia, northwest Florida, and Arkansas. Distinguished by leaves similar to *Quercus marilandica* but thinner, not as three lobed at the apex, and lacking rusty pubescence below, and by an acorn cap covering only the base of the nut.

Quercus marilandica fruit

Quercus marilandica twig

Quercus marilandica bark

Quercus marilandica leaves

Fagaceae

Quercus michauxii Nutt.

swamp chestnut oak, cow oak, basket oak

Quick Guide: *Leaves* alternate, simple, unlobed, obovate throughout the canopy, margin with round and callous-tipped teeth; *Acorn* large with a scaly cap; *Bark* gray-brown and scaly; *Habitat* well-drained bottomlands and stream edges.

Leaves: Simple, alternate, deciduous, obovate, 12 to 22 cm long, margin with rounded and callous-tipped teeth, base cuneate to rounded, apex may be abruptly acuminate, underside pubescent; autumn color yellow-brown to dull red.

Twigs: Stout, red-brown to gray-brown, glabrous to densely pubescent; leaf scar crescent shaped to oval with numerous bundle scars.

Buds: Terminal buds 3 to 6 mm long, ovoid; scales overlapping, red-brown, lightly pubescent.

Bark: Light gray to gray-brown, scaly.

Flowers: Imperfect, in spring with the leaves; staminate flowers in drooping catkins; pistillate spikes in leaf axils.

Fruit: Acorn, large, 2.5 to 4 cm long; cap bowl shaped with pubescent scales and a thin edge, covering one-third of the nut; maturing in one season.

Form: Up to 30 m (100 ft) in height with a long straight bole.

Habitat and Ecology: Shade intolerant; found in well-drained bottomlands and along edges of streams and swamps with *Acer rubrum, Carya cordiformis, Carya ovata, Fraxinus* spp., *Liquidambar styraciflua, Nyssa sylvatica, Pinus glabra, Quercus alba, Quercus lyrata, Quercus pagoda, Quercus phellos,* and *Quercus shumardii*.

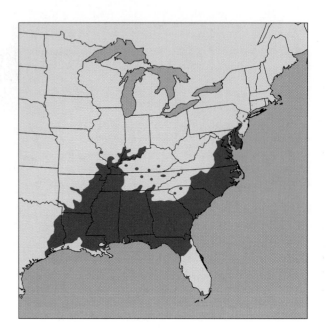

Uses: White oak lumber, was used in making baskets. See *Quercus alba* for wildlife uses.

Botanical Name: *Quercus* is Latin for "oak tree"; *michauxii* is for the French botanist F. Michaux.

Quercus michauxii fruit

Quercus michauxii leaves

Quercus michauxii bark

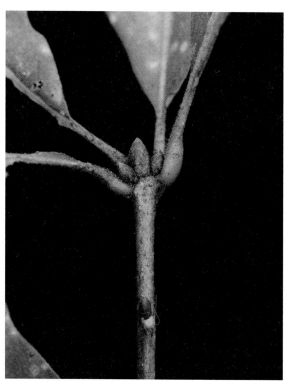

Quercus michauxii twig

Quercus muehlenbergii Engelm.

chinkapin oak, scrub chestnut oak, yellow oak, yellow chestnut oak, rock oak

Quick Guide: *Leaves* alternate, simple, unlobed, elliptical, margin with curved callous-tipped teeth; *Acorn* cap with appressed scales covering one-third of the nut; *Bark* gray-brown and scaly; *Habitat* alkaline soils.

Leaves: Simple, alternate, deciduous, 9 to 18 cm long, obovate in the lower canopy, elliptical to lanceolate and smaller in the upper canopy, margin with curved callous-tipped teeth but sometimes the margin is only wavy, apex often acuminate, base acute, underside pubescent; autumn color yellow-brown or dull red.

Twigs: Slender, red-brown to gray, glabrous or lightly pubescent; leaf scar crescent shaped to oval with numerous bundle scars.

Buds: Terminal buds 3 to 5 mm long, acute; scales overlapping, red-brown, with white pubescence on the margins, appearing striped.

Bark: Gray-brown, scaly.

Flowers: Imperfect, in spring with the leaves; staminate flowers in drooping catkins; pistillate spikes in leaf axils.

Fruit: Acorn, 1.3 to 2.5 cm long; nut brown to black; cap thin and bowl shaped with appressed pubescent scales and a thin edge covering one-third to one-half of the nut; maturing in one season.

Form: Up to 24 m (80 ft) in height and 1 m (3 ft) in diameter.

Habitat and Ecology: Shade intolerant; found on alkaline soils of bluffs and upland slopes, and along stream edges. Forest associates include *Acer rubrum, Carya* spp., *Fagus grandifolia, Fraxinus* spp., *Juniperus virginiana, Liriodendron tulipifera, Pinus echinata, Pinus rigida, Pinus virginiana, Quercus alba, Quercus coccinea, Quercus falcata,*

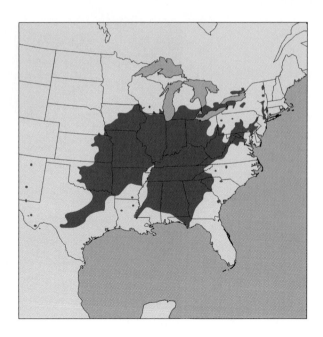

Quercus prinus, Quercus rubra, Quercus velutina, and *Ulmus alata.*

Uses: White oak lumber, construction, and fuel. See *Quercus alba* for wildlife uses.

Botanical Name: *Quercus* is Latin for "oak tree"; *muehlenbergii* is for the American botanist Muhlenberg.

Quercus muehlenbergii fruit

Quercus muehlenbergii leaves

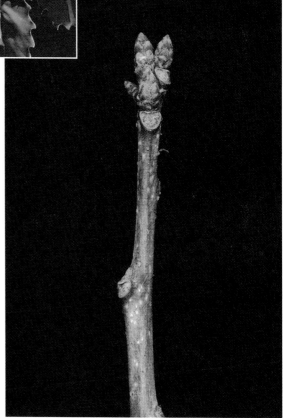
Quercus muehlenbergii twig

Quercus muehlenbergii bark

Fagaceae

Quercus myrtifolia Willd.
myrtle oak

Quick Guide: *Leaves* alternate, simple, leathery, variable in shape, margin entire and often cupped under, mostly glabrous; *Acorn* cap covering one-fourth of the nut; *Habitat* pine-oak scrub forests.

Leaves: Simple, alternate, evergreen, very leathery, dark green and shiny, variable in shape from oblong to oblanceolate, obovate, or oval, 2 to 8 cm long, apex sometimes bristle tipped, margin entire and revolute, mostly glabrous below when mature and with a yellow midrib.

Twigs: Gray-brown, with dense pubescence; leaf scar crescent shaped to oval with numerous bundle scars.

Buds: Terminal buds 3 mm long, ovoid-acute; scales red-brown and fringed with pubescence.

Bark: Gray-brown, smooth, becoming ridged on larger trees.

Flowers: Imperfect, in spring with the new leaves; staminate flowers in drooping catkins; pistillate spikes in leaf axils.

Fruit: Acorn, about 1.3 cm long, cap enclosing one-fourth of the nut; maturing in two seasons.

Form: Thicket forming shrub or gnarly formed tree up to 9 m (30 ft) in height.

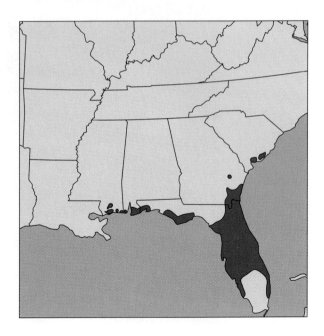

Habitat and Ecology: Found in pine-oak scrub forests with *Pinus clausa, Pinus elliottii, Pinus palustris, Quercus chapmanii, Quercus geminata, Quercus incana, Quercus laevis,* and *Quercus margaretta*.

Uses: See *Quercus alba* for wildlife uses.

Botanical Name: *Quercus* is Latin for "oak tree"; *myrtifolia* is for the myrtle like leaves.

Quercus myrtifolia fruit

Quercus myrtifolia leaves

Quercus myrtifolia bark

Quercus myrtifolia twig

Fagaceae

Quercus nigra L.
water oak, opossum oak, fiddle oak

Quick Guide: *Leaves* alternate, simple, spatulate, margin usually entire, some leaves may be one to three pronged, apex with a bristle tip; *Acorn* cap covering the base of the small, dark, flat-topped nut; *Bark* brown-gray and smooth or shallowly ridged; *Habitat* uplands to bottomlands.

Leaves: Simple, alternate, persistent, leathery, obovate to spatulate but very variable, 5 to 10 cm long, margin usually entire or may be lobed near the apex, apex obtuse to rounded and bristle tipped, base acute to cuneate, nearly sessile, underside mostly glabrous; autumn color yellow-brown. Leaves of seedlings, saplings, and root sprouts variable, often larger, oblong, and lobed.

Twigs: Slender, red-brown, glabrous; leaf scar crescent shaped to oval with numerous bundle scars.

Buds: Terminal buds 5 mm long, acute; scales overlapping, gray-brown to red-brown, glabrous or pubescent.

Bark: Brown-gray and smooth or shallowly ridged.

Flowers: Imperfect, in spring with the leaves; staminate flowers in drooping catkins; pistillate spikes in leaf axils.

Fruit: Acorn, 1.2 cm long; nut dark brown, pubescent, flat topped; cap flat with appressed pubescent scales, covering one-fourth of the nut; maturing in two seasons.

Form: Up to 36 m (120 ft) in height, possibly with mistletoe in the canopy.

Habitat and Ecology: Shade intolerant; found on a variety of sites including dry uplands and moderately wet bottomlands. Forest associates are numerous and include *Fagus grandifolia, Fraxinus* spp., *Liquidambar styraciflua, Quercus michauxii, Quercus pagoda, Quercus phellos, Pinus elliottii, Pinus palustris, Pinus taeda, Ulmus alata,* and *Ulmus americana.*

Uses: Red oak lumber, construction, and fuel. See *Quercus alba* for wildlife uses.

Botanical Name: *Quercus* is Latin for "oak tree"; *nigra* means "dark," referring to the nut.

Quercus nigra fruit

Quercus nigra twig

Quercus nigra bark

Quercus nigra leaves

Fagaceae

Quercus oglethorpensis Duncan
Oglethorpe oak

Quick Guide: *Leaves* alternate, simple, oblong to elliptical, margin entire, underside with yellow pubescence; *Bark* brown-gray to orange-brown and scaly; *Acorn* cap covering one-third of the nut; *Habitat* moist soils, rare.

Leaves: Simple, alternate, deciduous, oblong, elliptical, or oblanceolate, 4 to 17 cm long, apex acute to rounded and lacking a bristle tip, base cuneate, margin entire and sometimes wavy, underside with yellow pubescence.

Twigs: Gray-brown, densely hairy when new; leaf scar crescent shaped to oval with numerous bundle scars.

Buds: Terminal buds 2 to 5 mm long, ovoid, with red-brown scales.

Bark: Orange-brown to gray-brown and scaly.

Flowers: Imperfect, in spring with the leaves; staminate flowers in drooping catkins; pistillate spikes in leaf axils.

Fruit: Acorn, 1 to 2 cm long, cap covering one-third of the nut, maturing in one season.

Form: Up to 24 m (80 ft) in height.

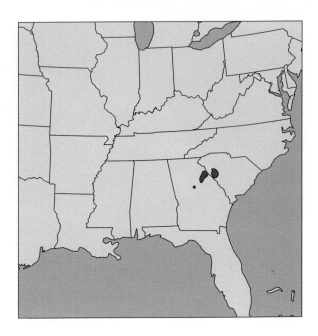

Habitat and Ecology: Rare, found in bottoms and next to streams.

Uses: See *Quercus alba* for wildlife uses.

Botanical Name: *Quercus* is Latin for "oak tree"; *oglethorpensis* refers to the county in Georgia where this tree is found.

Quercus oglethorpensis twig

Quercus oglethorpensis bark

Quercus oglethorpensis bark

Quercus oglethorpensis leaves

Fagaceae

Quercus pagoda Raf.

cherrybark oak, red oak, swamp red oak, bottomland red oak

Quick Guide: *Leaves* alternate, simple, lobes five to seven and bristle tipped, basal lobes prominent, underside densely hairy; *Acorn* cap covering up to one-half of the pubescent nut; *Bark* gray-brown, flaky or with scaly ridges; *Habitat* bottomlands and swamp edges.

Leaves: Simple, alternate, deciduous, 13 to 23 cm long, lobes mostly five to seven and bristle tipped, basal lobes prominent, base truncate to tapered, underside densely pubescent; autumn color yellow-brown. Sinuses shallow on shade leaves and deep on sun leaves with sun leaves similar to *Quercus falcata* except at the base.

Twigs: Green to red-brown with gray pubescence; leaf scar crescent shaped to oval with numerous bundle scars.

Buds: Terminal buds 5 mm long, ovoid, somewhat angled; scales overlapping, red-brown, pubescent.

Bark: On small trees smooth and on larger trees gray-brown and flaky, somewhat like black cherry; very large trees shallowly furrowed with scaly ridges.

Flowers: Imperfect, in spring with the leaves; staminate flowers in drooping catkins; pistillate spikes in leaf axils.

Fruit: Acorn, 1.3 cm long; nut pubescent; cap bowl-like with appressed pubescent scales, covering one-third to one-half of the nut; maturing in two seasons.

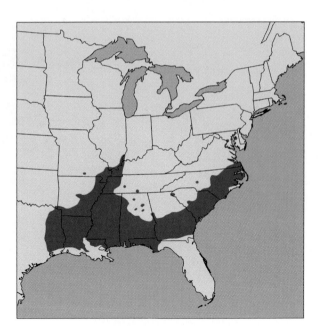

Form: Up to 39 m (130 ft) in height and 1.5 m (5 ft) in diameter with a long straight bole.

Habitat and Ecology: Shade intolerant; found in floodplain forests and on terraces, stream margins, and swamp edges with *Carya cordiformis, Carya ovata, Fraxinus* spp., *Liquidambar styraciflua, Pinus glabra, Quercus lyrata, Quercus michauxii, Quercus nigra, Quercus phellos, Quercus shumardii,* and *Ulmus americana.*

Uses: Red oak wood commercially important, used for trim, finish work, flooring, and furniture. See *Quercus alba* for wildlife uses.

Botanical Name: *Quercus* is Latin for "oak tree"; *pagoda* refers to the pagoda-shaped leaves.

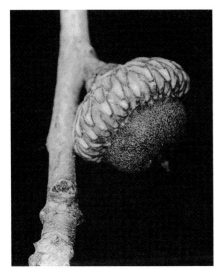
Quercus pagoda fruit

Quercus pagoda bark

Quercus pagoda twig

Quercus pagoda bark

Quercus pagoda shade leaves

Fagaceae

Quercus palustris Muenchh.

pin oak, swamp oak, swamp Spanish oak

Quick Guide: *Leaves* alternate, simple, lobes five to nine and bristle tipped, sinuses deep; *Acorn* small with a maroon-brown striped nut and cap covering the top of the nut; *Bark* gray-brown, smooth or lightly ridged; *Habitat* bottomlands.

Leaves: Simple, alternate, deciduous, 8 to 15 cm long, lobes five to nine and bristle tipped, U- or C-shaped sinuses deep and extending almost to the midrib, base truncate to acute, vein axils with tufts of hair; autumn color red.

Twigs: Slender, red-brown to gray-brown, glabrous; leaf scar crescent shaped to oval with numerous bundle scars; with short spurlike branches ("pins").

Buds: Terminal buds 3 mm long, ovoid, slightly angled; scales overlapping, red-brown, lightly pubescent.

Bark: Gray-brown, smooth or lightly ridged.

Flowers: Imperfect, in spring with the leaves; staminate flowers in drooping catkins; pistillate spikes in leaf axils.

Fruit: Acorn, 1.3 cm long; nut maroon-brown, often striped, pubescent, flat topped; cap flat with pubescent appressed scales, covering about one-fourth of the nut; maturing in two seasons.

Form: Up to 36 m (120 ft) in height and 1.5 m (5 ft) in diameter with branches forming 90° angles.

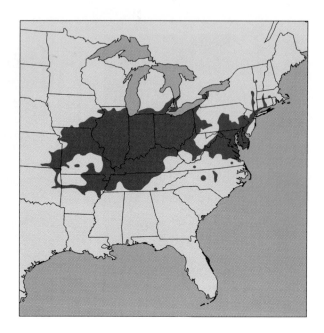

Habitat and Ecology: Shade intolerant; found in bottomland forests and moist uplands with *Acer rubrum, Fraxinus nigra, Liquidambar styraciflua, Nyssa* spp., *Quercus bicolor, Quercus lyrata, Quercus macrocarpa, Quercus nigra, Quercus phellos, Quercus texana,* and *Ulmus americana.*

Uses: Red oak lumber and fuel. A popular shade tree planted for the pyramidal crown and fall foliage but requires acidic, moist soils. See *Quercus alba* for wildlife uses.

Botanical Name: *Quercus* is Latin for "oak tree"; *palustris* means "swamp."

Quercus palustris fruit

Quercus palustris twig

Quercus palustris bark

Quercus palustris bark

Quercus palustris leaf

Fagaceae

Quercus phellos L.

willow oak, swamp willow oak, pin oak, peach oak

Quick Guide: *Leaves* alternate, simple, unlobed, lanceolate, thin, often drooping, margin entire, apex bristle tipped, midrib usually with tufts of yellow hair; *Acorn* cap covering the base of the flat-topped nut; *Bark* gray and smooth or shallowly ridged; *Habitat* bottomlands.

Leaves: Simple, alternate, deciduous, thin, "willowlike," linear to lanceolate, 5 to 12 cm long, apex acute with a bristle tip, base acute, margin entire or sometimes wavy, nearly sessile, midrib yellow possibly with tufts of hair; autumn color yellow-brown.

Twigs: Slender, gray-brown, glabrous; leaf scar crescent shaped to oval with numerous bundle scars.

Buds: Terminal buds 3 mm long, ovoid; scales overlapping, red-brown, glabrous.

Bark: Brown-gray and smooth, becoming shallowly ridged; very large trees with scaly or blocky ridges.

Flowers: Imperfect, in spring with the leaves; staminate flowers in drooping catkins; pistillate spikes in leaf axils.

Fruit: Acorn, 0.6 to 1.3 cm long; nut flat topped, pubescent; cap green-brown to maroon when young, flat, with appressed scales, covering one-fourth or less of the nut; maturing in two seasons.

Form: Up to 36 m (120 ft) in height and 1 m (3 ft) in diameter.

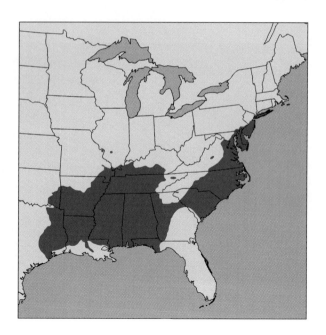

Habitat and Ecology: Shade intolerant; found on moist to poorly drained soils of bottomlands and along stream edges in association with *Acer rubrum, Celtis laevigata, Diospyros virginiana, Fraxinus* spp., *Liquidambar styraciflua, Pinus taeda, Populus deltoides, Quercus lyrata, Quercus michauxii, Quercus nigra, Quercus pagoda, Quercus palustris, Quercus shumardii, Quercus texana,* and *Ulmus americana.*

Uses: Red oak lumber, pulpwood, and fuel. A popular ornamental especially in the Southeast. See *Quercus alba* for wildlife uses.

Botanical Name: *Quercus* is Latin for "oak tree"; *phellos* means "cork tree."

Quercus phellos leaves

Quercus phellos twig

Quercus phellos bark

Quercus phellos bark

Quercus phellos fruit

Fagaceae

Quercus prinus L.

chestnut oak, tanbark oak, rock oak, rock chestnut oak

Quick Guide: *Leaves* alternate, simple, obovate, margin with rounded smooth teeth; *Acorn* cap thin with warty scales, covering one-third of the oblong nut; *Bark* dark gray and deeply grooved with orange in the grooves; *Habitat* dry or rocky uplands.

Leaves: Simple, alternate, deciduous, obovate to elliptical, 10 to 22 cm long, margin with rounded teeth lacking bristles or callous tips, base acute, underside pubescent or glabrous; autumn color yellow-brown to dull red.

Twigs: Stout, red-brown to orange-brown, glabrous; leaf scar crescent shaped to oval with numerous bundle scars.

Buds: Terminal buds 6 to 13 mm long, acute; scales overlapping, chestnut brown, lightly pubescent on the margins.

Bark: Gray and smooth or shallowly grooved on small trees; deeply furrowed with orange in the grooves on large trees.

Flowers: Imperfect, in spring with the leaves; staminate flowers in drooping catkins; pistillate spikes in leaf axils.

Fruit: Acorn, oblong, 2 to 4 cm long; nut shiny yellow-brown when young; cap bowl shaped with fused warty scales, covering one-third of the nut, cap thin on the edge and becoming loose on the nut when mature; maturing in one season.

Form: Up to 24 m (80 ft) in height and 1 m (3 ft) in diameter.

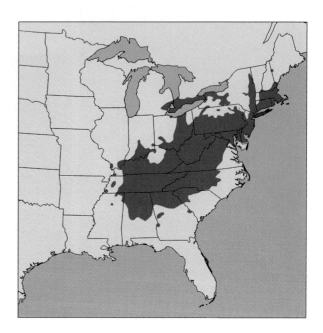

Habitat and Ecology: Intermediate shade tolerance; found primarily on sandy or rocky uplands with *Acer rubrum*, *Carya* spp., *Juniperus virginiana*, *Liquidambar styraciflua*, *Nyssa sylvatica*, *Oxydendrum arboreum*, *Pinus echinata*, *Pinus pungens*, *Pinus resinosa*, *Pinus rigida*, *Pinus strobus*, *Pinus virginiana*, *Quercus alba*, *Quercus coccinea*, *Quercus muehlenbergii*, *Quercus stellata*, and *Quercus velutina*.

Uses: White oak lumber, the thick bark was used as a source of tannin. Planted as a shade tree for the fall color and drought tolerance. See *Quercus alba* for wildlife uses.

Botanical Name: *Quercus* is Latin for "oak tree"; *prinus* is Greek for a European oak. Also named *Quercus montana* Willd.

Quercus prinus fruit

Quercus prinus leaves

Quercus prinus twig

Quercus prinus bark

Quercus prinus bark

Fagaceae

Quercus rubra L.

northern red oak, eastern red oak, red oak, gray oak

Quick Guide: *Leaves* alternate, simple, lobes seven to eleven and bristle tipped, sinuses shallow; *Buds* mostly glabrous and chestnut brown; *Acorn* cap beanielike; *Bark* gray-black with flat ridges; *Habitat* moist, upland forests.

Leaves: Simple, alternate, deciduous, 10 to 23 cm long, lobes seven to eleven and bristle tipped, sinuses shallow and extending at most half way to the midrib, base truncate to acute, vein axils with tufts of hair; autumn color red.

Twigs: Red-brown, glabrous; leaf scar crescent shaped to oval with numerous bundle scars.

Buds: Terminal buds 7 mm long, acute, slightly angled; scales overlapping, chestnut brown, with minute pubescence on the margins.

Bark: Gray-black and shallowly cracked or grooved with flattened gray or white ridges; inner bark orange-brown. The bark varies with range.

Flowers: Imperfect, in spring with the leaves; staminate flowers in drooping catkins; pistillate spikes in leaf axils.

Fruit: Acorn, 2–3 cm long; nut pubescent; cap with appressed pubescent or glabrous scales, beanielike with rolled edges, covering at most one-fourth of the nut; maturing in two seasons.

Form: Up to 30 m (100 ft) in height and 1 m (3 ft) in diameter with a straight bole.

Habitat and Ecology: Intermediate shade tolerance; found on moist, well-drained sites with

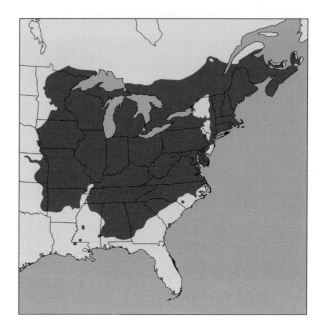

many species including *Acer pensylvanicum, Acer rubrum, Acer saccharum, Aesculus flava, Betula alleghaniensis, Betula papyrifera, Carya* spp., *Cercis canadensis, Fagus grandifolia, Fraxinus* spp., *Liriodendron tulipifera, Nyssa sylvatica, Pinus rigida, Pinus strobus, Pinus taeda, Prunus serotina, Quercus alba, Quercus velutina, Tilia americana, Tsuga canadensis,* and *Ulmus americana.*

Uses: Red oak wood red-brown, heavy, hard; commercially important, used for trim, flooring, cabinets, and furniture. Planted as an ornamental for fall color and site adaptability. See *Quercus alba* for wildlife uses.

Botanical Name: *Quercus* is Latin for "oak tree"; *rubra* means "red," referring to the autumn leaves or wood.

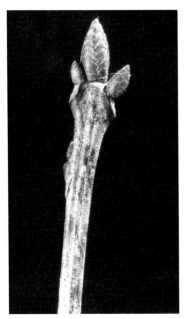

Quercus rubra twig

Quercus rubra bark

Quercus rubra bark

Quercus rubra fruit

Quercus rubra bark

Quercus rubra leaves

Fagaceae

Quercus shumardii Buckl.

Shumard oak, red oak, swamp red oak

Quick Guide: *Leaves* alternate, simple, lobes five to nine with many bristle tips, sinuses deep; *Acorn* cap covering the top of the nut; *Bark* gray-brown and smooth or shallowly fissured; *Habitat* bottomlands and limestone soils.

Leaves: Simple, alternate, deciduous, 10 to 20 cm long, lobes five to nine with many bristle tips, U-shaped sinuses extending halfway or more to the midrib, base truncate to obtuse, vein axils with tufts of hair; autumn color red.

Twigs: Brown to gray-brown, glabrous; leaf scar crescent shaped to oval with numerous bundle scars.

Buds: Terminal buds 6 mm long, acute, angled; scales overlapping, gray-brown, with gray pubescence.

Bark: Gray-brown and smooth on small trees; large trees shallowly grooved possibly with scaly ridges.

Flowers: Imperfect, in spring with the leaves; staminate flowers in drooping catkins; pistillate spikes in leaf axils.

Fruit: Acorn, 2–3 cm long; nut pubescent; cap with appressed pubescent or glabrous scales and rolled edges, covering up to one-third of the nut; maturing in two seasons.

Form: Up to 30 m (100 ft) in height and 1.2 m (4 ft) in diameter.

Habitat and Ecology: Shade intolerant; found next to streams and rivers, in bottomlands, and on limestone soils in association with *Carya aquatica, Carya cordiformis, Carya ovata, Fagus grandifolia, Fraxinus* spp., *Pinus glabra, Quercus lyrata, Quercus michauxii, Quercus pagoda, Quercus palustris, Quercus phellos, Quercus texana,* and *Ulmus americana*.

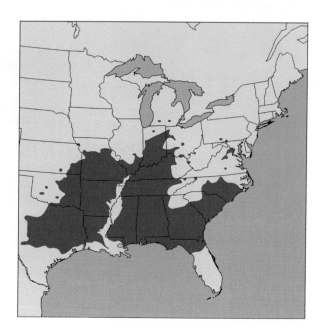

Uses: Red oak wood commercially important, used for trim, flooring, cabinets, and furniture. A popular landscape tree for its form and fall color. See *Quercus alba* for wildlife uses.

Botanical Name: *Quercus* is Latin for "oak tree"; *shumardii* is for the geologist Shumard.

Quercus shumardii fruit

Quercus shumardii leaf

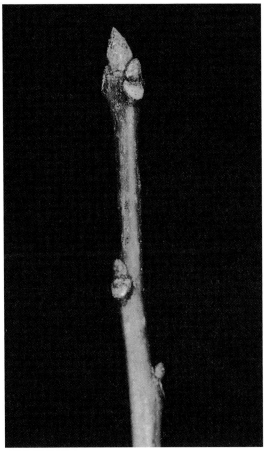
Quercus shumardii twig

Quercus shumardii bark

Quercus shumardii leaf

Quercus stellata Wangenh.

post oak, iron oak, turkey oak

Quick Guide: *Leaves* alternate, simple, leathery, pubescent above and below, lobes five and bristleless, central lobes may form a cross; *Twigs* pubescent; *Acorn* nut and cap downy pubescent; *Bark* gray-brown with scaly ridges; *Habitat* rocky, sandy soils.

Leaves: Simple, alternate, deciduous, leathery, 10 to 15 cm long, lobes five and bristleless and variable in shape, some leaves with central lobes squarish forming a cross, sinuses variable in depth and length, base cuneate, upper surface with stellate pubescence, densely pubescent below; autumn color yellow-brown.

Twigs: Stout, red-brown to gray and densely pubescent; leaf scar crescent shaped to oval with numerous bundle scars.

Buds: Terminal buds 3 mm long, ovoid, angled; scales overlapping, red-brown, glabrous or with yellow-brown pubescence.

Bark: Gray-brown with scaly ridges on small trees; large trees with more thickened irregular ridges; becoming deeply grooved on very large trees.

Flowers: Imperfect, in spring with the leaves; staminate flowers in drooping catkins; pistillate spikes in leaf axils.

Fruit: Acorn, 1.3 to 2 cm long; nut pubescent, dark; cap bowl shaped with pubescent, loosely appressed scales, covering one-third to one-half of the nut; maturing in one season.

Form: Up to 30 m (100 ft) in height and 1.2 cm (4 ft) in diameter.

Habitat and Ecology: Shade intolerant; found on a variety of dry sites but commonly on well-drained, rocky, sandy soils with *Carya* spp., *Juniperus virginiana*, *Liquidambar styraciflua*, *Nyssa sylvatica*, *Oxydendrum arboreum*, *Pinus echinata*, *Pinus rigida*, *Pinus strobus*, *Pinus taeda*, *Pinus virginiana*, *Quercus alba*, *Quercus coccinea*, *Quercus falcata*, *Quercus marilandica*, *Quercus muehlenbergii*, *Quercus prinus*, and *Quercus velutina*.

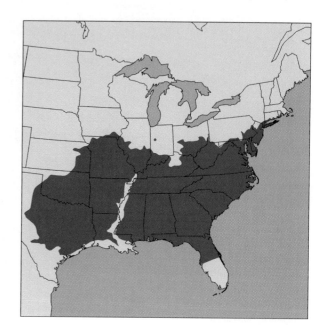

Uses: White oak lumber, construction, pulpwood, veneer, railroad ties, posts, and fuel. See *Quercus alba* for wildlife uses.

Botanical Name: *Quercus* is Latin for "oak tree"; *stellata* means "star" and refers to star-shaped clusters of hairs on the leaf or to the radiate terminal lobes.

Similar Species: Delta post oak or bottomland post oak (*Quercus similis* Ashe) is found in bottomlands in South Carolina, Mississippi, Arkansas, Louisiana, and Texas, and distinguished from post oak by habitat and leaves not as cruciform.

Quercus stellata fruit

Quercus stellata leaf

Quercus stellata bark

Quercus stellata leaves

Quercus stellata leaves

Quercus stellata twig

Quercus texana Buckl.

Nuttall oak, Texas red oak, striped oak, red river oak, pin oak, Nuttall's oak

Quick Guide: *Leaves* alternate, simple, lobes five to seven and bristle tipped, apical lobe elongated, sinuses deep and variable on the same leaf; *Acorn* nut maroon-brown and striped with the cap covering one-third to one-half of the nut; *Bark* gray-brown and smooth or shallowly grooved; *Habitat* bottomlands and wet clay soils.

Leaves: Simple, alternate, deciduous, 10 to 18 cm long, lobes five to seven and bristle tipped, parallel lobes often uneven, basal lobes perpendicular to the midrib and other lobes pointing toward the apex, apical lobe often elongated, sinuses variable on the same leaf and extending more than halfway to the midrib, base truncate to acute or inequilateral, vein axils with tufts of hair; autumn color red.

Twigs: Red-brown to gray-brown, glabrous; leaf scar crescent shaped to oval with numerous bundle scars.

Buds: Terminal buds 6 mm long, acute, slightly angled; scales overlapping, light gray-brown, pubescent or glabrous.

Bark: Gray-brown, smooth or shallowly grooved and ridged.

Flowers: Imperfect, in spring with the leaves; staminate flowers in drooping catkins; pistillate spikes in leaf axils.

Fruit: Acorn, about 3 cm long; nut dark, maroon-brown striped, pubescent; cap bowl-like with pubescent appressed scales, covering one-third to one-half of the nut; maturing in two seasons.

Form: Up to 36 m (120 ft) in height and 1 m (3 ft) in diameter.

Habitat and Ecology: Shade intolerant; found on wet clay flats and bottomlands with *Acer*

rubrum, Fraxinus pennsylvanica, Liquidambar styraciflua, Nyssa spp., *Quercus lyrata, Quercus macrocarpa, Quercus michauxii, Quercus nigra, Quercus palustris, Quercus phellos,* and *Ulmus americana.*

Uses: Red oak lumber, construction, railroad ties, and fuel. Planted as a shade tree. See *Quercus alba* for wildlife uses.

Botanical Name: *Quercus* is Latin for "oak tree"; *texana* refers to the geographic range. Previously named *Quercus nuttallii* Palmer.

Similar Species: Texas Shumard oak (*Quercus buckleyi* Nixon and Dorr) is found in central Texas and southeastern Oklahoma on limestone ridges and slopes and along stream edges. However, some consider it to be the same species as *Quercus texana* or *Quercus shumardii* and the taxonomy is unclear.

Quercus texana bark

Quercus texana leaf

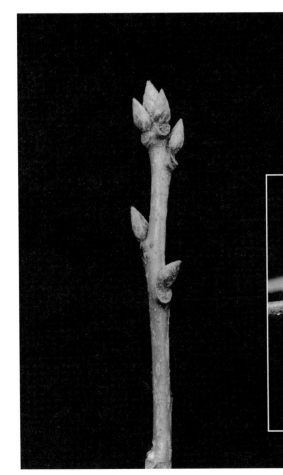
Quercus texana twig

Quercus texana fruit

Fagaceae

Quercus velutina Lam.

black oak, yellow oak, yellowbark oak, quercitron oak

Quick Guide: *Leaves* alternate, simple, variable in shape, lobes five to seven and bristle tipped; *Buds* angled and tawny pubescent; *Acorn* cap slightly fringed and pubescent covering one-third of the nut; *Bark* dark and ridged, inner bark yellow-orange; *Habitat* upland sites.

Leaves: Simple, alternate, deciduous, variable in shape from sun to shade leaves, 10 to 23 cm long, lobes five to seven and bristle tipped, base truncate to obtuse; autumn color yellow to dull red. Shade leaves leathery with shallow sinuses and pubescence on the underside. Sun leaves usually thinner with sinuses reaching halfway or more to the midrib.

Twigs: Red-brown, glabrous or pubescent; leaf scar crescent shaped to oval with numerous bundle scars.

Buds: Terminal buds up to 1.3 cm long, acute, angled; scales overlapping, covered with tawny pubescence.

Bark: Gray-black and rough on small trees; large trees more blocky and ridged, upper trunk may show white flattened ridges; inner bark yellow-orange.

Flowers: Imperfect, in spring with the leaves; staminate flowers in drooping catkins; pistillate spikes in leaf axils.

Fruit: Acorn, 1 to 2 cm long; nut pubescent; cap with pubescent scales, slightly fringed at the base, covering one-third to one-half of the nut; maturing in two seasons.

Form: Up to 30 m (100 ft) in height and 1.2 m (4 ft) in diameter.

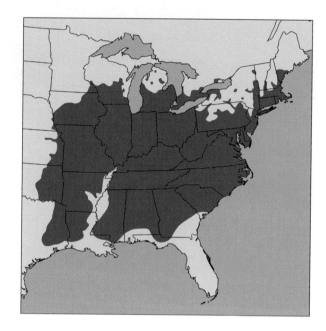

Habitat and Ecology: Intermediate shade tolerance; found on a variety of upland sites. Forest associates include *Acer rubrum, Acer saccharum, Carya* spp., *Diospyros virginiana, Fagus grandifolia, Nyssa sylvatica, Pinus echinata, Pinus strobus, Pinus taeda, Pinus virginiana, Quercus alba, Quercus coccinea, Quercus falcata, Quercus nigra, Quercus prinus, Quercus rubra, Quercus stellata, Ulmus alata,* and *Ulmus rubra*.

Uses: Red oak wood commercially important, used for trim, flooring, cabinets, and furniture. Once used as a source of tannin and yellow dye made from the inner bark was used in printing calicoes. See *Quercus alba* for wildlife uses.

Botanical Name: *Quercus* is Latin for "oak tree"; *velutina* means "velvety," referring to the underside of developing leaves.

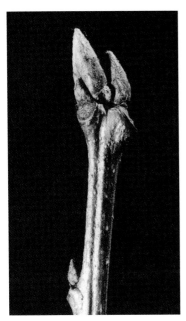

Quercus velutina twig

Quercus velutina fruit

Quercus velutina shade leaves

Quercus velutina bark

Quercus velutina sun leaf

Quercus velutina bark

Fagaceae

Quercus virginiana P. Mill.
live oak, Virginia live oak

Quick Guide: *Leaves* alternate, simple, unlobed, leathery, evergreen, oblong, margin with an occasional bristle-tipped tooth; *Acorn* cap long stalked; *Bark* gray-brown, deeply grooved, blocky; *Habitat* sandy soils.

Leaves: Simple, alternate, evergreen, shiny, leathery, oblanceolate to oblong or elliptical, 4 to 15 cm long, apex obtuse to acute with and without a rigid bristle tip, base acute to cuneate, margin entire or revolute, with an occasional rigid bristle-tipped tooth, nearly sessile, petiole and underside with pale or yellow-brown pubescence.

Twigs: Slender, gray-brown, glabrous or pubescent; leaf scar crescent shaped to oval with numerous bundle scars.

Buds: Terminal buds 2 mm long, round; scales overlapping, brown, lightly pubescent.

Bark: Red-brown to brown-gray and deeply fissured, with corky blocky ridges.

Flowers: Imperfect, in spring with the new leaves; staminate flowers in short drooping catkins; pistillate spikes in leaf axils.

Fruit: Acorn, 1 to 2.5 cm long; nut shiny dark brown to black; cap long stalked and buglelike, bowl shaped, with appressed pubescent scales, covering one-third to one-half of the nut; maturing in one growing season.

Form: Up to 23 m (75 ft) in height, open-grown trees with a short and heavy trunk. Crown of very old trees may span over 40 m and branches may rest on the ground.

Habitat and Ecology: Intermediate shade tolerance; found on sandy soils ranging from dry to wet and can withstand brief inundation and salt spray. Forests associates include *Acer rubrum, Chameacyparis thyoides, Magnolia virginiana, Nyssa biflora, Pinus elliottii, Pinus glabra, Pinus palustris, Quercus laurifolia, Quercus nigra, Taxodium distichum* var. *distichum* and var. *imbricarium*, and scrub oaks.

Uses: Wood yellow-brown to white, hard, heavy; once used in ship-building before the steel era. A popular ornamental because of the evergreen leaves, low crown, and Spanish moss draping the limbs but not cold hardy. See *Quercus alba* for wildlife uses.

Botanical Name: *Quercus* is Latin for "oak tree"; *virginiana* refers to the geographic range.

Quercus virginiana fruit

Quercus virginana leaves

Quercus virginana bark

Quercus virginana bark

Quercus virginana leaves

Fagaceae

Hamamelis virginiana L.
witch-hazel, American witch-hazel

Quick Guide: *Leaves* alternate, simple, obovate, lopsided, margin wavy, base inequilateral; *Buds* stalked, flattened, curved, and scurfy pubescent; *Flowers* threadlike and yellow in late fall; *Fruit* a bumpy capsule; *Habitat* moist, fertile soils.

Leaves: Simple, alternate, deciduous, obovate to oval, often lopsided, variable in size from 6 to 15 cm long, apex acute to rounded, base inequilateral to acute, margin wavy underside with pale pubescence; autumn color yellow.

Twigs: Slender, red-brown, with scurfy pale pubescence; leaf scar shield shaped with three bundle scars.

Buds: Terminal bud stalked, flattened, curved, up to 1.5 cm long, naked, yellow-brown, with scurfy pubescence.

Bark: Gray-brown, smooth, with lenticels.

Flowers: Perfect, with four stringy petals usually yellow but sometimes reddish, blooming in late fall.

Fruit: Capsule, woody, yellow-brown, urn shaped to nearly round, 1 cm long, containing shiny black seeds "ejected" in fall of the following year.

Form: Shrub or small understory tree up to 6 m (20 ft) in height.

Habitat and Ecology: Shade tolerant; found in the understory on fertile, moist soils.

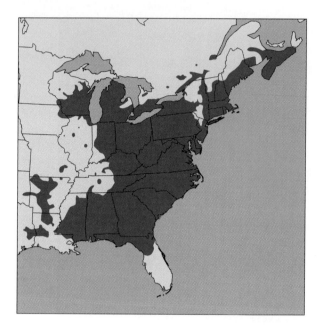

Uses: Wood white and heavy with brown-red heartwood. Inner bark used in witch-hazel astringent and branches used as "water diviners." An interesting specimen for naturalizing gardens. Browsed to a limited degree by white-tailed deer; seeds eaten by wild turkey, northern bobwhite, and gray squirrel; seeds, buds, and flowers used by ruffed grouse.

Botanical Name: *Hamamelis* is from a Greek word and refers to the fall blooming flowers; *virginiana* refers to the geographic range.

Hamamelis virginiana leaves

Hamamelis virginiana fruit

Hamamelis virginiana twig

Hamamelis virginiana bark

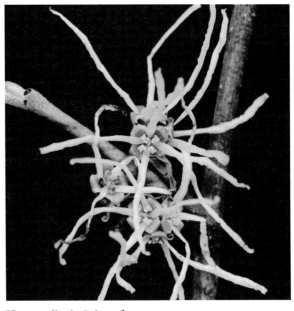
Hamamelis virginiana flowers

Hamamelidaceae

Liquidambar styraciflua L.

sweetgum, redgum, satin walnut, sapgum

Quick Guide: *Leaves* alternate, simple, star shaped, lobes five to seven and deeply palmate, margin finely serrate; *Branches* often with corky wings; *Fruit* a spiny ball of capsules; *Bark* gray-brown with flattened interlacing ridges; *Habitat* dry sandy soils to bottomlands.

Leaves: Simple, alternate, deciduous, star shaped, 8 to 15 cm wide, lobes five to seven and deeply palmate, apices acuminate, base truncate, margin finely serrate, palmately veined, petiole long; autumn color yellow or red.

Twigs: Moderately stout, yellow-brown to green-brown, shiny, with lenticels, often with corky wings; leaf scar shield shaped with three bundle scars.

Buds: Terminal bud up to 2 cm long, acute; scales overlapping, red-brown to purple-green, shiny, tipped with white pubescence.

Bark: Gray-brown with corky ridges on small trees; darker with flat interlacing ridges on large trees.

Flowers: Imperfect, with the leaves in spring; staminate flowers in yellow-green round heads in erect terminal clusters up to 6 cm long; pistillate flowers in small green heads on long drooping stalks.

Fruit: Capsules, in a spiny woody ball 3 cm wide ("gumballs"), maturing in early fall.

Form: Up to 36 m (120 ft) in height and 1.2 m (4 ft) in diameter; often forming thickets due to root sprouting.

Habitat and Ecology: Shade intolerant; an aggressive colonizer found on a variety of sites ranging from dry sandy soils to intermittently

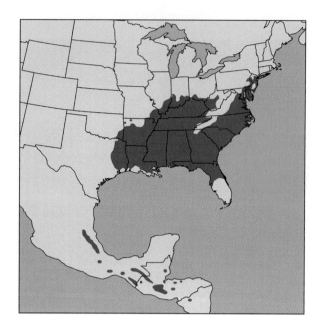

flooded bottomlands. Forest associates are numerous and include *Acer rubrum, Acer negundo, Betula nigra, Carya* spp.*, Celtis laevigata, Fagus grandifolia, Nyssa* spp.*, Pinus echinata, Pinus taeda, Pinus virginiana, Quercus* spp., and *Ulmus americana*.

Uses: Sapwood white, heartwood red-brown, moderately heavy, moderately hard; used for veneer, trim, boxes, pallets, and pulpwood, and heartwood used for furniture. The sweet sap was used in chewing gum and skin balms. Cultivars have been developed to enhance fall color and cold hardiness, and a fruitless cultivar is available. Seeds used by mallard ducks and eaten directly from seed balls by various finches and occasionally by gray squirrels; bark relished by beavers.

Botanical Name: *Liquidambar* refers to the brown-yellow sap; *styraciflua* refers to the fragrant, gummy resin (styrax) exuded from cut twigs.

Liquidambar styraciflua leaf

Liquidambar styraciflua bark

Liquidambar styraciflua staminate and pistillate flowers

Liquidambar styraciflua fruit

Liquidambar styraciflua twig

Hamamelidaceae

Hippocastanaceae Guide

1. A large tree; twigs with an unpleasant odor when cut; flowers yellow with stamens longer than the petals; capsule prickly or spiny. *Aesculus glabra*
2. A large tree; flowers yellow and petals longer than the stamens; capsule smooth. *Aesculus flava*
3. A small tree or shrub; flowers bright red; capsule smooth. *Aesculus pavia*
4. A small tree or shrub; flowers yellow-red-green; capsule smooth. *Aesculus sylvatica*
5. A small tree or shrub; flowers white and bottlebrush-like; capsule smooth. *Aesculus parviflora*

Aesculus sylvatica flowers

Aesculus flava leaf and flowers. ©Dorling Kindersley.

Aesculus pavia flowers

Aesculus glabra fruit

Aesculus parviflora flowers

Aesculus flava Ait.

yellow buckeye, big buckeye, sweet buckeye

Quick Guide: *Leaves* opposite, palmately compound, leaflets five; *Buds* stout, buff color, and glabrous; *Flower* petals longer than the stamens; *Fruit* a globose, smooth capsule with large maroon seeds; *Bark* smooth becoming loosely plated with "bull's-eye" lines; *Habitat* moist, cool sites.

Leaves: Palmately compound, opposite, deciduous, up to 35 cm long. Leaflets five, obovate to ovate or elliptical, 10 to 20 cm long, margin serrate, glabrous below; petiole long (up to 15 cm); autumn color yellow.

Twigs: Stout, yellow-brown, glabrous, with lenticels; leaf scar large, shield shaped, bundle scars forming a V shape.

Buds: Terminal bud stout, up to 2 cm long, acute; scales overlapping, buff color, glabrous.

Bark: Gray-brown and smooth with lenticels on small trees; large trees scaly and loosely plated with "bull's-eye" lines.

Flowers: Usually perfect, most petals longer than the stamens, in erect yellow panicles in spring after the leaves (see page 305).

Fruit: Capsule, 3 to 4 cm wide, round to pear shaped, brown, leathery, mostly smooth, containing one to three large, shiny maroon seeds with a pale spot (buck's eye); maturing in fall.

Form: Up to 26 m (85 ft) in height and 1.5 m (5 ft) in diameter.

Habitat and Ecology: Shade tolerant; found on moist, fertile soils of cool mountain slopes, coves,

river bottoms, and stream edges with *Acer saccharum, Betula alleghaniensis, Betula lenta, Fraxinus americana, Liriodendron tulipifera, Picea rubens, Pinus strobus, Quercus alba, Quercus rubra, Tilia americana,* and *Tsuga canadensis.*

Uses: Wood yellow-white to gray, light, very soft; used for pulpwood, novelty items, boxes, and crates. A paste made from the fruit was used in bookbinding. The smooth seed is carried as a good-luck piece. Young foliage and seeds are poisonous to humans and livestock; gray and fox squirrels and feral hogs will occasionally eat the seeds. Can be planted as an ornamental on moist, cool sites.

Botanical Name: *Aesculus* is a Latin name for "a mast-bearing tree"; *flava* means "golden yellow," referring to the flowers.

Aesculus flava bark

Aesculus flava fruit

Aesculus flava twig

Aesculus flava bark

Aesculus flava leaf

Hippocastanaceae

Aesculus glabra Willd.

Ohio buckeye, stinking buckeye, fetid buckeye

Quick Guide: *Leaves* opposite, palmately compound, leaflets five, leaves and twigs smelly when crushed; *Buds* with keeled scales; *Flowers* yellow with exserted stamens; *Fruit* a prickly or spiny capsule; *Bark* with scaly plates; *Habitat* moist sites.

Leaves: Palmately compound, opposite, deciduous, up to 30 cm long, emitting an unpleasant odor when crushed. Leaflets five, obovate to elliptical, 8 to 15 cm long, margin serrate, glabrous below; petiole long (up to 15 cm); autumn color yellow.

Twigs: Stout, yellow-brown to maroon, glabrous or pubescent, with lenticels and an unpleasant odor; leaf scar large, shield-shaped, bundle scars forming a V-shape.

Buds: Terminal bud stout, up to 1.5 cm long, acute; scales overlapping, maroon-brown to orangish, keeled.

Bark: Gray-brown to gray with corky ridges on small trees; large trees with scaly plates.

Flowers: Usually perfect, petals shorter than the stamens, in erect yellow panicles in spring after the leaves, unpleasant smelling.

Fruit: Capsule, up to 5 cm wide, round to pear-shaped, yellow-brown, leathery, with prickles or short spines, containing one large shiny maroon seed with a pale spot (buck's eye); maturing in fall.

Form: Up to 20 m (66 ft) in height and 61 cm (2 ft) in diameter.

Habitat and Ecology: Shade tolerant; found on moist sites such as stream banks and bottoms with

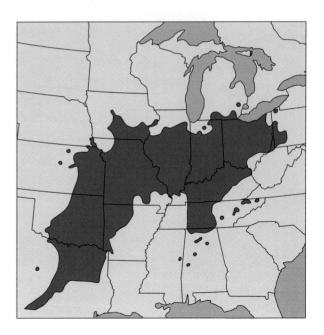

Acer saccharum, Fagus grandifolia, Fraxinus americana, Juglans nigra, Prunus serotina, Quercus macrocarpa, Tilia americana, and *Ulmus americana.*

Uses: Wood yellow-white to gray, light, very soft; used for pulpwood, novelty items, boxes, and crates, and was used in making artificial limbs. Young foliage and fruit poisonous to humans and livestock; squirrels and feral hogs will occasionally eat the seeds. Limited value as a landscape tree because of disease problems.

Botanical Name: *Aesculus* is Latin for "a mast-bearing tree"; *glabra* means "smooth," referring to the flower pedicels, which lack the hairs seen on flower pedicels of *Aesculus flava*.

Aesculus glabra twig

Aesculus glabra fruit

Aesculus glabra bark

Aesculus glabra leaf

Aesculus glabra bark

Aesculus glabra flowers

Hippocastanaceae

Aesculus pavia L.
red buckeye

Quick Guide: *Leaves* opposite, palmately compound with five leaflets; *Buds* stout, buff color, and glabrous; *Flowers* bright red; *Fruit* a round smooth capsule with large maroon seeds; *Habitat* moist, fertile soils.

Leaves: Palmately compound, opposite, deciduous, up to 30 cm long. Leaflets five, obovate to ovate or elliptical, 7 to 15 cm long, margin serrate; petiole long (up to 15 cm); autumn color yellow.

Twigs: Stout, yellow-brown, glabrous, with lenticels; leaf scar large, shield shaped, bundle scars forming a V shape.

Buds: Terminal bud stout, up to 1.5 cm long, acute; scales overlapping, buff color, glabrous.

Bark: Light brown, smooth, with lenticels or warts, becoming scaly on large individuals.

Flowers: Usually perfect, in erect bright red panicles after the leaves (see page 305).

Fruit: Capsule, up to 5 cm wide, round to pear shaped, smooth, brown, leathery, containing one to three large shiny maroon seeds with a pale spot (buck's eye); maturing in fall.

Form: Shrub, often forming thickets, or a small understory tree up to 6 m (20 ft) in height.

Habitat and Ecology: Found in the understory on moist, fertile soils.

Uses: Mass plantings are great for naturalizing a garden but require moist soils with some shade. Wildlife uses similar to *Aesculus flava*. Flowers visited by ruby-throated hummingbirds.

Botanical Name: *Aesculus* is a Latin name for "a mast-bearing tree"; *pavia* is for Petrus Pavius.

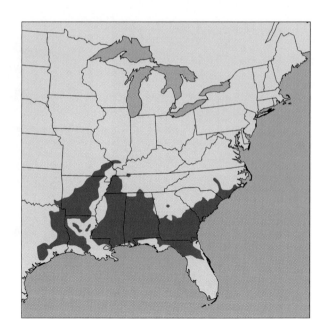

Similar Species: Painted buckeye (*Aesculus sylvatica* Bartr.) is found from Virginia to northern Alabama and has yellow-green-red flowers. Bottlebrush buckeye (*Aesculus parviflora* Walt.) is found mainly in central Alabama and Georgia, and has small white flowers with greatly exserted stamens in erect clusters creating a bottlebrush appearance, and small darker terminal buds with scurfy white pubescence. Both species (see page 305) are good for naturalizing gardens but require moist soils and partial shade for best growth.

Aesculus pavia seeds

Aesculus pavia fruit

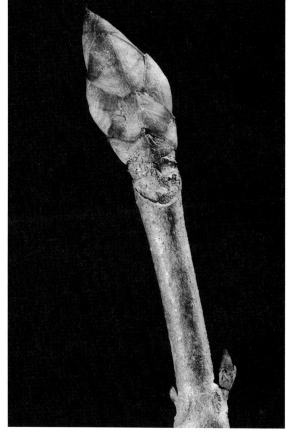
Aesculus pavia twig

Aesculus pavia bark

Aesculus pavia leaf

Hippocastanaceae

Illicium floridanum Ellis

anise-tree, Florida anise, stink-bush

Quick Guide: *Leaves* alternate, simple, evergreen, elliptical, margin entire, lateral veins obscure, anise odor when bruised; *Flowers* with bright red fringelike petals and anise smell; *Fruit* a star-shaped cluster of follicles persisting over winter; *Habitat* stream edges and ravines.

Leaves: Simple, alternate, evergreen, leathery, anise odor when cut, elliptical to lanceolate, 7 to 15 cm long, apex acute, base cuneate, margin entire, lateral veins obscure, underside glandular, petiole possibly red.

Twigs: Red to green or gray-brown, glabrous; leaf scar with one bundle scar.

Buds: Up to 1.5 cm long, acute, with green or red scales.

Bark: Brown-gray, smooth or lightly ridged.

Flowers: Perfect, star shaped, with up to 33 red fringelike petals, emitting an aniselike fragrance (that some find unpleasant), blooming in spring and early summer.

Fruit: Follicles, one seeded, in a star-shaped cluster about 3 cm wide, becoming dry when maturing in fall.

Form: Shrub or small tree to 9 m (30 ft).

Habitat and Ecology: Moist soils along streams and in ravines.

Uses: Can be planted as an ornamental but requires a moist, partially shaded site. Leaves toxic to livestock.

Botanical Name: *Illicium* means "alluring fragrance"; *floridanum* refers to the geographic range.

Illicium floridanum leaves

Illicium floridanum flowers

Illicium floridanum bark

Illicium floridanum fruit

Juglandaceae Guide

1. Leaflets five to seven, obovate to lanceolate, underside mostly glabrous, rachis glabrous; twigs glabrous; husk thin, shiny, often pear shaped, splitting halfway to the base; bark with interlacing ridges. *Carya glabra*
2. Similar to *Carya glabra* but petiole often red; husk rough and splitting to the base; bark with scaly ridges. *Carya ovalis*
3. Leaflets five to seven, mostly obovate, with minute tufts of hair on margin teeth; nut large and round in a thick husk with four deep sutures splitting to the base; bark shaggy with curved loose plates. *Carya ovata*
4. Similar to *Carya ovata* but buds smaller, nearly hairless, with shiny black outer scales; smaller fruit; slender blackish twigs. *Carya carolinae-septentrionalis*
5. Similar to *Carya ovata* but with seven larger leaflets; an orange-brown twig; a larger nut with four to six ridges. *Carya laciniosa*
6. Leaflets five to nine, underside with silvery blue or yellow scales, rachis pubescent; husk with yellow scales, splitting to the base; bark with interlacing ridges. *Carya pallida*
7. Leaflets seven to nine, rachis and underside densely pubescent; nut large and round in a thick husk splitting to the base; bark with interlacing ridges. *Carya tomentosa*
8. Similar to *Carya tomentosa* but leaves and buds with rusty scurfy pubescence; bark darker and rougher. *Carya texana*
9. Leaflets seven to eleven, obovate to lanceolate; buds sulfur yellow, valvate; husk yellow scurfy, with winged sutures above the middle; bark with tight interlacing ridges. *Carya cordiformis*
10. Leaflets mostly nine and yellow or silver scaly below; twigs with yellow scales; buds with buff color, scurfy, valvate scales; husk with keeled or winged sutures. *Carya myristiciformis*
11. Leaflets mostly 7 to 13, falcate, underside rusty scurfy pubescent; buds yellow-brown, valvate, scurfy pubescent; husk flattened, sutures winged; bark scaly. *Carya aquatica*
12. Leaflets 9 to 17, falcate, underside with pale pubescence; buds valvate, pubescent; nut ellipsoidal with thinly winged sutures; bark scaly. *Carya illinoinensis*
13. Leaflets 11 to 17, ovate, with a terminal leaflet, rachis very hairy; leaf scar with a fuzzy mustache; nut sharply ridged and corrugated in a sticky pubescent indehiscent husk; bark ash-gray with flattened ridges. *Juglans cinerea*
14. Leaflets 8 to 24, ovate, often lacking the terminal; nut ridged and corrugated in a yellow-green indehiscent husk; bark dark brown and grooved. *Juglans nigra*

Carya tomentosa staminate flowers

Carya myristiciformis staminate flowers

Carya illinoinensis staminate flowers

Carya aquatica (Michx. f.) Nutt.
water hickory, pecan hickory, bitter pecan

Quick Guide: *Leaves* alternate, pinnately compound, leaflets 7 to 13 and falcate, underside with some rusty scurfy pubescence; *Buds* yellow-brown, valvate, scurfy pubescent; *Fruit* a nut in a flattened, winged husk; *Bark* gray and shaggy; *Habitat* wet sites.

Leaves: Pinnately compound, alternate, deciduous, up to 40 cm long. Leaflets usually 7 to 13 (sometimes 5 to 17), lanceolate to falcate, 5 to 12 cm long, sessile, margin finely serrate, underside rusty scurfy pubescent at least along veins, rachis scurfy pubescent; autumn color yellow.

Twigs: Red-brown, mottled, glabrous or scurfy-pubescent, with lenticels; leaf scar heart shaped with numerous bundle scars.

Buds: Terminal bud up to 1cm long, conical; scales valvate, yellow-brown, scurfy-pubescent.

Bark: Gray-brown and smooth on small trees; large trees shaggy.

Flowers: Imperfect, after the leaves in spring; staminate flowers in drooping catkins; pistillate flowers in terminal spikes.

Fruit: Nut, 3 cm long, flattened; husk thin, with four thinly winged sutures, splitting to the base; nut ridged, kernel bitter; maturing in fall.

Form: Up to 30 m (100 ft) in height and 1 m (3 ft) in diameter.

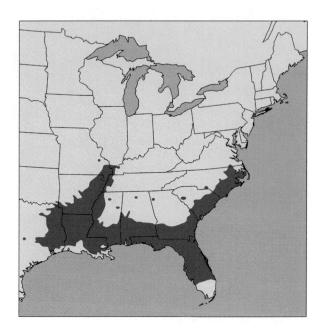

Habitat and Ecology: Intermediate shade tolerance; found in swamps, bottoms, wet clay flats, and shallow sloughs with *Carya cordiformis, Celtis laevigata, Fraxinus pennsylvanica, Fraxinus profunda, Gleditsia aquatica, Liquidambar styraciflua, Nyssa biflora, Pinus glabra, Quercus laurifolia, Quercus lyrata, Quercus pagoda, Quercus texana, Taxodium distichum* var. *distichum*, and *Ulmus americana*.

Uses: Wood white to red-brown, hard, heavy, suffers from longitudinal separations or cracks. Nuts very bitter but eaten by some wildlife species.

Botanical Name: *Carya* is derived from a Greek word meaning "walnut tree"; *aquatica* refers to the wet habitat.

Carya aquatica leaf

Carya aquatica twig

Carya aquatica bark

Carya aquatica bark

Carya aquatica fruit

Juglandaceae

Carya cordiformis (Wangenh.) K. Koch

bitternut hickory, pecan hickory, swamp hickory

Quick Guide: *Leaves* alternate, pinnately compound, leaflets 7 to 11; *Buds* sulfur yellow and valvate; *Fruit* husk with winged sutures above the middle; *Bark* gray-brown with tight interlacing ridges forming a diamond pattern; *Habitat* bottoms to dry uplands.

Leaves: Pinnately compound, alternate, deciduous, up to 25 cm long. Leaflets usually 7 to 9 but sometimes 11, obovate to lanceolate, 5 to 15 cm long, sessile, margin serrate, underside and rachis lightly pubescent; autumn color yellow.

Twigs: Gray-brown, lightly pubescent, with lenticels; leaf scar heart shaped with numerous bundle scars.

Buds: Terminal bud 1.5 cm long, falcate, flattened; scales valvate, sulfur yellow, with scurfy pubescence.

Bark: Gray-brown and smooth on small trees; large trees fissured with tight interlacing ridges forming a diamond pattern.

Flowers: Imperfect, after the leaves in spring; staminate flowers in drooping catkins; pistillate flowers in terminal spikes.

Fruit: Nut, nearly round, pointed, 2.5 cm long; husk thin, with four thinly winged sutures above the middle and splitting to the base; nut lightly ridged, kernel bitter; maturing in fall.

Form: Up to 30 m (100 ft) in height and 1 m (3 ft) in diameter.

Habitat and Ecology: Shade intolerant; found on a variety of sites from bottomlands to dry uplands but restricted to moist sites in the South. Forest associates include *Acer nigrum*, *Acer saccharum*, *Carya ovata*, *Celtis laevigata*, *Fraxinus* spp., *Liriodendron tulipifera*, *Pinus taeda*, *Populus deltoides*, *Quercus alba*, *Quercus lyrata*, *Quercus michauxii*, *Quercus pagoda*, *Quercus rubra*, *Quercus shumardii*, *Quercus velutina*, *Tilia americana*, and *Ulmus americana*.

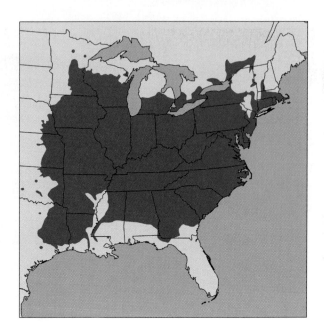

Uses: Wood white to red-brown, hard, heavy; used for pulpwood, handles, furniture, flooring, novelty items, and fuel. Nuts very bitter but eaten by some wildlife species.

Botanical Name: *Carya* is derived from a Greek word meaning "walnut tree"; *cordiformis* means "heart shaped," referring to the fruit.

Carya cordiformis fruit

Carya cordiformis bark

Carya cordiformis leaf

Carya cordiformis twig

Juglandaceae

Carya glabra (Mill.) Sweet

pignut hickory, sweet pignut, smooth hickory, white hickory

Quick Guide: *Leaves* alternate, pinnately compound, leaflets five to seven, underside mostly glabrous, rachis glabrous; *Twigs* slender and glabrous; *Fruit* a smooth nut in a thin husk splitting halfway to the base; *Bark* gray-brown with interlacing ridges forming a diamond pattern; *Habitat* upland forests.

Leaves: Pinnately compound, alternate, deciduous, up to 30 cm long. Leaflets usually five but sometimes seven, obovate to lanceolate or falcate, 8 to 15 cm long, sessile, margin serrate, underside glabrous or sparsely pubescent, rachis slender and glabrous; autumn color yellow.

Twigs: Slender, red-brown, glabrous, with lenticels; leaf scar heart shaped with numerous bundle scars.

Buds: Terminal bud up to 1 cm long, ovoid; outer scales red-brown, loose, glabrous, inner scales paler with silky hairs.

Bark: Smooth or lightly grooved on small trees; large trees gray-brown and fissured with interlacing ridges forming a diamond pattern, becoming scaly.

Flowers: Imperfect, after the leaves in spring; staminate flowers in drooping catkins; pistillate flowers in terminal spikes.

Fruit: Nut, nearly round to egg or pear shaped (like a pig's snout), 2.5 to 5 cm long; husk dark red-brown, thin, splitting halfway to the base; nut smooth, kernel sweet; maturing in fall.

Form: Up to 36 m (120 ft) in height and 1.2 m (4 ft) in diameter.

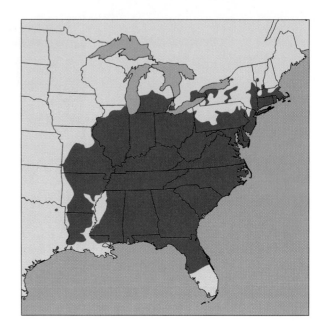

Habitat and Ecology: Intermediate shade tolerance; found on a variety of sites but common on dry slopes and ridges of upland forests. Associates are numerous and include *Acer rubrum, Carya tomentosa, Pinus echinata, Pinus elliottii, Pinus strobus, Pinus taeda, Pinus virginiana,* and *Quercus* spp.

Uses: Wood white to red-brown, hard, heavy; used for pulpwood, handles, furniture, flooring, novelty items, and fuel. Nuts eaten by squirrels and other small mammals.

Botanical Name: *Carya* is derived from a Greek word meaning "walnut tree"; *glabra* means "hairless," and refers to the mostly hairless twig, rachis, and leaf underside.

Carya glabra fruit

Carya glabra twig

Carya glabra bark

Carya glabra leaf

Juglandaceae

Carya illinoinensis (Wangenh.) K. Koch

pecan, pecan hickory, sweet pecan

Quick Guide: *Leaves* alternate, pinnately compound, leaflets 9 to 17 and falcate; *Buds* gray-brown, valvate, and pubescent; *Fruit* an ellipsoidal sweet nut; *Bark* scaly and silvery gray; *Habitat* well-drained soils.

Leaves: Pinnately compound, alternate, deciduous, up to 50 cm long. Leaflets 9 to 17, lanceolate or falcate, 8 to 15 cm long, sessile, margin serrate or doubly serrate, underside pale pubescent; rachis glabrous or pubescent; autumn color yellow.

Twigs: Moderately stout, red-brown, pubescent, with lenticels; leaf scar heart shaped with numerous bundle scars.

Buds: Terminal bud 1 cm long, ovoid; scales valvate, gray-brown to yellowish, pubescent; lateral buds divergent.

Bark: Gray-brown to silvery gray, shallowly ridged, scaly, often with woodpecker holes.

Flowers: Imperfect, after the leaves in spring; staminate flowers in drooping catkins (see page 315); pistillate flowers in terminal spikes.

Fruit: Nut, ellipsoidal, 2.5 to 5 cm long; husk thin with thinly winged sutures; nut smooth or lightly ridged, kernel sweet; maturing in fall.

Form: Up to 45 m (150 ft) in height or sometimes taller (the largest of the native hickories).

Habitat and Ecology: Shade intolerant; originally in well-drained bottomlands with *Acer negundo*,

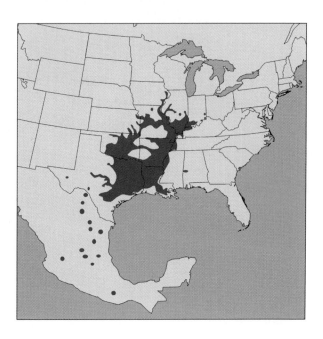

Acer saccharinum, Liquidambar styraciflua, Populus deltoides, Platanus occidentalis, and *Ulmus americana.* Planted extensively throughout the Southeast for nut production; range has extended eastward. Now found on a variety of sites.

Uses: Wood white to red-brown, hard, heavy; used for pulpwood, handles, furniture, paneling, flooring, novelty items, and fuel. A commercially important nut producer and popular ornamental despite brittle limbs. Because of the sweet meat and thin shell, the nuts are favored by all squirrels, some birds (in particular crows), and small mammals.

Botanical Name: *Carya* is derived from a Greek word meaning "walnut tree"; *illinoinensis* refers to the geographic range.

Carya illinoinensis bark

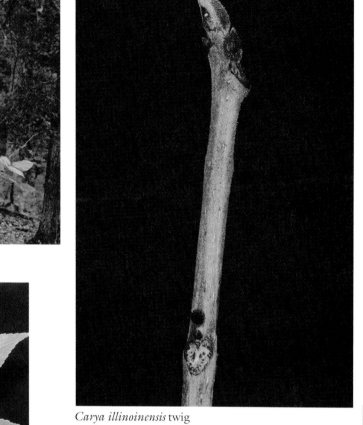
Carya illinoinensis twig

Carya illinoinensis leaf

Carya illinoinensis fruit

Carya laciniosa (Michx. f.) G. Don

shellbark hickory, big shagbark hickory, bigleaf shagbark hickory, kingnut hickory

Quick Guide: Similar to *Carya ovata* but *Leaves* with seven larger leaflets; *Twigs* orange-brown; *Fruit* a larger nut.

Leaves: Pinnately compound, alternate, deciduous, up to 60 cm long. Leaflets usually seven but from five to nine, obovate to oblanceolate, 5 to 22 cm long, sessile, margin serrate and possibly ciliate, underside with dense pale or yellow-maroon pubescence; rachis glabrous or densely pubescent; autumn color yellow.

Twigs: Stout, orange-brown to yellowish, glabrous or densely pubescent, with lenticels; leaf scar heart shaped with numerous bundle scars.

Buds: Terminal bud up to 3 cm long, ovoid, outer scales pubescent and loose; lateral buds divergent.

Bark: Similar to *Carya ovata;* smaller trees with scaly ridges; large trees with large, loose, shaggy plates.

Flowers: Imperfect, after the leaves in the spring; staminate flowers in drooping catkins; pistillate flowers in terminal spikes.

Fruit: Nut, egg-shaped to nearly round, up to 6 cm long (the largest of the hickories); husk depressed at the apex, brown, very thick (12 mm), with four deep sutures splitting to the base, inner husk pale; nuts four to six ribbed; kernel sweet; maturing in fall.

Form: Up to 30 m (100 ft) in height and 1 m (3 ft) in diameter.

Habitat and Ecology: Shade tolerant; found on moist to periodically inundated, loamy bottoms and occasionally on drier sites. Forest associates include *Carya ovata, Fraxinus americana, Liquidambar styraciflua, Populus deltoides, Quercus macrocarpa, Quercus michauxii, Quercus shumardii,* and *Ulmus americana.*

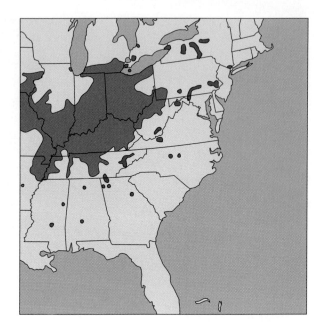

Uses: Wood white to red-brown, hard, heavy, flexible; used for pulpwood, handles, furniture, sporting goods, novelty items, and fuel. Nuts eaten by humans, squirrels, turkey, waterfowl, and small mammals.

Botanical Name: *Carya* is derived from a Greek word meaning "walnut tree"; *laciniosa* is Latin for "shred or torn," referring to the shaggy bark.

Carya laciniosa fruit

Carya laciniosa fruit

Carya laciniosa bark

Carya laciniosa leaf

Carya laciniosa twig

Juglandaceae

Carya myristiciformis (Michx. f.) Nutt.

nutmeg hickory, swamp hickory

Quick Guide: *Leaves* alternate, pinnately compound, leaflets nine with yellow scales below; *Twigs* and buds scurfy; *Fruit* with winged sutures; *Bark* scaly on large trees; *Habitat* rich, moist soils.

Leaves: Pinnately compound, alternate, deciduous, up to 35 cm long. Leaflets usually nine but can be five to eleven, 5 to 15 cm long, sessile, margin serrate, underside with yellow or silver scales; petiole and rachis scurfy; autumn color yellow.

Twigs: Brown-gray, scurfy; leaf scar heart shaped with numerous bundle scars (forming a monkey face).

Buds: Terminal bud up to 1 cm long, ovoid; scales valvate, yellowish and scurfy; lateral buds divergent.

Bark: Red-brown to brown-gray and smooth on young trees; becoming scaly like *Carya ovata* on large trees.

Flowers: Imperfect, after the leaves in the spring; staminate flowers in drooping catkins (see page 315); pistillate flowers in terminal spikes.

Fruit: Nut, ellipsoidal, up to 4 cm long; husk thin with yellow scales and four winged sutures splitting almost to the base; kernel sweet; maturing in fall.

Form: Up to 30 m (100 ft) in height.

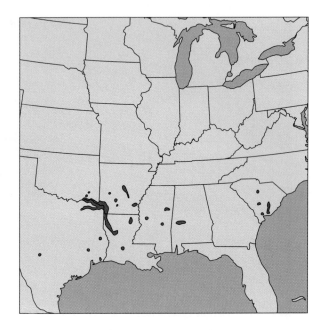

Habitat and Ecology: Considered the rarest of hickories, found on moist soils of riverbanks, swamp margins, bottoms, and limestone ridges with *Fraxinus americana*, *Nyssa sylvatica*, *Quercus michauxii*, *Quercus pagoda*, *Quercus shumardii*, and other bottomland hickories.

Uses: The sweet nuts are enjoyed by wildlife and used in cooking.

Botanical Name: *Carya* is derived from a Greek word meaning "walnut tree"; *myristiciformis* refers to the similarity of the nut to nutmeg (*Myristica fragrans*).

Carya myristiciformis twig

Carya myristiciformis leaf and fruit

Carya myristiciformis bark

Carya myristiciformis fruit

Juglandaceae

Carya ovalis (Wangenh.) Sarg.
red hickory, false pignut hickory

Quick Guide: Similar to *Carya glabra*, distinguished by a red petiole, a rough husk on the nut splitting to the base, and more scaly bark.

Leaves: Pinnately compound, alternate, deciduous, up to 30 cm long. Leaflets five to seven, sometimes nine, obovate to lanceolate or sometimes falcate, 8 to 15 cm long, sessile, margin serrate, underside glabrous or sparsely pubescent, rachis slender and mostly glabrous, petiole often red, strongly spicy aromatic when crushed; autumn color yellow.

Twigs: Slender, red-brown, glabrous or lightly pubescent, with lenticels; leaf scar heart shaped with numerous bundle scars.

Buds: Terminal bud 8 mm long, ovoid; outer scales red-brown, loose, glabrous or pubescent, inner scales paler with silky hairs.

Bark: Smooth or lightly grooved on small trees; large trees gray-brown and fissured with scaly, loose, interlacing ridges.

Flowers: Imperfect, after the leaves in spring; staminate flowers in drooping catkins; pistillate flowers in terminal spikes.

Fruit: Nut, nearly round, 2.5 cm long; husk rough, dark red-brown, thin, splitting to the base; nut not ribbed, kernel sweet; maturing in fall.

Form: Similar to *Carya glabra*.

Habitat and Ecology: Intermediate shade tolerance; found in moist hardwood forests, often intergrading with *Carya glabra*.

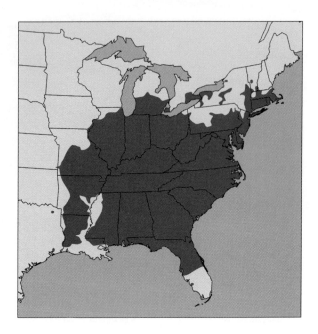

Uses: Similar to *Carya glabra*.

Botanical Name: *Carya* is derived from a Greek word meaning "walnut tree"; *ovalis* means "oval or egg shaped," referring to the nut or leaves.

Carya ovalis fruit

Carya ovalis leaf and fruit

Carya ovalis bark

Carya ovalis bark

Juglandaceae

Carya ovata (P. Mill.) K. Koch

shagbark hickory, scalybark hickory, little shellbark hickory

Quick Guide: *Leaves* alternate, pinnately compound, leaflets five; *Fruit* a large round nut enclosed by a dark thick husk with four deep sutures splitting to the base; *Bark* gray and shaggy with curved loose plates; *Habitat* moist soils.

Leaves: Pinnately compound, alternate, deciduous, up to 40 cm long. Leaflets usually five but sometimes seven, variable in shape and size from widely obovate with an abruptly acuminate apex to elliptical or lanceolate, 8 to 18 cm long, sessile, margin serrate with fine tufts of hair from the teeth (see page 331), underside glabrous or pubescent; rachis stout and pubescent or glabrous; autumn color yellow.

Twigs: Stout, red-brown, glabrous or pubescent, with lenticels; leaf scar heart shaped to semicircular with numerous bundle scars.

Buds: Terminal bud up to 2.0 cm long, ovoid; outer scales pubescent and very loose, inner scales paler with silky hairs.

Bark: Smooth or lightly grooved on small trees; large trees gray and shaggy with curved loose plates.

Flowers: Imperfect, after the leaves in spring; staminate flowers in drooping catkins; pistillate flowers in terminal spikes.

Fruit: Nut, egg shaped or nearly round, up to 5 cm wide, husk depressed at the apex, almost black, shiny, very thick (12 mm), with four deep sutures splitting to the base, inner husk pale; nut four ridged, white to light brown, kernel sweet; maturing in fall.

Form: Up to 39 m (130 ft) in height and 1.2 m (4 ft) in diameter.

Habitat and Ecology: Intermediate shade tolerance; in its southern range found on moist upland sites, swamp edges, and in well-drained bottomlands, and farther north found on moist slopes and ridges. On more upland sites associated with *Acer saccharum, Liriodendron tulipifera, Pinus strobus, Quercus alba, Quercus rubra,* and *Quercus velutina*. On bottomland sites associated with *Carya cordiformis, Liquidambar styraciflua,*

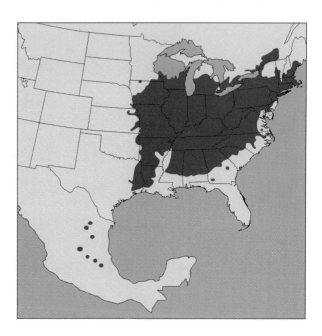

Quercus lyrata, Quercus palustris, Quercus michauxii, Quercus pagoda, and *Quercus palustris.*

Uses: Wood white to red-brown, hard, heavy; used for pulpwood, handles, furniture, paneling, flooring, novelty items, and fuel. Planted as an ornamental because of interesting bark. Nuts eaten by all squirrels and other small mammals.

Botanical Name: *Carya* is derived from a Greek word meaning "walnut tree"; *ovata* means "egg shaped," referring to the leaves or fruit.

Similar Species: Southern shagbark hickory (*Carya carolinae-septentrionalis* Engelm. & Graebn.) is found in bottomlands and moist woods and on limestone soils in the Piedmont and distinguished by smaller buds with shiny reddish or black outer scales, smaller fruit, and very slender blackish twigs.

Carya ovata fruit

Carya ovata fruit

Carya ovata leaf margin

Carya ovata twig

Carya ovata bark

Carya ovata leaf

Juglandaceae 331

Carya pallida (Ashe) Engelm. & Graebn.

sand hickory, pale-leaf hickory, pale hickory

Quick Guide: *Leaves* alternate, pinnately compound, leaflets five to nine and underside with silvery blue or yellow scales, rachis pubescent; *Fruit* a round nut with a yellow scurfy husk splitting to the base; *Bark* brown-gray with interlacing ridges; *Habitat* dry uplands.

Leaves: Pinnately compound, alternate, deciduous, up to 40 cm long. Leaflets five to nine but usually seven, obovate to lanceolate or falcate, 8 to 14 cm long, sessile, margin serrate, underside pale with pubescence and silvery blue or yellow scales; rachis slender and pubescent; autumn color yellow.

Twigs: Slender, red-brown, with yellow-brown scales and lenticels; leaf scar heart shaped with numerous bundle scars.

Buds: Terminal bud 8 mm long, ovoid; scales overlapping, yellow-scurfy, fringed with pubescence.

Bark: Smooth or lightly grooved on small trees; large trees gray-brown and fissured with somewhat scaly interlacing ridges.

Flowers: Imperfect, after the leaves in spring; staminate flowers in drooping catkins; pistillate flowers in terminal spikes.

Fruit: Nut, nearly round to egg shaped, up to 4 cm wide; husk, yellow scurfy, splitting to the base; nut ridged, kernel sweet; maturing in fall.

Form: Up to 30 m (100 ft) tall.

Habitat and Ecology: Found on dry slopes and ridges. Forest associates in the sandhills of the Coastal Plain include *Pinus palustris, Quercus incana, Quercus laevis, Quercus margaretta,* and *Quercus marilandica*. In the Piedmont associates include *Carya glabra, Carya tomentosa, Quercus coccinea,* and *Quercus stellata*.

Uses: Wood white to red-brown, hard, heavy, used primarily for fuel. Nuts eaten by squirrels and other small mammals.

Botanical Name: *Carya* is derived from a Greek word meaning "walnut tree"; *pallida* means "pale," and refers to the pale leaf underside.

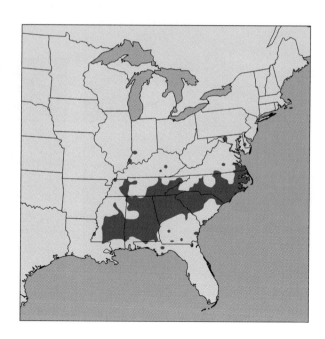

Carya pallida fruit

Carya pallida leaf

Carya pallida bark

Carya pallida twig

Juglandaceae

Carya texana Buckl.

black hickory, red hickory, Texas hickory

Quick Guide: *Leaves* alternate, pinnately compound, leaflets seven and rusty scurfy pubescent below; *Buds* with rusty scurfy pubescence; *Fruit* a round nut with the husk splitting to the base; *Bark* dark and furrowed; *Habitat* dry, rocky soils.

Leaves: Pinnately compound, alternate, deciduous, up to 35 cm long. Leaflets usually five or seven but up to nine, obovate, oblanceolate, or diamond shaped, up to 15 cm long, sessile, base of some leaflets cuneate, margin serrate; underside and rachis with rusty scurfy pubescence; autumn color yellow.

Twigs: Slender, brown-gray, glabrous or with rusty scurfy pubescence; leaf scar heart shaped with numerous bundle scars.

Buds: Terminal bud up to 1 cm long, ovoid; scales overlapping with yellow-brown or rusty scurfy pubescence.

Bark: Brown-gray to nearly black and furrowed with thick, scaly ridges.

Flowers: Imperfect, after the leaves in the spring; staminate flowers in drooping catkins; pistillate flowers in terminal spikes.

Fruit: Nut, round to pear shaped, up to 5 cm long; husk 3 to 4 mm thick with four sutures splitting to the base; kernel sweet; maturing in fall.

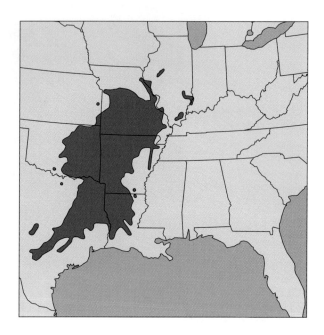

Form: Up to 30 m (100 ft) in height but usually much smaller.

Habitat and Ecology: Found on dry, sandy slopes and ridges in association with *Nyssa sylvatica, Quercus falcata, Quercus marilandica, Quercus stellata,* and *Quercus velutina*.

Uses: Wood used for fuel, charcoal, fence posts and tool handles, and was used for barrel hoops. Fruit eaten by squirrels, peccary, opossum, and turkey.

Botanical Name: *Carya* is derived from a Greek word meaning "walnut tree"; *texana* refers to the geographic range.

Carya texana leaf

Carya texana bark

Carya texana bark

Carya texana leaves

Juglandaceae

Carya tomentosa (Poir. ex Lam.) Nutt.

mockernut hickory, big bud hickory, white hickory

Quick Guide: *Leaves* alternate, pinnately compound, leaflets seven to nine, rachis and underside pubescent; *Twigs* stout and hairy; *Fruit* a large nut with a thick husk splitting to the base; *Bark* gray-brown with interlacing ridges; *Habitat* a variety of sites.

Leaves: Pinnately compound, alternate, deciduous, up to 43 cm long. Leaflets seven to nine but sometimes five, obovate to oblanceolate or lanceolate, 5 to 18 cm long, sessile, margin serrate, underside pubescent; rachis stout and pubescent; autumn color yellow.

Twigs: Stout, red-brown, pubescent, with lenticels; leaf scar heart shaped with numerous bundle scars.

Buds: Terminal bud up to 2.0 cm long, ovoid; outer scales tomentose and loose, inner scales paler with silky hairs.

Bark: Blue-gray and smooth or lightly grooved on small trees; large trees gray-brown and fissured with interlacing ridges forming a diamond pattern.

Flowers: Imperfect, after the leaves in spring; staminate flowers in drooping catkins (see page 315); pistillate flowers in terminal spikes.

Fruit: Nut, elliptical to nearly round, up to 5 cm long; husk with four deep sutures, dark red-brown, thick (5 mm), shiny, splitting to the base; nut four ridged, kernel sweet; maturing in fall.

Form: Up to 30 m (100 ft) in height and 1 m (3 ft) in diameter.

Habitat and Ecology: Shade intolerant; found on a variety of sites from bottomlands to fertile uplands and dry, rocky slopes in association with many species.

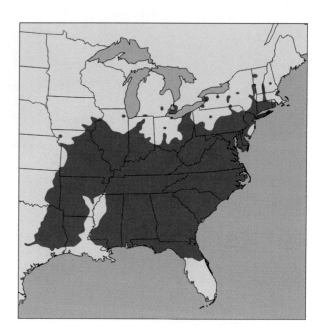

Uses: Wood white to red-brown, hard, heavy; used for pulpwood, handles, furniture, paneling, flooring, novelty items, and fuel. Nuts a favorite of squirrels and small mammals.

Botanical Name: *Carya* is derived from a Greek word meaning "walnut tree"; *tomentosa* refers to the tomentose twig, rachis, and leaf underside. Also named *Carya alba* (L.) Nutt. ex Ell.

Carya tomentosa fruit

Carya tomentosa leaf

Carya tomentosa bark

Carya tomentosa bark

Carya tomentosa twig

Juglans cinerea L.
butternut, white walnut, oil nut

Quick Guide: *Leaves* alternate, pinnately compound, leaflets 11 to 17 and ovate, usually with a terminal leaflet, rachis very hairy; *Leaf scars* with a fuzzy mustache; *Fruit* an ellipsoidal, prominently ridged, corrugated nut encased in a sticky pubescent husk; *Bark* gray and grooved with flattened ridges; *Habitat* moist, fertile soils.

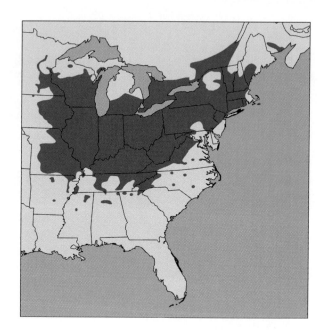

Leaves: Pinnately compound, alternate, deciduous, up to 75 cm long. Leaflets 11 to 17, ovate to oblong-lanceolate, usually with a terminal leaflet, 5 to 12 cm long, nearly sessile, margin serrate, underside pubescent; autumn color yellow. Very hairy on the rachis and petiole.

Twigs: Stout, brown, glabrous or pubescent, with lenticels; pith dark chocolate brown, chambered. Leaf scar three lobed, heart shaped, with a hairy fringe or "mustache" above the scar, bundle scars in groups of three.

Buds: Terminal bud elongated, conical, about 1.5 cm long; scales overlapping, white or gray-brown, scurfy pubescent; lateral buds often superposed.

Bark: Gray to light brown with interlacing ridges on small trees; large trees gray to gray-brown and fissured with flat interlacing ridges, inner bark chocolate brown.

Flowers: Imperfect, after the leaves in late spring or early summer; staminate flowers in drooping catkins; pistillate flowers in terminal spikes.

Fruit: Nut, ellipsoid to oblong or ovoid, about 6 cm long; husk yellow-green to green-brown, pubescent, sticky, thick, indehiscent; nut corrugated and sharply ridged (see photos for *Juglans nigra*), kernel oily and sweet; maturing in fall.

Form: Up to 30 m (100 ft) in height and 1 m (3 ft) in diameter.

Habitat and Ecology: Shade intolerant; found on streambanks and moist upland slopes, and in coves. Forest associates include *Acer rubrum*, *Acer saccharum*, *Betula alleghaniensis*, *Betula lenta*, *Fagus grandifolia*, *Fraxinus* spp., *Juglans nigra*, *Liriodendron tulipifera*, *Platanus occidentalis*, *Prunus serotina*, *Quercus alba*, *Quercus rubra*, *Tilia americana*, and *Ulmus americana*. Distribution greatly reduced because of butternut canker, a fungus that causes stem-girdling cankers.

Uses: Wood white to chestnut brown, coarse grained, light, soft; used for plywood, boxes, trim work, cabinets, and furniture. Dye was made from the husks and bark, and immature nuts were pickled to make an oilnut relish.

Botanical Name: *Juglans* is Latin for "walnut tree"; *cinerea* means "ash," referring to the ash-gray bark.

Juglans cinerea bark

Juglans cinerea twig

Juglans cinerea bark

Juglans cinerea leaf

Juglans cinerea fruit

Juglandaceae

Juglans nigra L.
black walnut, American walnut

Quick Guide: *Leaves* alternate, pinnately compound, leaflets 8 to 22 and ovate, often lacking the terminal on mature trees; *Fruit* a round, ridged, corrugated nut encased in a yellow-green juicy husk; *Bark* dark brown and grooved; *Habitat* moist, fertile soils.

Leaves: Pinnately compound, alternate, deciduous, up to 60 cm long. Leaflets 8 to 24 on mature trees due to an aborted terminal leaflet (young trees often with a terminal leaflet), ovate to ovate-lanceolate, 5 to 13 cm long, nearly sessile, margin serrate, underside pubescent; rachis lightly pubescent; autumn color yellow.

Twigs: Stout, brown, pubescent, with lenticels; pith light brown, chambered. Leaf scar three lobed, heart shaped, bundle scars in groups of three ("monkey-faced leaf scar").

Buds: Terminal bud conical to round, 1 cm long; scales overlapping, brown-gray, pubescent; lateral buds often superposed.

Bark: Dark gray-brown to almost black with interlacing ridges, inner bark chocolate brown.

Flowers: Imperfect, after the leaves in late spring or early summer; staminate flowers in drooping catkins; pistillate flowers in terminal spikes.

Fruit: Nut, round, up to 7 cm wide; husk yellow-green, juicy, indehiscent, thick; nut ridged and corrugated, kernel sweet; maturing in fall.

Form: Up to 39 m (130 ft) in height and 2.4 m (8 ft) in diameter.

Habitat and Ecology: Shade intolerant; found on moist, fertile soils in association with *Acer saccharum, Carya cordiformis, Carya ovata, Fagus grandifolia, Fraxinus americana, Gymnocladus dioicus, Liriodendron tulipifera, Prunus serotina, Quercus alba, Quercus rubra, Tilia americana,* and *Ulmus americana.*

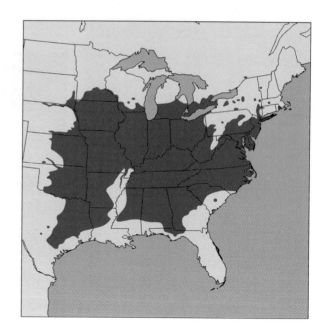

Uses: Sapwood white, heartwood red-brown, dense, fine textured, strong, hard; wood commercially valuable, used for furniture, cabinetry, and veneer. Nuts eaten by fox, gray and red squirrels, and people; beaver will occasionally use the bark. Leaves were once used to repel insects and dye cloth and hair.

Botanical Name: *Juglans* is Latin for "walnut tree"; *nigra* refers to the dark bark or dark wood.

Juglans nigra (left) and *Juglans cinerea* (right) nuts

Juglans nigra bark

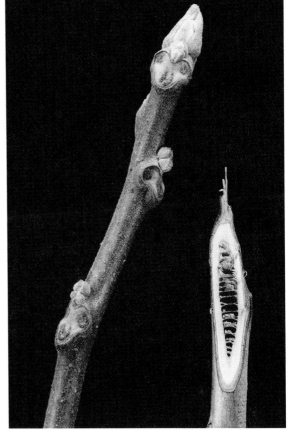
Juglans nigra twig

Juglans nigra leaf

Juglans nigra fruit

Juglandaceae

Lauraceae Guide

1. Leaves deciduous, unlobed or with two or three lobes, venation prominent; fruit on bright red stalks. *Sassafras albidum*
2. Leaves evergreen, unlobed, often disfigured, underside pale; fruit on green stalks. *Persea borbonia*
3. Leaves similar to *Persea borbonia* but midrib with rusty pubescence; twigs with long brown hairs; habitat wetter. *Persea palustris*

Sassafras albidum staminate flowers

Sassafras albidum bark

Sassafras albidum flowers

Sassafras albidum pistillate flowers

Persea borbonia (L.) Spreng.
redbay

Quick Guide: *Leaves* alternate, simple, elliptical, often disfigured, evergreen, spicy aromatic, underside pale; *Fruit* a round blue drupe; *Bark* gray-brown to reddish with flattened, scaly ridges; *Habitat* mesic to xeric woods.

Leaves: Simple, alternate, evergreen, leathery, elliptical to oblong or oblanceolate, often disfigured from insect galls, 7 to 15 cm long, apex and base acute to rounded, margin entire, petiole stout and gray-brown, underside pale with minute pubescence, spicy aromatic when crushed.

Twigs: Slender, slightly angled, mottled green-brown, pubescent; leaf scar crescent shaped with one linear bundle scar.

Buds: Terminal bud naked, rusty pubescent, 5 mm long.

Bark: Red-brown to gray, with flattened ridges; large trees more furrowed with scaly ridges. Inner bark red-brown.

Flowers: Perfect, yellow-white, 5 mm wide, in spring with the new leaves.

Fruit: Drupe, blue, nearly round, 1 cm wide, maturing in early fall.

Form: Up to 18 m (60 ft) tall and 61 cm (2 ft) in diameter.

Habitat and Ecology: Shade tolerant; found on moist to dry sandy soils. Commonly found on sandy bluffs and ridges, and in mesic woods.

Uses: Wood bright red, heavy, hard; used for furniture, cabinets, boats, and trim work. Leaves used as a substitute for bay leaves in cooking. Intermediate value as a deer browse; fruit eaten by songbirds, northern bobwhite, and small mammals.

Botanical Name: *Persea* is an ancient Greek name for an Oriental tree; *borbonia* is an old name for Persea.

Persea borbonia fruit

Persea borbonia bark

Persea borbonia bark

Persea borbonia leaves

Lauraceae

Persea palustris (Raf.) Sarg.
swamp redbay, swampbay

Quick Guide: Similar to *Persea borbonia* but a smaller tree; *Leaves* with rusty hairs on the midrib and veins; *Twigs* with long brown hairs; *Habitat* wetter.

Leaves: Simple, alternate, evergreen, leathery, oblong or elliptical, often disfigured from insect galls, 5 to 20 cm long, apex and base generally acute, margin entire, petiole stout with long brown hairs, underside with rusty pubescence on the midrib and veins, spicy aromatic when crushed.

Twigs: Slender, angled, covered with wooly, long, brown hairs; leaf scar crescent shaped with one linear bundle scar.

Buds: Terminal bud naked, with rusty pubescence, 5 mm long.

Bark: Gray-brown and shallowly grooved, becoming more furrowed with flat, scaly ridges.

Flowers: Perfect, yellow-white, 5 mm wide, long-stalked, in spring with the new leaves.

Fruit: Drupe, blue, nearly round, 1 cm wide, maturing in early fall.

Form: Up to 12 m (40 ft) in height.

Habitat and Ecology: Found in the same range as *Persea borbonia* but on wetter sites in swamps, bays, and marsh edges with *Chamaecyparis thyoides*, *Cyrilla racemiflora*, *Magnolia ashei*, *Myrica heterophylla*, *Nyssa biflora*, and *Taxodium distichum* var. *imbricarium*.

Uses: Uses as for *Persea borbonia*.

Botanical Name: *Persea* is an ancient Greek name for an Oriental tree; *palustris* means "swamp."

Persea palustris leaves

Persea palustris bark

Persea palustris bark

Persea palustris immature fruit

Lauraceae

Sassafras albidum (Nutt.) Nees
sassafras, cinnamon wood, golden elm

Quick Guide: *Leaves* alternate, simple, elliptical, margin entire, some two to three lobed; leaves, twigs, and bark aromatic; *Twigs* mottled yellow-green; *Fruit* a dark blue and red-stalked drupe; *Bark* brown to red-brown and thickly ridged; *Habitat* well-drained soils.

Leaves: Simple, alternate, deciduous, elliptical to ovate, obovate or oval, 8 to 15 cm long, aromatic, margin entire, some two or three lobed, veins prominent on the leaf surface, leaf underside paler with pubescence along the veins; autumn color red, orange, or yellow.

Twigs: Slender, yellow-green, mottled, often curved, aromatic, finely pubescent, with lenticels; leaf scar half-round with one linear bundle scar.

Buds: Terminal bud ovoid, yellow-green, up to 1 cm long; scales three to four, overlapping, keeled.

Bark: Gray to red-brown, thick, irregularly fissured, with broad ridges, inner bark red-brown, fragrant when cut.

Flowers: Imperfect, species is dioecious, in spring before (staminate) or with (pistillate) the leaves in yellow clusters on branch terminals (see page 343).

Fruit: Drupe, dark blue, oblong to nearly round, up to 1.0 cm long, born on bright red swollen stalks in early fall.

Form: Usually only up to 15 m (50 ft) in height, but sometimes larger. Can form thickets.

Habitat and Ecology: Shade intolerant; found on a variety of soils ranging from moist to dry. Forest associates include *Acer saccharum, Carya* spp., *Diospyros virginiana, Fagus grandifolia,*

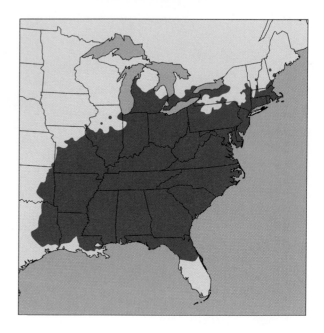

Juniperus virginiana, Liquidambar styraciflua, Liriodendron tulipifera, Nyssa sylvatica, Oxydendrum arboreum, Pinus echinata, Pinus palustris, Pinus taeda, Pinus virginiana, Quercus ilicifolia, and *Ulmus alata*.

Uses: Sapwood yellow, heartwood orange-brown, moderately heavy, moderately hard; used for furniture, specialty items, millwork, window frames, cabinets and fence posts, and was used for fishing rods, boat construction, and ox yokes. Oil of sassafras is extracted from the roots for perfumes and teas. Leaves used in flavoring gumbo soups. A browse for white-tailed deer, black bear, beaver, rabbits, and woodchuck; the fruit is of moderate value to a variety of birds and mammals.

Botanical Name: *Sassafras* comes from a Native American name; *albidum* means "whitish," referring to the leaf underside.

Sassafras albidum bark

Sassafras albidum twig

Sassafras albidum fruit

Sassafras albidum leaves

Sassafras albidum bark

Magnoliaceae Guide

1. Leaves broadly lobed, deciduous, apex broadly notched; flower tuliplike; fruit a cone of samaras; bark ash-gray and deeply furrowed. *Liriodendron tulipifera*
2. Leaves unlobed, deciduous, elliptical, apex acuminate, underside silky pubescent; flower yellow-green; fruit conelike, often curved like a cucumber, with red follicles; bark scaly. *Magnolia acuminata*
3. Leaves unlobed, deciduous, obovate or rhombic, earlike at the base, white and glabrous below; flower yellow or white; fruit conelike with horny carpels and shiny red follicles; bark smooth. *Magnolia fraseri*
4. Similar to *Magnolia fraseri* but found in the southern Coastal Plain. *Magnolia pyramidata*
5. Leaves unlobed, deciduous, extremely large, earlike at the base, underside with silvery white pubescence; flower large and white; fruit conelike with shiny red follicles; bark smooth. *Magnolia macrophylla*
6. Similar to *Magnolia macrophylla* but a smaller tree found in northwest Florida. *Magnolia ashei*
7. Leaves unlobed, deciduous, obovate, base acute, underside with white pubescence, clustered at branch terminals; flower white; fruit conelike with shiny red follicles; bark mostly smooth. *Magnolia tripetala*
8. Leaves unlobed, semi-evergreen or evergreen, elliptical, underside prominently silvery blue; flower white; fruit conelike with shiny red follicles; bark mostly smooth. *Magnolia virginiana*
9. Leaves unlobed, evergreen, leathery, elliptical, underside rusty pubescent; flower white and very fragrant; fruit conelike with shiny red follicles; bark smooth or flaky. *Magnolia grandiflora*

Magnolia fraseri twig

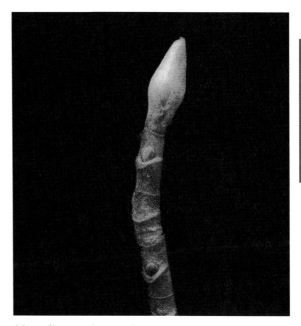
Magnolia acuminata twig

Magnolia macrophylla twig

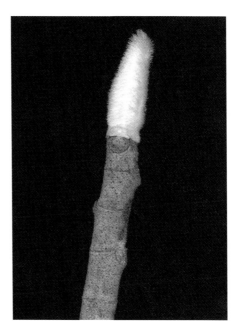
Magnolia grandiflora twig

Magnoliaceae

Liriodendron tulipifera L.

yellow-poplar, tulip-poplar, tulip tree, white-poplar, whitewood

Quick Guide: *Leaves* alternate, simple, broadly lobed, tulip shaped; *Bud* shaped like a duck's bill; *Flower* tuliplike, petals yellow-green with orange; *Fruit* a persistent cone of samaras; *Bark* ash-gray and deeply grooved; *Habitat* moist soils.

Leaves: Simple, alternate, deciduous, lobes mostly four and broad, 13 to 25 cm long, apex broadly notched to truncate, base truncate, margin entire, petiole long; autumn color bright yellow.

Twigs: Moderately stout, red-brown to purple-maroon, glabrous, stipular scars encircle the twig; leaf scar round, raised, with numerous bundle scars.

Buds: Terminal bud yellow-green to dark purple, up to 2 cm long, flattened, with two valvate scales, resembling a duck's bill.

Bark: Gray or gray-green and smooth with light vertical grooves and black branch scars on small trees; large trees ash-gray, thick, and deeply grooved.

Flowers: Perfect, tulip like, petals yellow-green with orange at the base, up to 10 cm wide, blooming in late spring or early summer.

Fruit: Samara, 3.5 cm long, in a conelike aggregation up to 8 cm long, maturing in early autumn and falling through winter.

Form: Tall tree up to 46 m (150 ft) in height and 1.5 m (5 ft) in diameter with a straight clear bole.

Habitat and Ecology: Shade intolerant; found on streambanks and cool slopes, and in well-drained bottomlands, coves, and ravines. Forest associates include *Acer rubrum, Acer saccharum, Diospyros virginiana, Fagus grandifolia, Liquidambar styraciflua, Magnolia virginiana, Nyssa* spp., *Pinus

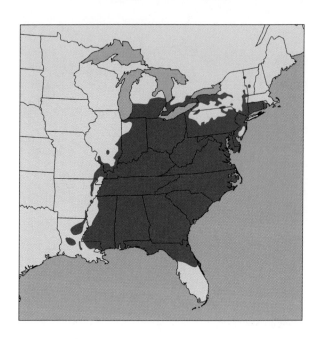

echinata, Pinus strobus, Pinus taeda, Prunus serotina, Quercus alba, Quercus rubra, Quercus velutina, Robinia pseudoacacia,* and *Tsuga canadensis.*

Uses: Wood white to green-brown with green to black stripes, light, soft, straight grained; used for pulpwood, veneer, furniture, paneling, trim, framing, pallets, musical instruments, and boxes. A popular shade tree due to fast growth, interesting leaves, attractive flowers, and good form but requires a large area, plenty of sun, and moist, well-drained soil. Flowers popular with bees. A preferred white-tailed deer browse when succulent in spring and summer; seeds eaten by a wide variety of songbirds and small mammals; beavers will cut young stems growing near water.

Botanical Name: *Liriodendron* means "lily tree"; *tulipifera* means "tuliplike," referring to the flowers.

Liriodendron tulipifera fruit

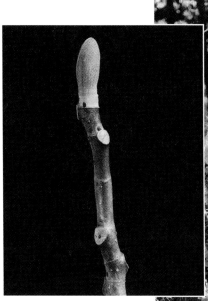
Liriodendron tulipifera twig

Liriodendron tulipifera bark

Liriodendron tulipifera bark

Liriodendron tulipifera leaves and flower

Magnoliaceae

Magnolia acuminata (L.) L.

cucumbertree, cucumber magnolia, yellow flower magnolia, mountain magnolia

Quick Guide: *Leaves* alternate, simple, elliptical, silky pubescent below; *Bud* with white silky pubescence; *Flower* yellow-green; *Fruit* conelike, bulbous, often curved like a cucumber, with shiny red follicles; *Bark* red-brown-gray and scaly; *Habitat* moist, deep soils.

Leaves: Simple, alternate, deciduous, elliptical to ovate, obovate, or oval, 13 to 25 cm long, apex acute to acuminate, base acute to rounded, margin entire, underside with soft white pubescence; autumn color yellow.

Twigs: Moderately stout, red-brown, glabrous, stipular scars encircle the twig; leaf scar U-shaped with numerous bundle scars (see page 351).

Buds: Terminal bud 1.5 cm long, ovoid, the single scale covered with silky white hairs.

Bark: Red-brown to gray-brown with flattened scaly ridges.

Flowers: Perfect, yellow-green, 6 to 8 cm long, blooming in late spring or early summer.

Fruit: Follicles, 13 mm long, shiny red; in a bulbous conelike aggregation up to 8 cm long, often curved, red, maturing in early fall. Green immature fruit looks like a cucumber.

Form: Up to 30 m (100 ft) in height and 1.2 m (4 ft) in diameter.

Habitat and Ecology: Intermediate shade tolerance; found on moist, deep soils of bottomlands, coves, and cool slopes. Forest associates include *Acer rubrum, Acer saccharum, Aesculus*

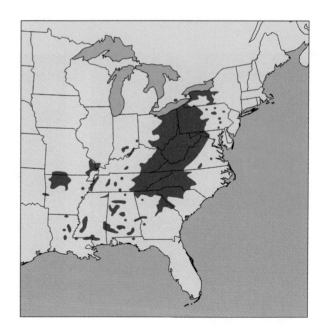

flava, Betula alleghaniensis, Betula lenta, Fraxinus, spp., *Halesia tetraptera, Juglans cinerea, Liriodendron tulipifera, Magnolia grandiflora, Magnolia macrophylla, Magnolia virginiana, Pinus strobus, Prunus serotina, Quercus alba, Quercus rubra, Tilia americana, Tsuga canadensis,* and *Ulmus americana.*

Uses: Wood sold as yellow-poplar, white to pale green-brown, heavy, hard, straight grained; used for pulpwood, veneer, furniture, paneling, trim, framing, pallets, and boxes. Planted as an ornamental for the interesting leaves, yellow flowers, and unusual fruit. Overall use of the magnolias by wildlife relatively low, but seeds eaten by some birds and small mammals.

Botanical Name: *Magnolia* refers to the botanist Peter Magnol; *acuminata* refers to the leaf apex.

Magnolia acuminata fruit

Magnolia acuminata leaves

Magnolia acuminata bark

Magnolia acuminata flower

Magnolia fraseri Walt.

Fraser magnolia, mountain magnolia, ear-leaved magnolia

Quick Guide: *Leaves* alternate, simple, large, earlike at the base, white and glabrous below; *Bud* elongated, purple-green, and glabrous; *Flowers* creamy white to yellow; *Fruit* conelike with scarlet horny carpels and shiny red follicles; *Bark* gray-brown and mostly smooth; *Habitat* moist soils of the Appalachians.

Leaves: Simple, alternate, deciduous, obovate or rhombic, 23 to 45 cm long, often clustered at the ends of shoots, apex acute, base prominently auriculate, margin entire, underside pale and glabrous; autumn color yellow.

Twigs: Moderately stout, green or purplish, glabrous, stipular scars encircle the twig; leaf scar nearly round with numerous bundle scars.

Buds: Terminal bud elongated, conical, up to 4 cm long, with one green or purple glabrous scale (see page 351).

Bark: Mottled gray-brown and smooth with lenticels and warts, becoming scaly at the base of large trees.

Flowers: Perfect, creamy white to pale yellow, up to 23 cm wide, blooming in spring after the leaves.

Fruit: Follicles, 13 mm long, bright red; in a conelike aggregation up to 12 cm long, carpels scarlet with long horny tips, maturing in early fall.

Form: Up to 20 m (65 ft) in height and 1 m (3 ft) in diameter, leaves arranged like an umbrella.

Habitat and Ecology: Intermediate shade tolerance; found on moist, rich soils of coves and cool slopes in the Appalachians with *Acer pensylvanicum, Acer saccharum, Acer spicatum, Aesculus flava, Betula alleghaniensis, Betula lenta, Halesia tetraptera, Magnolia acuminata, Magnolia tripetala, Pinus strobus, Prunus serotina, Tilia americana,* and *Tsuga canadensis.*

Uses: Lumber not commercially important, mixed with other species. Overall use of the magnolias by wildlife relatively low, but seeds eaten by some birds and small mammals.

Botanical Name: *Magnolia* refers to the botanist Peter Magnol; *fraseri* refers to the botanist John Fraser.

Similar Species: Pyramid magnolia (*Magnolia pyramidata* Bartr.) is very similar but is found in the Coastal Plain and has smaller flowers and fruit.

Magnolia fraseri flower © Dorling Kindersley

Magnolia fraseri leaves

Magnolia fraseri bark

Magnolia pyramidata leaves

Magnolia fraseri fruit

Magnoliaceae

Magnolia grandiflora L.

southern magnolia, great laurel magnolia, evergreen magnolia, bull bay

Quick Guide: *Leaves* alternate, simple, evergreen, leathery, elliptical, rusty pubescent below; *Flower* large, white, and very fragrant; *Fruit* conelike with shiny red follicles; *Bark* brown and smooth or scaly; *Habitat* moist soils.

Leaves: Simple, alternate, evergreen, leathery, elliptical to oblong or oval, 13 to 25 cm long, apex acute to rounded, base acute or obtuse, margin entire, shiny dark green above, underside rusty pubescent.

Twigs: Stout, red-brown, with rusty or white hair, stipular scars encircle the twig; leaf scar nearly round with numerous bundle scars (see page 351).

Buds: Terminal vegetative bud up to 4 cm long, conical, the one scale yellow-green with rusty hair; flower bud covered with pale wooly hair.

Bark: Brown to green-brown, smooth or with loose flakes; large trees scaly.

Flowers: Perfect, very showy, with large creamy white petals, very fragrant, up to 20 cm wide, blooming in late spring.

Fruit: Follicles, 13 mm long, shiny bright red, in a red conelike aggregation up to 10 cm long, maturing in fall.

Form: Up to 24 m (80 ft) in height and 1 m (3 ft) in diameter.

Habitat and Ecology: Shade tolerant; found on moist soils of bottomlands, stream and swamp edges, and low uplands with *Fagus grandifolia, Liquidambar styraciflua, Liriodendron tulipifera, Magnolia pyramidata, Magnolia virginiana, Nyssa

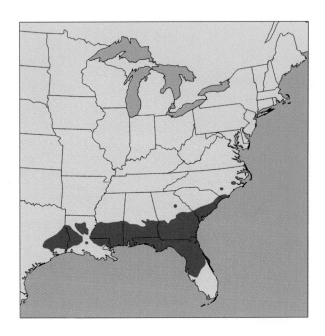

biflora, Pinus palustris, Pinus taeda, Quercus pagoda,* and *Quercus virginiana.*

Uses: Wood sold as yellow-poplar, white to dark green-brown, heavy, hard, straight grained; used for pulpwood, veneer, furniture, paneling, trim, framing, pallets, and boxes. Overall use of the magnolias by wildlife relatively low, but seeds eaten by some birds, in particular cardinals, and small mammals including gray squirrel. Dense foliage offers roosting, loafing, and nesting cover for a variety of birds. A popular ornamental because of the evergreen foliage, wonderfully fragrant flowers and bright red fruit, but the fruit and leaves can be messy and the roots may block septic lines.

Botanical Name: *Magnolia* refers to the botanist Peter Magnol; *grandiflora* refers to the large showy flowers.

Magnolia grandiflora leaves

Magnolia grandiflora fruit

Magnolia grandiflora bark

Magnolia grandiflora bark

Magnolia grandiflora flower

Magnolia macrophylla Michx.
bigleaf magnolia, large-leaved cucumbertree

Quick Guide: *Leaves* alternate, simple, extremely large, with an earlike base and white pubescent underside; *Flowers* large and white; *Fruit* conelike with shiny red follicles; *Bark* gray and smooth with corky lenticels; *Habitat* moist, well-drained soils.

Leaves: Simple, alternate, deciduous, mostly obovate, very large, 50 to 80 cm long, apex acute, base auriculate, margin entire, leaf underside with white pubescence; autumn color yellow.

Twigs: Stout, mottled green-black, pubescent, stipular scars encircle the twig; leaf scar nearly round with numerous bundle scars.

Buds: Terminal bud elongated, flattened, up to 6 cm long, the one bud scale covered with pale pubescence (see page 351).

Bark: Light gray-brown and smooth with corky lenticels or warts; large trees scaly at the base.

Flowers: Perfect, petals creamy white with a purple blotch at the base, large, up to 40 cm wide, blooming in spring after the leaves.

Fruit: Follicles, 13 mm long, shiny red, in a red conelike nearly round aggregation up to 8 cm long, maturing in early fall.

Form: Up to 15 m (50 ft) in height.

Habitat and Ecology: Shade tolerant; found in coves and on streambanks with *Acer barbatum, Asimina triloba, Betula nigra, Liriodendron tulipifera, Platanus occidentalis,* and *Tilia americana.*

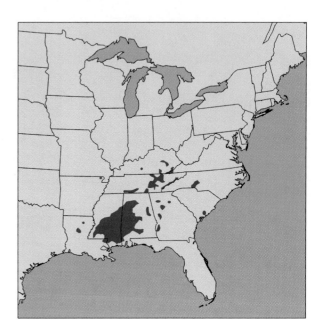

Uses: Wood similar to *Magnolia acuminata* but not commercially important due to limited availability. Overall use of the magnolias by wildlife relatively low, but seeds eaten by some birds and small mammals. Planted as an ornamental for the large leaves and flowers but requires moist, well-drained soils.

Botanical Name: *Magnolia* refers to the botanist Peter Magnol; *macrophylla* means "large leaf."

Similar Species: Ashe magnolia (*Magnolia ashei* Weatherby) is very similar but is a small tree found in the Florida panhandle.

Magnolia macrophylla flower

Magnolia ashei leaves

Magnolia macrophylla immature fruit

Magnolia macrophylla bark

Magnolia macrophylla leaves

Magnolia tripetala (L.) L.
umbrella magnolia, umbrella tree

Quick Guide: *Leaves* alternate, simple, obovate, clustered at branch terminals, base acute, underside with soft white pubescence; *Bud* glabrous with a single green-purple scale; *Flower* creamy white; *Fruit* conelike with shiny red follicles; *Bark* gray-brown and mostly smooth; *Habitat* mountain slopes and along streams.

Leaves: Simple, alternate, deciduous, obovate, 25 to 60 cm long, apex acute, base acute to cuneate, margin entire, underside with soft white pubescence; autumn color yellow. Leaves clustered at the branch terminal like an umbrella.

Twigs: Moderately stout, green-brown to purplish, mostly glabrous, stipular scars encircle the twig, swollen at start of each year's growth; leaf scar nearly round with numerous bundle scars.

Buds: Terminal bud elongated, conical, curved, up to 5 cm long, with one purple glabrous scale.

Bark: Mottled gray-brown, smooth, with lenticels and warts.

Flowers: Perfect, creamy white, up to 25 cm wide, with an unpleasant fragrance, blooming in spring after the leaves.

Fruit: Follicles, 13 mm long, shiny red, in a red conelike aggregation up to 12 cm long, maturing in early fall.

Form: Up to 12 m (40 ft) in height, branches tend to curve upward emphasizing the umbrella appearance.

Habitat and Ecology: Intermediate shade tolerance; found mainly in coves and ravines, and on cool slopes and stream edges. In the Appalachian mountains often found with *Magnolia fraseri*.

Uses: Not commercially important or a popular ornamental because of the unpleasant odor of the flowers. Overall use of the magnolias by wildlife relatively low, but seeds eaten by some birds and small mammals.

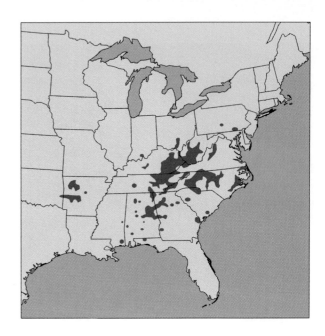

Botanical Name: *Magnolia* refers to the botanist Peter Magnol; *tripetala* means "three petalled," referring to the three green-white outer petals.

Magnolia tripetala flower ©Dorling Kindersley

Magnolia tripetala bud

Magnolia tripetala fruit

Magnolia tripetala bark

Magnolia tripetala leaves

Magnoliaceae

Magnolia virginiana L.

sweetbay magnolia, white bay, swamp bay, sweet bay, swamp magnolia

Quick Guide: *Leaves* alternate, simple, elliptical, underside prominently silvery blue; *Flowers* with creamy white petals and fragrant; *Fruit* conelike with shiny red follicles; *Bark* mottled gray-brown and smooth; *Habitat* stream edges and bottomlands.

Leaves: Simple, alternate, semi-evergreen or evergreen, leathery, elliptical to oblong, 8 to 17 cm long, apex acute to obtuse, base acute to cuneate, margin entire, shiny green above, underside silvery blue-white, aromatic when crushed.

Twigs: Slender, mottled green, with silky pubescence, stipular scars encircle the twig; leaf scar nearly round with numerous bundle scars.

Buds: Terminal bud up to 2.0 cm long, conical, flattened; the one green bud scale with silky, white-gray pubescence.

Bark: Mottled gray-brown, smooth, with lenticels; large trees scaly at the base.

Flowers: Perfect, with creamy white petals, up to 8 cm wide, fragrant, blooming in late spring or early summer.

Fruit: Follicles, 13 mm long, shiny, dark red, in a conelike aggregation up to 5 cm long, maturing in early fall.

Form: Up to 30 m (100 ft) in height and 1 m (3 ft) in diameter.

Habitat and Ecology: Intermediate shade tolerance; found on moist to wet but not deeply

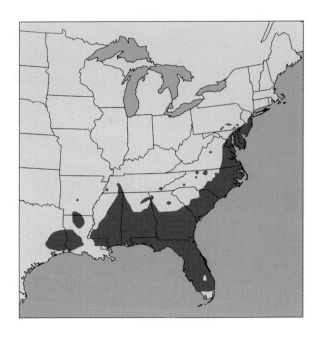

flooded soils of stream edges, swamps, and bottomlands with many species including *Acer rubrum, Chamaecyparis thyoides, Fagus grandifolia, Gordonia lasianthus, Liquidambar styraciflua, Magnolia grandiflora, Nyssa biflora, Pinus elliottii, Pinus palustris, Pinus serotina, Pinus taeda,* and *Taxodium distichum* spp.

Uses: White to pale green-brown, soft, straight grained; used for pulpwood, veneer, pallets, and boxes. Overall use of the magnolias by wildlife relatively low, but seeds eaten by small mammals and some birds, including the red-cockaded woodpecker; bark eaten by beaver.

Botanical Name: *Magnolia* refers to the botanist Peter Magnol; *virginiana* refers to the geographic range.

Magnolia virginiana flower

Magnolia virginiana immature fruit

Magnolia virginiana bark

Magnolia virginiana leaves

Melia azedarach L.
chinaberry, Chinese umbrella tree, Pride of India

Quick Guide: *Leaves* alternate, bi- or tripinnately compound, leaflets coarsely serrate and some basally one to two lobed; *Flowers* in purple fragrant clusters in spring; *Fruit* a yellow drupe in long-stalked clusters persisting over winter; *Bark* brown-gray with loose interlacing ridges; *Habitat* open areas.

Leaves: Bi- or tripinnately compound, alternate, deciduous, up to 50 cm long. Leaflets ovate to obovate, 3 to 8 cm long, margin coarsely serrate, sometimes one to two lobed at the base, apex acuminate; autumn color bright yellow.

Twigs: Stout, purple-brown to green-brown, with brown pubescence and orange lenticels; leaf scar raised, three lobed, heart shaped, with three bundle scars.

Buds: True terminal bud lacking; lateral buds round, fuzzy, covered with dense buff-colored hairs.

Bark: Small stems purple-green to maroon, shiny, smooth, with lenticels; large trees brown-gray and thick with loose interlacing ridges.

Flowers: Perfect, fragrant, in pink-purple clusters, individual flowers resemble purple firecrackers, blooming in spring after the leaves.

Fruit: Drupe, leathery, yellow, round, 1 to 2 cm wide, maturing in early fall, wrinkled clusters of fruit persisting over winter.

Form: Up to 15 m (50 ft) in height with an open low crown.

Habitat and Ecology: Shade intolerant; originally from Asia but naturalized in the South. Found in thickets, fencerows, fields, and along roadsides.

Uses: Once planted as an ornamental and a popular tree for tree houses. Fruit is poisonous to livestock but eaten by some birds, in particular cedar waxwings and American robins. The leaves were once used in bedding to repel fleas and lice, and seeds were used as beads. Fruits used in sling shots and popguns.

Botanical Name: *Melia* means "like an ash tree," referring to the leaves; *azedarach* means "noble tree."

Melia azedarach leaf

Melia azedarach fruit

Melia azedarach flowers

Melia azedarach bark

Melia azedarach twig

Melia azedarach fruit

Meliaceae

Albizia julibrissin Durazz.
mimosa, silk-tree

Quick Guide: *Leaves* alternate, bipinnately compound, leaflets small and scalloped; *Flowers* pink, resembling pom-poms, fragrant, blooming throughout the summer; *Fruit* a legume; *Bark* smooth and gray; *Habitat* open sites.

Leaves: Bipinnately compound, alternate, deciduous, up to 45 cm long. Leaflets scalloped, small, 1 cm long, margin entire; autumn color yellow.

Twigs: Stout, mottled gray-brown to black, glabrous, with lenticels; leaf scar raised, three lobed, with three bundle scars.

Buds: True terminal bud lacking; lateral buds minute, round; scales overlapping, brown, glabrous.

Bark: Pale gray and smooth.

Flowers: Perfect, very fragrant, with silky bright pink filaments, in erect pom-poms throughout summer.

Fruit: Legume, brown, up to 20 cm long, maturing in fall.

Form: Usually only up to 9 m (30 ft) in height with a flat-topped and open crown.

Habitat and Ecology: Shade intolerant; originally from Eurasia but naturalized throughout the South. Found in fields and along fencerows, roads, and wood edges.

Uses: Planted as an ornamental but susceptible to disease. Recently has attracted attention as a potentially important deer browse because of high protein content. This tree can be very invasive, and is considered an exotic pest plant in Florida and Tennessee.

Botanical Name: *Albizia* is for an Italian naturalist, Filippo Albizzi; who brought this tree to Italy; *julibrissin* is from a Persian name. Also grouped in Fabaceae.

Albizia julibrissin flowers

Albizia julibrissin twig

Albizia julibrissin leaf

Albizia julibrissin bark

Albizia julibrissin leaves, flowers, and fruit

Mimosaceae

Moraceae Guide

1. Leaves alternate, opposite or whorled, lobed or unlobed, margin serrate, with rough hairs on top and velvety pubescence below; twigs with long hairs. *Broussonetia papyrifera*
2. Leaves alternate, unlobed or lobed, margin serrate, scabrous above and hairy below; twigs mostly glabrous. *Morus rubra*
3. Similar to *Morus rubra* but leaves shiny and glabrous; bark ridged rather than scaly. *Morus alba*
4. Leaves alternate, unlobed, margin entire, glabrous; twigs glabrous and with spines. *Maclura pomifera*

Maclura pomifera bark

Morus rubra staminate flowers

Morus rubra pistillate flowers

Broussonetia papyrifera (L.) L'Her. ex Vent.

paper-mulberry, tapa cloth tree

Quick Guide: *Leaves* alternate, opposite or whorled, simple, ovate, unlobed or lobed, underside velvety pubescent, petiole with white sap; *Twigs* with long hairs; *Bark* tan and shallowly grooved; *Habitat* open, disturbed sites.

Leaves: Simple, alternate, opposite or whorled, sometimes all three on the same plant, deciduous, ovate or heart shaped, unlobed or with up to five lobes, 8 to 20 cm long, apex acuminate, base rounded or cordate or inequilateral, margin serrate, upper surface rough with hair, underside with velvety pubescence, petiole with long hairs and exuding milky sap when cut; autumn color yellow.

Twigs: Moderately stout, zig-zag, brittle, brown, rough, with corky lenticels and long hairs, pith large and white; leaf scar nearly round, raised, with numerous bundle scars.

Buds: True terminal bud lacking; lateral buds ovoid, 3 mm long, with two to three brown pubescent scales.

Bark: Tan or greenish and smooth when young, becoming shallowly grooved and gray-brown.

Flowers: Species is dioecious; staminate flowers in catkins; pistillate flowers with elongated styles in a dense ball; appearing in late spring.

Fruit: A red-brown ball of drupes 2 to 3 cm wide, maturing in late summer.

Form: Up to 15 m (50 ft) in height, often forming large thickets from root sprouting.

Habitat and Ecology: Originally from China and Japan and brought to the United States as an ornamental; naturalized throughout the eastern United States on disturbed, open sites.

Uses: In China the inner bark is used to make paper, umbrellas, and lanterns. Tapa cloth of the Polynesian islands is also made from the inner bark.

Botanical Name: *Broussonetia* is for the French naturalist Auguste Broussonet; *papyrifera* refers to the use of the fibrous inner bark in making paper.

Broussonetia papyrifera leaf

Broussonetia papyrifera pistillate flowers

Broussonetia papyrifera pistillate flowers and twig

Broussonetia papyrifera bark

Broussonetia papyrifera young twig

Broussonetia papyrifera staminate flowers

Moraceae

Maclura pomifera (Raf.) Schneid.

osage-orange, bowwood, hedge-apple, bodark

Quick Guide: *Leaves* alternate, simple, shiny, ovate, apex acuminate, petiole with white sap; *Twigs* often with a stout spine at the node; *Fruit* a large, juicy, green, brainlike ball; *Bark* gray-brown and thick with interlacing ridges, inner bark orange; *Habitat* pastures and fencerows.

Leaves: Simple, alternate, deciduous, shiny, often clustered in spur shoots, ovate, 8 to 15 cm long, apex acuminate, base obtuse or rounded, margin entire, petiole long and exuding white sap when cut; autumn color yellow.

Twigs: Moderately stout, green to orange-brown, glabrous, with warty lenticels and some with moundlike spur shoots, often with a stout spine at the node; leaf scar nearly round with numerous bundle scars.

Buds: True terminal bud lacking; lateral buds small, round, sunken.

Bark: Gray-brown to orange-brown with thick interlacing ridges and orange inner bark (see page 370).

Flowers: Imperfect, species is dioecious; staminate flowers in yellow-green racemes; pistillate flowers in round hairy heads; appearing in late spring.

Fruit: Seeds in a large (up to 15 cm wide), round, green, juicy, sticky, brainlike ball, maturing in early fall. Reported to be toxic.

Form: Shrub often forming thickets or tree up to 9 m (30 ft) in height with a scraggly crown.

Habitat and Ecology: Shade intolerant; originally found only in Texas, Oklahoma, and Arkansas, mostly in bottomlands. Now found on a range of sites throughout the United States including roadsides, abandoned pastures, and fencerows.

Uses: Wood yellow to bright orange, very hard and heavy, was used for bows by Native Americans (including Osage Native Americans) and was planted extensively for dyes, fence posts, and hedges. Now used in waste site reclamation. Seeds occasionally used by squirrels and northern bobwhite. Thickets provide nesting substrates for birds and excellent cover for other wildlife.

Botanical Name: *Maclura* refers to the geologist William Maclure; *pomifera* means "apple bearing."

Maclura pomifera pistillate flowers

Maclura pomifera fruit

Maclura pomifera staminate flowers

Maclura pomifera twig

Maclura pomifera bark

Maclura pomifera leaves

Morus alba L.

white mulberry, Chinese mulberry, common mulberry, Russian mulberry, silkworm mulberry

Quick Guide: *Leaves* alternate, simple, ovate, shiny, glabrous, unlobed or lobed, margin serrate, petiole with white sap; *Fruit* red, white, or purple blackberrylike drupes; *Bark* ridged rather than scaly; *Habitat* open sites.

Leaves: Simple, alternate, deciduous, ovate to orbicular, dark green and shiny above, 5 to 15 cm long, unlobed or lobed, apex acute or acuminate, base cordate to truncate, margin serrate, petiole exuding milky sap when cut, underside glabrous or pubescent along veins; autumn color yellow.

Twigs: Slender, zig-zag, orangish to gray-brown, shiny; leaf scar round and sunken with numerous bundle scars.

Buds: True terminal bud lacking; lateral buds round, 3 mm long, red-brown, appressed to the twig.

Bark: Orange-brown to yellow-brown and smooth when young; becoming shallowly and irregularly ridged and gray-brown with age.

Flowers: Imperfect, usually dioecious, flowers similar to *Morus rubra*.

Fruit: Drupes, sweet and juicy, in white, pink, or red-purple clusters up to 2.5 cm long; maturing in summer.

Form: Up to 15 m (50 ft) tall with a wide, round crown.

Habitat and Ecology: Introduced from China for use in the silk industry during colonial times, now naturalized in open disturbed areas throughout the United States.

Uses: Wood orange-brown and soft, used for fence posts. Fruit used in jams and baking and also enjoyed by songbirds and small mammals. Fruitless cultivars with attractive autumn color and fast growth are available for landscaping, but this species is considered invasive.

Botanical Name: *Morus* is Latin for "mulberry tree"; *alba* means "white," referring to the fruit.

Morus alba leaf

Morus alba fruit

Morus alba leaf

Morus alba leaf

Moraceae

Morus rubra L.
red mulberry

Quick Guide: *Leaves* alternate, simple, scabrous above, ovate, unlobed or lobed, apex acuminate, margin serrate, petiole with white sap; *Fruit* red-purple, juicy, blackberrylike drupes; *Bark* gray-brown and scaly; *Habitat* moist soils.

Leaves: Simple, alternate, deciduous, ovate or orbicular, may be unlobed or lobed, 10 to 25 cm long, but variable in size, veins prominent, apex acuminate, base rounded or cordate, margin serrate, scabrous above, underside hairy, petiole exuding milky sap when cut; autumn color yellow.

Twigs: Moderately stout, zig-zag, yellow-brown to red-brown, black dotted, mostly glabrous, with lenticels; leaf scar round, sunken, with numerous bundle scars.

Buds: True terminal bud lacking; lateral buds ovoid, 6 mm long, with glabrous scales varying in color from maroon to green-brown, margins darker.

Bark: Gray to red-brown with raised ridges on small trees; large trees gray-brown with scaly or loose ridges.

Flowers: Imperfect, usually dioecious; staminate flowers in long yellow-green catkins; pistillate flowers in shorter catkins; appearing in spring (see page 371).

Fruit: Drupes, in red-purple clusters up to 3 cm long, sweet, juicy, maturing in summer.

Form: Up to 21 m (70 ft) in height and 1 m (3 ft) in diameter.

Habitat and Ecology: Shade tolerant; found on a variety of sites including coves, stream edges, bottomlands, pastures, and forest edges. Forest

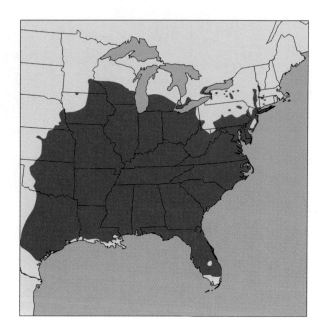

associates include *Acer negundo, Acer rubrum, Acer saccharinum, Celtis laevigata, Gleditsia triacanthos, Liquidambar styraciflua, Quercus nigra,* and *Ulmus americana.*

Uses: Wood yellow to orange-brown, heavy, hard, of limited commercial use, used mainly for fence posts. Was used in building boats; fiber was used in making rope and coarse cloth by Native Americans. Fruit used in jams and wines. Fruit relished by a wide variety of song and game birds including northern bobwhite and wild turkey, and also relished by mammals such as foxes, squirrels, opossum, and raccoon; foliage browsed by white-tailed deer while succulent and rabbits eat the bark.

Botanical Name: *Morus* is Latin for "mulberry tree"; *rubra* means "red," referring to the fruit.

Morus rubra bark

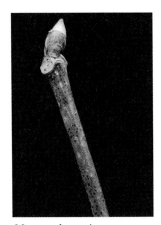

Morus rubra twig

Morus rubra leaf

Morus rubra leaves

Morus rubra fruit

Morus rubra bark

Moraceae

Myricaceae Guide

1. Leaves spicy aromatic with yellow resin dots on both sides. *Myrica cerifera*
2. Similar to *Myrica cerifera* but leaves with yellow resin dots on the underside and found only in a wet habitat. *Myrica heterophylla*
3. Similar to *Myrica cerifera* but leaves deciduous with yellow resin dots below; fruit larger. *Myrica pensylvanica*
4. Similar to *Myrica cerifera* but leaves odorless with an entire margin; fruit larger and brown rather than blue. *Myrica inodora*

Myrica cerifera bark

Myrica cerifera staminate flowers

Myrica cerifera pistillate flowers

Myrica cerifera L.
southern bayberry, wax-myrtle, candleberry

Quick Guide: *Leaves* alternate, simple, oblanceolate, spicy aromatic, evergreen, yellow resin dots above and below; *Fruit* a small, waxy, blue-gray drupe; *Bark* smooth and gray with corky lenticels; *Habitat* a variety of sites.

Leaves: Simple, alternate, evergreen, mostly oblanceolate, 3 to 10 cm long, spicy aromatic when crushed, apex acute, base cuneate, margin entire or with several coarse teeth above the middle, yellow resin dots above and below.

Twigs: Slender, green or brown, with brown pubescence and a spicy aroma when broken; leaf scar crescent shaped.

Buds: Terminal bud minute, yellow scurfy.

Bark: Gray-brown, smooth, with warty lenticels.

Flowers: Imperfect, species is dioecious, pistillate and staminate flowers in oblong catkins in early spring (see page 381).

Fruit: Drupe, round, blue-gray, waxy, 3 mm wide, in compact clusters along the branch, maturing in fall.

Form: Shrub or small tree with multiple trunks (see page 380) up to 12 m (40 ft) in height.

Habitat and Ecology: Found on a range of sites including flatwoods, swamp edges, old fields, fence rows, bogs, dry sandy soils and mixed woodlands.

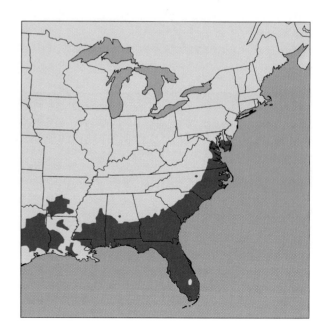

Uses: Planted as an ornamental for the evergreen, fragrant foliage. Fruit was used in candle making and leaves were used to repel insects and flavor soups. Fruit eaten by many songbirds, northern bobwhite, red-cockaded woodpecker, wild turkey, and foxes.

Botanical Name: *Myrica* means "fragrant," referring to the leaves; *cerifera* means "wax bearing," referring to the fruit. Also named *Morella cerifera* (L.) Small.

Myrica cerifera leaves

Myrica cerifera bark

Myrica cerifera fruit

Myrica heterophylla Raf.

swamp candleberry, evergreen bayberry

Quick Guide: Similar to *Myrica cerifera* but *Leaves* lacking yellow resin dots on the upper surface; *Twigs* with shaggy, brown pubescence; *Habitat* swamps and bays.

Leaves: Simple, alternate, evergreen, mostly oblanceolate, 3 to 12 cm long, spicy aromatic when crushed, apex obtuse to rounded, base cuneate, margin with a few coarse teeth above the middle, yellow resin dots lacking above and dense on the underside.

Twigs: Slender, gray-brown, with brown shaggy pubescence and lenticels.

Buds: Minute, red-brown, pubescent.

Bark: Mottled gray-brown to gray-black, with lenticels.

Flowers: Similar to *Myrica cerifera*.

Fruit: Drupe, round, up to 4.5 mm wide, pale blue to gray sometimes brown, waxy, in clusters along the branch, maturing in fall.

Form: Shrub or small tree to 4.5 m (15 ft) in height.

Habitat and Ecology: Found in bays, bogs, flatwoods, and swamps.

Uses: See *Myrica cerifera*.

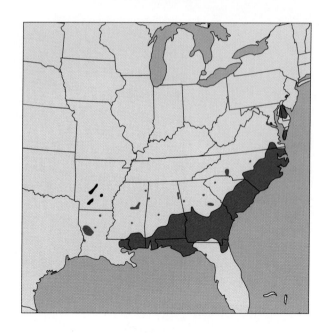

Botanical Name: *Myrica* means "fragrant," referring to the leaves; *heterophylla* means "with variously formed leaves." Also named *Morella caroliniensis* (P. Mill.) Small.

Similar Species: Odorless bayberry (*Myrica inodora* Bartr. or *Morella inodora* (Bartr.) Small) is found on wet sites in the lower Coastal Plain and has odorless leaves with an entire margin, glabrous twigs, and larger brown fruit with a thin waxy coating.

Myrica heterophylla leaves

Myrica heterophylla bark

Myrica heterophylla fruit

Myrica pensylvanica Loisel.

northern bayberry, candleberry, waxmyrtle

Quick Guide: *Leaves* alternate, simple, deciduous, oblanceolate, spicy aromatic, margin entire or with teeth near the apex, yellow resin dots below; *Twigs* pubescent; *Fruit* a waxy, blue drupe; *Habitat* a variety of sites.

Leaves: Simple, alternate, deciduous to semi-evergreen, oblanceolate to obovate or oblong, 3 to 10 cm long, spicy aromatic when crushed, apex acute to obtuse, base cuneate to obtuse, margin entire or with a few coarse teeth above the middle, yellow resin dots only on the underside.

Twigs: Slender, red-brown to gray, with pubescence hairs and lenticels.

Buds: Round, red-brown.

Bark: Gray-brown with corky lenticels.

Flowers: Similar to *Myrica cerifera*.

Fruit: Drupe, round, 3 to 6 mm wide, dark blue, waxy, in clusters along the branch, maturing in fall.

Form: Usually only a shrub but can be up to 9 m (30 ft) in height, can form thickets.

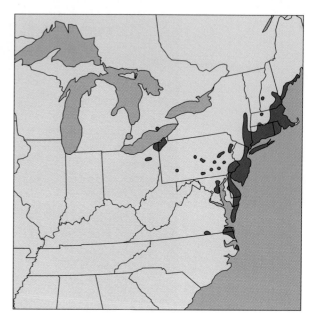

Habitat and Ecology: Found in bogs, swamps, marshes and woodlands, and on dry sand flats and dunes.

Uses: See *Myrica cerifera;* a popular landscape plant except in the extreme South.

Botanical Name: *Myrica* means "fragrant," referring to the leaves; *pensylvanica* refers to the geographic range. Also named *Morella pensylvanica* (Mirbel) Kartesz, comb. nov. ined.

Myrica pensylvanica leaves

Myrica pensylvanica bark

Myrica pensylvanica fruit

Oleaceae Guide

1. Leaves simple, up to 3 cm long. *Ligustrum sinense*
2. Leaves simple, greater than 5 cm long, deciduous, margin finely serrate above the middle. *Forestiera acuminata*
3. Leaves simple, greater than 6 cm long, evergreen, leathery; bark mottled and smooth. *Osmanthus americanus*
4. Leaves simple, greater than 10 cm long, deciduous, petiole purplish; bark grooved or scaly. *Chionanthus virginicus*
5. Leaves pinnately compound; leaf scar shield shaped with the lateral bud above the scar; samara wing narrow and extending halfway down the seed; bark with diamond-shape ridges. *Fraxinus pennsylvanica*
6. Leaves pinnately compound; leaf scar crescent shaped with the lateral bud sitting within the scar; samara wing extending to the top of the seed; bark with diamond-shape ridges. *Fraxinus americana*
7. Leaves pinnately compound; leaf scar shield shaped with the lateral bud above the scar; samara elliptical to diamond shaped and often three-winged, the wing extending to the base of the flattened seed; bark scaly. *Fraxinus caroliniana*
8. Leaves pinnately compound, leaflets with dense pubescence especially along the midrib; leaf scar crescent shaped with the lateral bud sitting within the scar; samara oblong to oblanceolate with the wing extending only to the base of the seed; trunk enlarged and pumpkinlike at the base. *Fraxinus profunda*
9. Leaves pinnately compound, leaflets with a serrate margin and acuminate apex; buds blue-black; samara oblong with the wide wing extending around the flattened seed; bark with corky soft ridges or scaly. *Fraxinus nigra*
10. Leaves pinnately compound; twigs four-angled and winged; samara with the broad wing extending to the base of the seed; bark scaly. *Fraxinus quadrangulata*

Fraxinus pennsylvanica staminate flowers

Fraxinus pennsylvanica staminate flowers

Fraxinus americana pistillate flowers

Fraxinus pennsylvanica pistillate flowers

Chionanthus virginicus L.

fringe-tree, old-man's-beard, gray-beard, white fringetree, flowering ash

Quick Guide: *Leaves* opposite or subopposite, simple, elliptical, deciduous, with a slightly winged and purplish petiole; *Flowers* white, four-petalled, and fringelike; *Twigs* with enlarged nodes and raised semicircular leaf scars; *Bark* scaly; *Habitat* moist soils.

Leaves: Simple, opposite or subopposite, deciduous, elliptical to lanceolate or oblanceolate, 10 to 20 cm long, apex acute to rounded, base cuneate, margin entire, underside glabrous or pubescent, petiole somewhat winged and purplish; autumn color yellow.

Twigs: Moderately stout, red-brown to gray, glabrous or pubescent, with lenticels and enlarged nodes; leaf scar raised, semicircular, concave within, with one linear or U-shaped bundle scar.

Buds: Terminal bud ovoid, 6 mm long, and scales overlapping, brown, keeled, fringed with white; lateral buds moundlike, embedded.

Bark: Gray-brown and shallowly grooved on small trees; large trees with scaly ridges.

Flowers: Imperfect, species is mostly dioecious, flowers creamy white, four-petalled, in fringelike clusters in spring.

Fruit: Drupe, dark blue, nearly round, up to 2 cm long, maturing in late summer or fall.

Form: Up to 9 m (30 ft) in height.

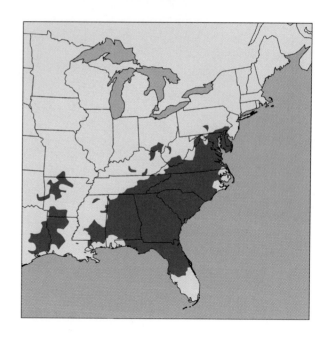

Habitat and Ecology: Found on fertile moist soils of upland forests, and along edges of streams and bogs.

Uses: Planted as an ornamental because of the attractive flowers, which appear even on seedlings. Leaves were once used as a folk remedy for fevers and "yaws." Fruit eaten by birds and mammals including wild turkey, northern bobwhite, and white-tailed deer.

Botanical Name: *Chionanthus* is from Greek meaning "snow flower"; *virginicus* refers to the geographic range.

Chionanthus virginicus leaves and flowers

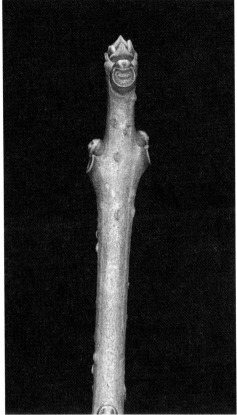
Chionanthus virginicus twig

Chionanthus virginicus leaves and immature fruit

Chionanthus virginicus bark

Forestiera acuminata (Michx.) Poir.
swamp-privet, eastern swampprivet

Quick Guide: *Leaves* opposite, simple, lanceolate, with an acuminate apex and long petiole; *Leaf scar* with one bundle scar; *Fruit* a purplish, wrinkled drupe; *Habitat* wet sites.

Leaves: Simple, opposite, deciduous, lanceolate to elliptical, 5 to 10 cm long, apex acuminate, base acute to cuneate, margin minutely serrate above the middle, underside glabrous or pubescent, petiole up to 2 cm long; autumn color yellow.

Twigs: Slender, gray-brown, glabrous, with warty white lenticels; leaf scar nearly round with one U-shaped bundle scar.

Buds: Terminal and lateral buds with overlapping reddish scales.

Bark: Gray-brown with warty lenticels.

Flowers: Species is dioecious; flowers lacking petals, in yellow-green clusters before the leaves.

Fruit: Drupe, purplish, wrinkled, ellipsoidal, up to 2 cm long, with a ridged stone, maturing in late summer.

Form: A shrub or small tree up to 9 m (30 ft) tall, occasionally larger.

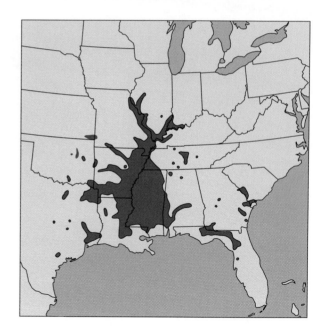

Habitat and Ecology: Found in swamps, sloughs, and margins of streams and ponds.

Uses: Sometimes planted on wet soils; fruit eaten by waterfowl and small mammals.

Botanical Name: *Forestiera* is for the French naturalist Charles Le-Forestier; *acuminata* refers to the leaf apex.

Forestiera acuminata leaf

Forestiera acuminata flowers

Forestiera acuminata bark

Forestiera acuminata leaves and twig

Oleaceae

Fraxinus americana L.
white ash

Quick Guide: *Leaves* opposite, pinnately compound, leaflets usually seven and ovate; *Lateral bud* sitting within the crescent shaped leaf scar; *Fruit* a single samara with the broad wing extending to the top of the seed; *Bark* brown-gray with interlacing ridges forming a diamond pattern; *Habitat* moist, fertile soils.

Leaves: Pinnately compound, opposite, deciduous, up to 35 cm long. Leaflets five to nine, elliptical to ovate, 5 to 13 cm long, margin entire or remotely serrate, pubescent or glabrous below; autumn color yellow to dull red.

Twigs: Moderately stout, gray-brown to green-brown, glabrous or pubescent, with lenticels and flattened nodes; leaf scar crescent shaped with bundle scars forming a U shape.

Buds: Terminal bud round, up to 8 mm long, dark chocolate brown, with suedelike scalloped scales; lateral buds smaller, sitting within the leaf scar.

Bark: Brown-gray with interlacing ridges forming a diamond pattern; large trees deeply furrowed and sometimes blocky.

Flowers: Imperfect, species is dioecious; staminate flowers in red-green, short stalked, dense, compact clusters; pistillate flowers are vase shaped with an elongated style in long-stalked, slender clusters; appearing before the leaves (see page 389).

Fruit: Samara, single, up to 5 cm long, elliptical to oblong, wing fairly broad and mostly extending only to the top of the seed, maturing in fall.

Form: Up to 30 m (100 ft) in height and 1 m (3 ft) in diameter.

Habitat and Ecology: Shade intolerant; found on moist, fertile soils of uplands, bottomlands, and along edges of streams and rivers. Forest associates are numerous and include *Abies balsamea, Acer rubrum, Acer saccharum, Betula alleghaniensis, Fagus grandifolia, Fraxinus nigra, Liriodendron tulipifera, Pinus strobus, Quercus alba, Quercus rubra, Tsuga canadensis,* and *Ulmus americana*.

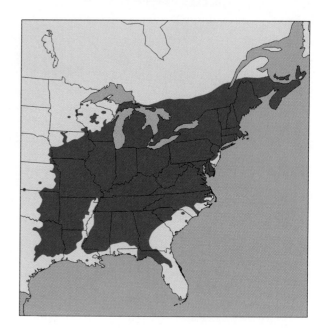

Uses: Wood white to light brown, heavy, hard, stiff, straight grained; used for handles, veneer, furniture, crates, pallets, and sports equipment such as baseball bats (Louisville Slugger). Planted as a shade tree. Seed eaten by a variety of birds and small mammals. White-tailed deer will lightly browse the foliage and beaver will eat the bark.

Botanical Name: *Fraxinus* is Latin for "ash tree"; *americana* refers to the New World.

Fraxinus americana fruit

Fraxinus americana leaf scar

Fraxinus americana bark

Fraxinus americana leaf

Fraxinus caroliniana P. Mill.

Carolina ash, water ash, swamp ash, pop ash

Quick Guide: *Leaves* opposite, pinnately compound, leaflets five to seven; *Fruit* a single samara with a wide wing surrounding the seed, some diamond shaped and three-winged; *Bark* with scaly ridges; *Habitat* wet sites.

Leaves: Pinnately compound, opposite, deciduous, up to 30 cm long. Leaflets five to seven, lanceolate to oval, 2.5 to 10 cm long, margin serrate or entire, midrib lightly pubescent; autumn color yellow.

Twigs: Slender, green-brown, glabrous or pubescent, with lenticels and flattened nodes; leaf scar shield shaped with bundle scars forming a U shape.

Buds: Terminal bud ovoid-acute, brown, with suedelike scales; lateral buds rounder and sitting mostly above the leaf scar.

Bark: Gray-brown, small trees shallowly grooved, larger trees with scaly ridges.

Flowers: Imperfect, species is dioecious; flower clusters appearing before the leaves.

Fruit: Samara, single, up to 5 cm long, variable in shape but often with the widest point in the middle, some diamond shaped and three-winged, the wing widely surrounding the seed; maturing in fall.

Form: Up to 18 m (60 ft) in height but usually smaller, may be multistemmed.

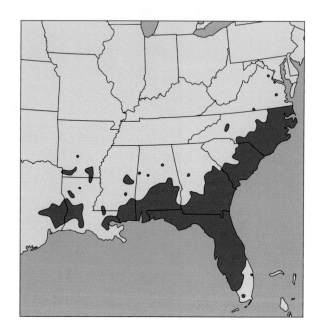

Habitat and Ecology: Found in inundated areas of the Coastal Plain in association with *Fraxinus profunda, Gleditisia aquatica, Nyssa aquatica, Planera aquatica, Populus heterophylla, Quercus laurifolia,* and *Taxodium distichum* var. *distichum*.

Uses: Wood similar to other *Fraxinus* species but not commercially important; wildlife uses as for *Fraxinus americana* plus seed eaten by waterfowl.

Botanical Name: *Fraxinus* is Latin for "ash tree"; *caroliniana* refers to the geographic range.

Fraxinus caroliniana fruit

Fraxinus caroliniana twig

Fraxinus caroliniana leaf

Fraxinus caroliniana bark

Fraxinus caroliniana leaf

Fraxinus nigra Marsh.

black ash, swamp ash, water ash, hoop ash, basket ash

Quick Guide: *Leaves* opposite, pinnately compound, leaflets usually nine, sessile, with an acuminate apex; *Buds* with blue-black scales; *Fruit* a single samara with the wing extending to the bottom of the seed; *Bark* gray with corky or scaly ridges; *Habitat* moist to poorly drained soils.

Leaves: Pinnately compound, opposite, deciduous, up to 40 cm long. Leaflets seven to eleven, usually nine, dark green, sessile, lanceolate to ovate or elliptical, 8 to 15 cm long, apex acuminate, margin serrate, underside with pale or brown pubescence or glabrous; autumn color yellow.

Twigs: Stout, green to gray, with lenticels; leaf scar nearly round with bundle scars forming a C shape.

Buds: Terminal bud conical, 6 mm long, dark brown to almost black or blue-black, with suedelike scalloped scales; lateral buds round, sitting above the leaf scar, first pair of lateral buds sitting below the terminal (rather than immediately adjacent).

Bark: Gray with soft, corky ridges on small trees; larger trees with rough or scaly ridges.

Flowers: Perfect or dioecious, in purplish panicles before the leaves.

Fruit: Samara, single, up to 4 cm long, elliptical to oblong, some twisted, the wide wing extending to the base of the seed, maturing in fall.

Form: Up to 21 m (70 ft) in height and 61 cm (2 ft) in diameter.

Habitat and Ecology: Shade intolerant; found on moist to poorly drained soils of swamps, bogs, and stream edges in association with *Abies balsamea, Acer rubrum, Betula alleghaniensis, Larix laricina, Picea mariana, Thuja occidentalis, Tsuga canadensis,* and *Ulmus americana.*

Uses: Wood with pale sapwood and dull brown or gray heartwood, moderately heavy, moderately hard; used for trim, furniture, cabinets, and because it splits easily along annual rings, splints and baskets. Wildlife uses as for *Fraxinus americana* plus seed eaten by waterfowl.

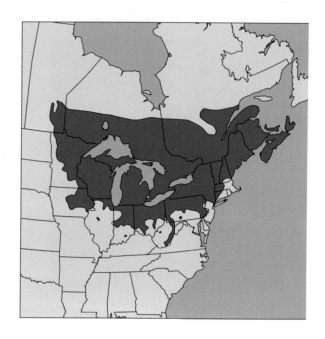

Botanical Name: *Fraxinus* is Latin for "ash tree"; *nigra* means "dark," referring to the leaves or buds.

Fraxinus nigra fruit

Fraxinus nigra terminal bud

Fraxinus nigra fruit

Fraxinus nigra leaf

Fraxinus nigra bark

Oleaceae

Fraxinus pennsylvanica Marsh.

green ash, swamp ash, water ash, red ash

Quick Guide: *leaves* opposite, pinnately compound, leaflets usually seven and elliptical to lanceolate; *Lateral bud* sitting above the shield shaped leaf scar; *Fruit* a single samara with the wing extending halfway down the seed; *Bark* brown-gray with interlacing ridges forming a diamond pattern; *Habitat* moist to wet soils.

Leaves: Pinnately compound, opposite, deciduous, up to 30 cm long. Leaflets seven to nine, elliptical to lanceolate, 5 to 15 cm long, but may be smaller, margin entire or remotely serrate, underside glabrous or pubescent; autumn color yellow.

Twigs: Moderately stout, gray-brown, glabrous or pubescent, with lenticels and flattened nodes; leaf scar shield shaped with bundle scars forming a U shape.

Buds: Terminal bud round, up to 8 mm long, dark chocolate brown, with suedelike scalloped scales; lateral buds smaller, sitting above the leaf scar.

Bark: Brown-gray with interlacing ridges forming a diamond pattern; large trees deeply furrowed.

Flowers: Imperfect, species is dioecious; staminate flowers in red-green, short-stalked, dense, compact clusters; pistillate flowers vase shaped with an elongated style in long-stalked slender clusters; appearing before the leaves (see page 389).

Fruit: Samara, single, up to 5 cm long, spatulate to oblanceolate, wing narrowing at the seed and extending about halfway down the seed, maturing in fall.

Form: Up to 36 m (120 ft) in height and 1 m (3 ft) in diameter.

Habitat and Ecology: Shade intolerant; found on moist uplands, periodically inundated

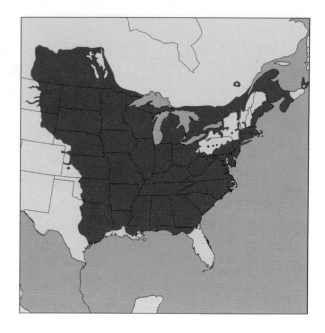

bottomlands, and along edges of streams and rivers. Associated with many species including *Acer negundo, Acer rubrum, Acer saccharinum, Acer saccharum, Carya aquatica, Celtis laevigata, Fraxinus americana, Liquidambar styraciflua, Populus deltoides, Quercus lyrata, Quercus michauxii, Quercus nigra, Quercus pagoda, Quercus phellos, Quercus rubra, Quercus velutina, Salix nigra, Tilia americana,* and *Ulmus americana.*

Uses: Wood white to light brown, heavy, hard, stiff, straight grained, quality not as high as for *Fraxinus americana;* used for handles, veneer, furniture, crates, pallets, and sports equipment such as baseball bats. Planted as a shade tree. Wildlife uses as for *Fraxinus americana.*

Botanical Name: *Fraxinus* is Latin for "ash tree"; *pennsylvanica* refers to the geographic range.

Oleaceae

Fraxinus pennsylvanica leaflets

Fraxinus pennsylvanica twig

Fraxinus pennsylvanica bark

Fraxinus pennsylvanica fruit

Oleaceae   401

Fraxinus profunda (Bush) Bush
pumpkin ash, red ash

Quick Guide: *Leaves* opposite, pinnately compound, leaflets seven to nine, underside with white velvety hair; *Twigs* densely pubescent; *Fruit* a large samara with the wing extending narrowly to the bottom of the seed; *Bark* gray-brown and deeply furrowed; *Habitat* deep swamps and bottoms.

Leaves: Pinnately compound, opposite, deciduous, up to 40 cm long. Leaflets seven to nine, ovate to elliptical, 8 to 20 cm long, apex acute or acuminate, margin entire or irregularly serrate, underside with dense white pubescence that is velvety or wooly, especially along the midrib; rachis with white hair; autumn color yellow.

Twigs: Moderately stout, gray-brown, densely pubescent, with lenticels and flattened nodes; leaf scar shield shaped with bundle scars forming a U shape.

Buds: Terminal bud round, 5 mm long, brown, with suedelike scalloped scales; lateral buds smaller, sitting above or shallowly within the leaf scar.

Bark: Gray-brown with interlacing ridges; large trees deeply furrowed and sometimes blocky.

Flowers: Imperfect, species is dioecious; clusters appearing before the leaves.

Fruit: Samara, single, the largest of the ashes, up to 8 cm long, oblong to oblanceolate, the wide wing extending narrowly around the seed, maturing in fall.

Form: Up to 36 m (120 ft) in height and 1 m (3 ft) in diameter but can be larger on good sites. The base of the trunk is often swollen like a pumpkin.

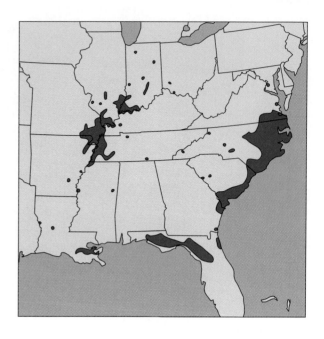

Habitat and Ecology: Intermediate to shade intolerant; found on wet to very wet sites in association with *Fraxinus caroliniana*, *Gleditsia aquatica*, *Nyssa aquatica*, *Nyssa biflora*, *Planera aquatica*, *Populus heterophylla*, and *Taxodium distichum* var. *distichum* and var. *imbricarium*.

Uses: Wood similar to other *Fraxinus* species. Wildlife uses as for *Fraxinus americana* plus seed eaten by waterfowl.

Botanical Name: *Fraxinus* is Latin for "ash tree"; *profunda* means "thick, dense," possibly referring to the trunk base or bark.

Fraxinus profunda leaf

Fraxinus profunda bark

Fraxinus profunda bark

Fraxinus profunda twig

Fraxinus quadrangulata Michx.
blue ash

Quick Guide: *Leaves* opposite, pinnately compound, leaflets usually nine; *Twigs* four-angled and winged; *Fruit* a samara with the broad wing extending to the bottom of the seed; *Bark* gray with scaly ridges; *Habitat* dry, upland sites.

Leaves: Pinnately compound, opposite, deciduous, up to 30 cm long. Leaflets usually nine but from seven to eleven, ovate to lanceolate, 6 to 13 cm long, apex acuminate, margin serrate, underside pubescent or glabrous; autumn color yellow.

Twigs: Stout, green to blue-gray, four angled, winged or ridged, with lenticels and exuding bluish sap when cut; leaf scar crescent shaped with bundle scars forming a U shape.

Buds: Terminal bud ovoid, 5 mm long, gray or brown, pubescent, with suedelike scalloped scales; lateral buds more rounded, sitting somewhat in the leaf scar.

Bark: Gray-blue and smooth on small trees; large trees gray with scaly ridges; inner bark blue when cut.

Flowers: Perfect; in panicles before the leaves.

Fruit: Samara, single, up to 5 cm long, oblong to obovate, the broad wing extending to the base of the seed, maturing in fall.

Form: Up to 24 m (80 ft) in height.

Habitat and Ecology: Found on dry upland sites and limestone soils. Forest associates include *Juniperus virginiana*, *Quercus muehlenbergii*, *Quercus shumardii*, *Quercus stellata*, and *Quercus velutina*.

Uses: Wood yellow-brown, heavy, hard; used for flooring, tool handles, and crates. A blue dye was made from the inner bark. Planted as an ornamental for its tolerance of dry soils. Wildlife uses as for *Fraxinus americana*.

Botanical Name: *Fraxinus* is Latin for "ash tree"; *quadrangulata* refers to the four angled twigs.

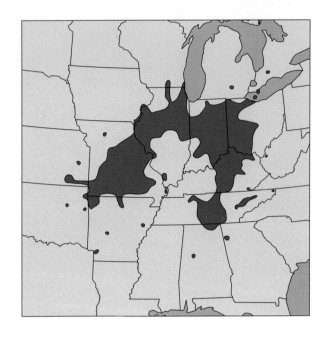

Fraxinus quadrangulata fruit

Fraxinus quadrangulata fruit

Fraxinus quadrangulata leaf

Fraxinus quadrangulata bark

Fraxinus quadrangulata twig

Ligustrum sinense Lour.
Chinese privet

Quick Guide: Usually a shrub or multistem small tree; *Leaves* opposite, simple, small, and nearly round; *Flowers* in terminal, malodorous, white clusters; *Fruit* a small blue drupe; *Bark* smooth and brown-gray; *Habitat* open areas.

Leaves: Simple, opposite, persistent to evergreen, elliptical to round, up to 3 cm long, apex and base acute to rounded, margin entire, underside pubescent, petiole short and pubescent.

Twigs: Slender, gray-brown, pubescent, with lenticels; leaf scar small, semicircular, raised, with one bundle scar.

Buds: Terminal bud minute with overlapping, green-brown, pubescent scales.

Bark: Brown-gray, mottled, smooth, with lenticels.

Flowers: Perfect, in white malodorous terminal clusters after the leaves.

Fruit: Drupe, blue, round, up to 5 mm wide, in drooping clusters maturing in early fall and persisting through winter.

Form: Usually a shrub but sometimes a small tree up to 6 m (20 ft).

Habitat and Ecology: Imported from Asia and naturalized throughout the eastern United States, found aggressively colonizing open and disturbed areas and considered an invasive pest plant.

Uses: Heavily browsed by white-tailed deer and eaten by beaver; abundant fruit eaten by a variety of songbirds. Dense thickets near streams provide cover for foraging American woodcock.

Botanical Name: *Ligustrum* is Latin for "privet"; *sinense* means "from China."

Ligustrum sinense flowers

Ligustrum sinense leaves and flowers

Ligustrum sinense bark

Ligustrum sinense fruit

Osmanthus americanus (L.) Benth. & Hook. f. ex Gray

devilwood, wild-olive, fragrant olive

Quick Guide: *Leaves* opposite, simple, leathery, evergreen, and elliptical; *Flowers* fragrant, white; *Fruit* an olivelike dark blue drupe; *Bark* brown-gray to black, small trees smooth and warty; *Habitat* sandy ridges to edges of swamps and streams.

Leaves: Simple, opposite, evergreen, leathery, elliptical to lanceolate or obovate, 6 to 15 cm long, apex acute, base cuneate, margin entire and sometimes revolute, glabrous below, petiole stout.

Twigs: Slender, green to gray-brown, smooth or scaly, with lenticels; leaf scar widely crescent shaped, raised, with one crescent shaped bundle scar.

Buds: Terminal bud acute, 1 cm long, with two brown valvate scales.

Bark: Brown-gray to black, often mottled, smooth with lenticels or warts; large trees darker and rougher.

Flowers: Perfect and imperfect, small and white, in fragrant clusters in spring. Immature flower clusters obvious in fall and winter.

Fruit: Drupe, dark blue, olivelike, round, up to 2.0 cm long, maturing in fall.

Form: Up to 9 m (30 ft) in height, can form thickets.

Habitat and Ecology: Found near streams, in hammocks, on sandy ridges, and in dune forests.

Uses: Wood hard and difficult to work. The fragrant flowers and evergreen foliage make this species an attractive specimen for landscaping.

Botanical Name: *Osmanthus* is from Greek for "fragrant flower"; *americanus* refers to the New World.

Osmanthus americanus leaves

Osmanthus americanus fruit and immature flowers

Osmanthus americanus flowers

Osmanthus americanus bark

Osmanthus americanus bark

Platanus occidentalis L.

sycamore, American sycamore, American planetree, buttonwood, buttonball-tree

Quick Guide: *Leaves* alternate, simple, large, fan shaped, lobes three to five, margin coarsely toothed; *Bud* conical with one smooth scale; *Fruit* a long-stalked ball of achenes; *Bark* brown and scaly exposing white smooth bark; *Habitat* riverbanks and bottomlands.

Leaves: Simple, alternate, deciduous, fan shaped, 13 to 20 cm wide, lobes three to five, sinuses shallow, apex acuminate, base truncate, margin wavy and coarsely toothed, underside pubescent, petiole stout and hollow at the base hiding the lateral bud; autumn color dull yellow.

Twigs: Moderately stout, zig-zag, orangish or yellow-brown, glabrous, with stipular scars; leaf scar thin and nearly encircling the bud.

Buds: True terminal bud lacking; lateral buds divergent, 8 mm long, conical, caplike, with one maroon glabrous scale.

Bark: Brown, flaky, and scaly revealing smooth green-white bark; upper trunk smooth and white.

Flowers: Imperfect, in long-stalked round heads from leaf axils; pistillate flowers red and on longer stalks; staminate flowers smaller, yellow-green, and on shorter stalks; appearing in spring with the leaves.

Fruit: Achenes, in a ball-like cluster up to 3.0 cm wide on a long stalk, maturing in fall.

Form: Very large tree up to 42 m (140 ft) in height and 3 m (10 ft) in diameter.

Habitat and Ecology: Intermediate shade tolerance; found on banks of streams and rivers, and in bottomlands with *Acer negundo, Acer rubrum,*

Acer saccharinum, Betula nigra, Celtis laevigata, Celtis occidentalis, Fraxinus spp., *Liquidambar styraciflua, Populus deltoides,* and *Salix nigra.*

Uses: Wood blond to light brown, close grained, coarse, moderately hard; used for veneer, pulpwood, inexpensive furniture, paneling, trim, butcher blocks, pallets, and boxes. Planted in short-rotation plantations due to fast growth and coppicing. The bark and leaves make for an interesting landscape tree but this species requires moist soils and plenty of space. Core of the fruit was used for buttons. Minimal wildlife value, although some finches will eat the seeds. Probably the most important use is as a den tree for birds and mammals.

Botanical Name: *Platanus* is Latin for "plane-tree," referring to the broad leaves; *occidentalis* refers to the Western Hemisphere.

Platanus occidentalis bark

Platanus occidentalis twig

Platanus occidentalis fruit

Platanus occidentalis pistillate flowers

Platanus occidentalis bark

Platanus occidentalis leaves and stipules

Platanaceae

Rhamnus caroliniana Walt.

Carolina buckthorn, Indian cherry, yellow buckthorn

Quick Guide: *Leaves* alternate, simple, elliptical, shiny, with prominent parallel venation, margin obscurely serrate; *Buds* naked and wooly; *Fruit* a red to black drupe; *Bark* gray, mottled, and mostly smooth; *Habitat* near streams and on limestone soils.

Leaves: Simple, alternate, deciduous, shiny green or yellow-green, oblong to elliptical, 5 to 15 cm long, apex acute, base rounded or tapered, parallel venation prominent and curving at the margin, margin obscurely serrate, petiole possibly red; autumn color yellow.

Twigs: Slender, red-brown, pubescent; leaf scar elliptical with three bundle scars.

Buds: Terminal bud 6 mm long, naked, with wooly pale hair.

Bark: Gray, mottled, smooth, becoming shallowly grooved.

Flowers: Perfect, yellowish or white, small, with five petals, in small clusters in late spring or early summer.

Fruit: Drupe, red to black, round, up to 8 mm wide, maturing in late summer or fall.

Form: Shrub or tree up to 12 m (40 ft) in height.

Habitat and Ecology: Found in moist woods, on streambanks, and on dry limestone soils.

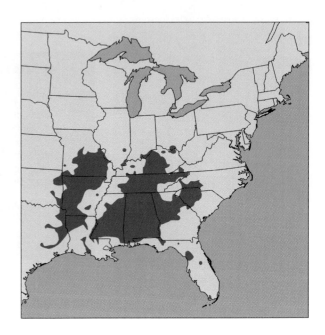

Uses: Bark was used in making a yellow dye. Charcoal from the wood was used in making gunpowder. Sometimes planted as an ornamental for the attractive foliage. Fruit occasionally eaten by birds.

Botanical Name: *Rhamnus* is from Greek for "buckthorn"; *caroliniana* refers to the geographic range. Also named *Frangula caroliniana* (Walt.) Gray.

Rhamnus caroliniana fruit and twig

Rhamnus caroliniana bark

Rhamnus caroliniana bark

Rhamnus caroliniana leaves and fruit

Rosaceae Guide

1. Leaves simple, ovate, margin serrate, petiole pubescent; flowers in white racemes before the leaves; buds long pointed; fruit a red or purple pome; bark smooth and gray with vertical stripes. *Amelanchier arborea*
2. Similar to *Amelanchier arborea* but leaves glabrous even when young; fruit blue-black when mature. *Amelanchier laevis*
3. Leaves simple, variable in shape, margin serrate to dentate, sometimes lobed; twigs with thorns; fruit a pome; bark scaly or peeling. *Crataegus* spp.
4. Leaves simple, elliptical, margin finely serrate, base sometimes lobed; twigs with thorns; flowers in showy, fragrant, white-pink terminal racemes with the leaves; fruit a sour pome; bark scaly. *Malus angustifolia*
5. Similar to *Malus angustifolia* but leaves more ovate and margin with larger teeth. *Malus coronaria*
6. Leaves simple, ovate to elliptical, apex acuminate, base rounded, margin serrate, petiole with and without glands; flowers in white umbels with the leaves; fruit a red sour drupe; bark scaly. *Prunus americana*
7. Leaves simple, ovate, margin serrate and glandular, petiole with dark glands; flowers in white-pink umbels with the leaves; fruit a red or yellow drupe; bark scaly. *Prunus nigra*
8. Leaves simple, lanceolate, often folding upward, margin with glandular teeth, petiole with glands; flowers in white umbels before the leaves; twigs with thorns; fruit a yellow-red sweet drupe; young bark red-black and shiny. *Prunus angustifolia*
9. Similar to *Prunus angustifolia* but leaf margin teeth without red-tipped glands and leaf midrib downy. *Prunus umbellata*
10. Leaves simple, evergreen, elliptical, margin entire or with an occasional hooked red tooth; flowers in short-stalked, white racemes after the leaves; fruit a blue-black drupe persisting over winter; bark gray-brown and smooth. *Prunus caroliniana*
11. Leaves simple, lanceolate, apex with a long-tapered tip, margin finely serrate, petiole bright red with one to two glands near the blade; flowers in white umbels after the leaves; fruit a bright red drupe; bark red to red-brown and lustrous on young trees and branches. *Prunus pensylvanica*
12. Leaves simple, oblong-lanceolate or oval, margin serrate, midrib with tawny pubescence, petiole with one or two glands near the blade; flowers in white racemes after the leaves; fruit a purple-black, juicy drupe; mature bark with black scales. *Prunus serotina*
13. Leaves simple, ovate, margin sharply serrate, underside densely pubescent; flowers in white umbels; twigs pubescent; fruit a large red drupe; bark scaly. *Prunus mexicana*
14. Leaves simple, oval to obovate, margin sharply and finely serrate, petiole glandular; flowers in white racemes; fruit a red to purplish drupe; twigs with a strong odor when cut. *Prunus virginiana*
15. Leaves simple, ovate, margin serrate, underside densely white pubescent; twigs very hairy; fruit an apple. *Malus pumila*
16. Leaves compound, leaflets 9 to 17, lanceolate, margin sharply serrate, rachis bright red; flowers in white flat-topped clusters; drupes in orange-red clusters; bark gray-brown becoming scaly. *Sorbus americana*
17. Similar to *Sorbus americana* but leaves not arching; flowers and fruits larger in more open clusters. *Sorbus decora*

Prunus serotina twig

Prunus angustifolia twig

Amelanchier arborea twig

Amelanchier laevis twig

Amelanchier arborea (Michx. f.) Fern.
downy serviceberry, juneberry, shadbush

Quick Guide: *Leaves* alternate, simple, ovate, margin serrate, petiole pubescent; *Flowers* in white racemes before the leaves; *Buds* long pointed; *Fruit* a red or purple pome; *Bark* with vertical stripes; *Habitat* well-drained soils.

Leaves: Simple, alternate, deciduous, ovate to oval, 4 to 10 cm long, apex acute, base rounded to cordate, margin serrate, underside downy pubescent at unfolding, petiole pubescent; autumn color yellow to red.

Twigs: Slender, red-brown, glabrous or pubescent, with white lenticels; leaf scar crescent shaped with three bundle scars.

Buds: Terminal bud acute, long pointed, about 1 cm long; scales five, overlapping, glabrous or with white pubescence on the margins, green-red; lateral buds appressed to the twig (see page 415).

Bark: Green to gray-brown, smooth, with vertical stripes or streaks sometimes twisting around the tree; large trees darker and grooved, appearing cracked.

Flowers: Perfect; in short-stalked, white racemes in early spring before the leaves.

Fruit: Pome, red to purple, nearly round, up to 1 cm wide, maturing in summer.

Form: Shrub or tree up to 15 m (50 ft) in height.

Habitat and Ecology: Found in the understory on a variety of sites including moist woods, dry limestone soils, and oak-pine forests.

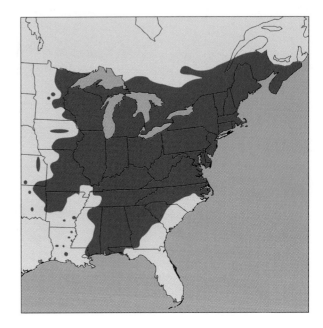

Uses: Wood red-brown and hard, used for tool handles. An attractive ornamental because of early spring flowers, bark, autumn foliage, and tolerance of a range of conditions. Serviceberries are considered valuable wildlife foods; fruit eaten by a variety of song and game birds, white-tailed deer, raccoon, foxes, chipmunk, bear, squirrel, and rabbit; white-tailed deer browse the foliage and twigs, and beaver eat the bark. Of all the serviceberries, *Amelanchier arborea* is the least edible.

Botanical Name: *Amelanchier* is a French name; *arborea* means "treelike."

Amelanchier arborea flowers

Amelanchier arborea leaf

Amelanchier arborea bark

Amelanchier arborea bark

Amelanchier arborea fruit

Amelanchier laevis Wieg.
Allegheny serviceberry, smooth juneberry, smooth serviceberry

Quick Guide: *Leaves* alternate, simple, ovate, margin serrate, mostly glabrous; *Flowers* in white racemes after the leaves; *Buds* long pointed; *Fruit* a blue-black juicy pome; bark smooth and gray with vertical stripes; *Habitat* moist woods.

Leaves: Simple, alternate, deciduous, ovate, 4 to 10 cm long, apex acute to abruptly acuminate, base rounded to cordate, margin serrate, underside glabrous, petiole glabrous, unfolding leaves bronze to purple and mostly glabrous; autumn color yellow.

Twigs: Slender, red-brown, glabrous or pubescent, with white lenticels; leaf scar crescent shaped with three bundle scars.

Buds: Terminal bud long pointed, about 1 cm long; scales five, overlapping, green-red, fringed with white pubescence; lateral buds appressed to the twig (see page 415).

Bark: Green to gray-brown, smooth, with vertical stripes; large trees darker and grooved with flattened ridges.

Flowers: Perfect; in short-stalked, white racemes in spring before or with the purplish unfolding leaves.

Fruit: Pome, red becoming blue-black when mature, sweet, juicy, nearly round, up to 1 cm wide, maturing in early summer.

Form: Shrub or small tree up to 9 m (30 ft) in height.

Habitat and Ecology: Found in moist woods in the eastern United States from Maine to Georgia and Alabama.

Uses: Wood red-brown and hard, used for tool handles. An attractive ornamental because of early spring flowers, bark, autumn foliage, and tolerance of a range of conditions. See *Amelanchier arborea* for wildlife uses.

Botanical Name: *Amelanchier* is a French name; *laevis* means "smooth," referring to the glabrous leaves.

Amelanchier laevis bark

Amelanchier laevis flowers

Amelanchier laevis bark

Amelanchier laevis leaves and immature fruit

Rosaceae

Crataegus spp.
hawthorns, haws

Leaves: Simple, alternate, deciduous, margin serrate to dentate, often lobed.

Twigs: Often with thorns; leaf scar crescent shaped with three bundle scars.

Buds: Round to ovoid with overlapping scales.

Bark: Green to maroon, brown or black, mottled, smooth; scaly or peeling on large trees.

Flowers: Perfect, showy, white or pink, with five petals, blooming in spring.

Fruit: Pome, round, 0.6 to 1.5 cm wide, usually maturing in fall.

Form: Shrub or small tree to 6 m (20 ft), some species with a fluted trunk.

Habitat and Ecology: Found in a variety of habitats throughout the East.

Uses: Wood hard and strong, used for tool handles. Many cultivars are available for landscaping. Fruit used in jellies and preserves. Fruit eaten by white-tailed deer, black bear, foxes, coyotes, song and game birds, and small mammals; lightly browsed by white-tailed deer. Thickets provide excellent nesting cover for a variety of songbirds.

Botanical Name: *Crataegus* is derived from a Greek work for "strength," referring to the hard wood.

Similar Species: A large genus and because of the confusion in classification, only basic characteristics are presented.

Crataegus marshallii leaves and flowers

Crataegus viridis leaves

Crataegus uniflora leaves and fruit

Crataegus marshallii bark

Malus angustifolia (Ait.) Michx.

southern crabapple, narrow-leaved crabapple, southern crab

Quick Guide: *Leaves* alternate, simple, elliptical, margin finely serrate to entire, base sometimes lobed; *Twigs* often with thorns; *Flowers* in showy, fragrant, white-pink racemes with the leaves; *Fruit* a sour pome; *Bark* scaly and gray-brown; *Habitat* a variety of sites.

Leaves: Simple, alternate, deciduous, elliptical to ovate or oblong, 2.5 to 8 cm long, margin ranging from finely serrate to crenate or entire, sometimes lobed on spur shoots, underside pubescent only when young, petiole pubescent; autumn color yellow.

Twigs: Red-brown, pubescent; leaf scar crescent shaped with three bundle scars. Spur shoots often ending in thorns.

Buds: Terminal bud small, 3 mm long, with red-brown overlapping and pubescent scales.

Bark: Gray-brown to red-brown and scaly.

Flowers: Perfect, white-pink, with five petals, fragrant, in terminal racemes with the leaves.

Fruit: Pome, yellow-green to red, round, 3 cm wide, sour, maturing in fall.

Form: A shrub or small tree up to 9 m (30 ft) in height, often forming thickets.

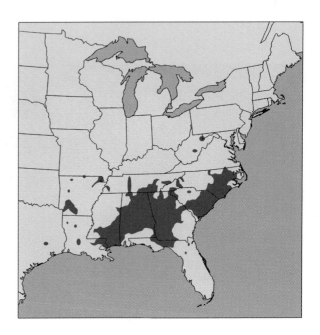

Habitat and Ecology: Found in moist woodlands, along fencerows, and on stream edges.

Uses: A wide selection of cultivars with many fruit and flower colors available for landscaping. Fruit used in jellies and preserves. Fruit eaten by white-tailed deer, wild turkey, ruffed grouse, various songbirds, and small mammals. Thickets provide cover.

Botanical Name: *Malus* is Latin for "apple tree", *angustifolia* means "narrow leaf."

Malus angustifolia fruit

Malus angustifolia flower

Malus angustifolia twig

Malus angustifolia leaves

Malus angustifolia bark

Malus coronaria (L.) P. Mill.

sweet crabapple, wild crabapple, American crabapple

Quick Guide: *Leaves* alternate, simple, ovate, margin coarsely serrate and nearly lobed at the base on some leaves; *Twigs* often ending in thorns; *Flowers* white-pink and very fragrant; *Fruit* a green, sour pome; *Bark* gray to red-brown with scaly ridges; *Habitat* open woods and fencerows.

Leaves: Simple, alternate, deciduous, ovate, 4 to 8 cm long, apex acute to abruptly acuminate, base rounded, margin coarsely serrate or doubly serrate, nearly lobed at the base on some leaves, underside glabrous when mature, petiole red and sometimes pubescent; autumn color yellow.

Twigs: Red-maroon to gray, glaucous, spur branches often ending in thorns; leaf scar crescent shaped with three bundle scars.

Buds: Up to 7 mm long, ovoid; scales overlapping, red, fringed with white pubescence.

Bark: Gray, shiny on small stems; large trees red-brown to gray with scaly ridges or plates.

Flowers: Perfect, white-pink, with five petals, very fragrant, appearing in spring.

Fruit: Pome, yellow-green, nearly round, up to 4 cm wide, fragrant when rubbed, maturing in fall.

Form: Up to 9 m (30 ft) in height, often forming thickets.

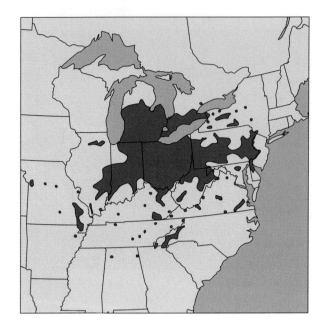

Habitat and Ecology: In open woods and along fences and road edges.

Uses: The hard, heavy wood is used for specialty items such as tool handles and was used for gear wheels. Can be planted as an ornamental but thicket forming and susceptible to rust diseases. The sour fruit is used in preserves and jellies. White-tailed deer, skunk, foxes, squirrels, and other small mammals eat the fruit.

Botanical Name: *Malus* is Latin for "apple tree"; *coronaria* means "fit for crowns and garlands," referring to the beautiful flowers in spring.

Malus coronaria bark

Malus coronaria leaf

Malus coronaria fruit

Malus coronaria flowers

Malus coronaria bark

Malus pumila P. Mill.
common apple, apple

Quick Guide: *Leaves* alternate, simple, ovate, margin serrate, underside and petiole densely white-pubescent; *Twigs* with hair but lacking thorns; *Buds* with wooly hairs; *Fruit* a green or red apple, misshapen on wild trees; *Bark* gray-brown and scaly; *Habitat* a variety of sites, especially old home sites.

Leaves: Simple, alternate, deciduous, ovate, oval, or elliptical, 5 to 10 cm long, apex acute to abruptly acuminate, base rounded to obtuse, margin serrate or crenate, underside and petiole with dense white pubescence; autumn color yellow.

Twigs: Maroon to brown, pubescent to tomentose, spur branches lacking thorns; leaf scar with three bundle scars.

Buds: Up to 6 mm long, rounded; scales overlapping and red with white wooly hair.

Bark: Brown-gray, thin, and scaly.

Flowers: Perfect, white, red, or pink, with five petals, appearing in spring.

Fruit: The apple (pome), yellow, green or red, nearly round, up to 10 cm wide, sweet or sour in taste, smaller and misshapen on wild trees, maturing in fall.

Form: Up to 12 m (40 ft) tall with a low crown; wild trees often crooked.

Habitat and Ecology: A native of Europe but naturalized in forests, old fields, and along fences and road edges, particularly near old home sites.

Uses: Wood hard and heavy, used for specialty items such as tool handles. The apple is a favorite fruit of humans and white-tailed deer, bear, fox, raccoon, and small mammals.

Botanical Name: *Malus* is Latin for "apple tree"; *pumila* means "dwarf." Also known as *Malus sylvestris* (L.) Mill.

Malus pumila leaves

Malus pumila bark

Malus pumila bark

Malus pumila fruit

Prunus americana Marsh.

American plum, wild plum, red plum, yellow plum

Quick Guide: *Leaves* alternate, simple, margin serrate, petiole with and without glands; *Twigs* with thornlike branches; *Flowers* in white umbels; *Fruit* a yellow to red drupe; *Habitat* a variety of open sites.

Leaves: Simple, alternate, deciduous, elliptical to ovate and lanceolate, 6 to 10 cm long, apex acuminate, base acute to rounded, margin serrate without glands, upper surface possibly scabrous, petiole with and without glands; autumn color yellow.

Twigs: Red-brown to gray, with lenticels and thornlike spur branches, odorless when cut; leaf scar semicircular with three bundle scars.

Buds: Acute, up to 8 mm long; scales overlapping, red-brown or gray.

Bark: Red-brown to gray or black, possibly lustrous, with lenticels on branches and small trees; large trees with scaly plates.

Flowers: Perfect, in white umbels, appearing with the leaves.

Fruit: Drupe, yellow turning to orange and red, glaucous, nearly round, 2.5 cm long, sweet or sour; maturing in late summer.

Form: Up to 9 m (30 ft) tall with a low crown, forming thickets.

Habitat and Ecology: Found on a variety of open sites.

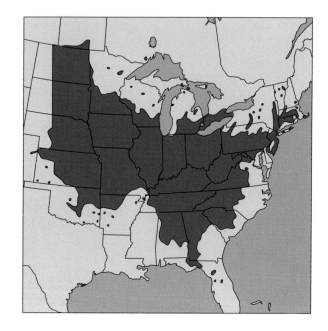

Uses: Flowers popular with bees and fruit eaten by many birds and small mammals. Fruit used in "plum butter" and in preserves.

Botanical Name: *Prunus* is Latin for "plum tree"; *americana* refers to the New World.

Similar Species: Canada plum (*Prunus nigra* Ait.) is found on a variety of sites from New England to Iowa and has ovate leaves with a serrate or doubly serrate margin and glandular teeth, petiole with large dark glands, and black twigs with thorny spur shoots.

Prunus americana flowers

Prunus americana leaf

Prunus americana fruit

Prunus americana bark

Prunus angustifolia Marsh.
chickasaw plum, wild plum

Quick Guide: *Leaves* alternate, simple, lanceolate, often folding upward, margin with gland-tipped teeth, petiole red; *Flowers* in white umbels before the leaves; *Twigs* shiny or glaucous, red-brown, and with thorns; *Habitat* fencerows and old fields.

Leaves: Simple, alternate, deciduous, elliptical to lanceolate, 3 to 8 cm long, often folding upward, apex acuminate, base acute, margin serrate and teeth tipped with yellow or red glands, petiole red with one or two glands near the blade; autumn color yellow.

Twigs: Slender, red-brown, shiny or glaucous, with lenticels and thorns; leaf scar small, semicircular, with three bundle scars.

Buds: True terminal bud lacking; lateral buds acute, 5 mm long, with red-brown overlapping scales (see page 415).

Bark: Red-brown and shiny with horizontal lenticels on small trees; large trees scaly.

Flowers: Perfect, in showy white umbels before the leaves.

Fruit: Drupe, yellow or red, round, up to 2.5 cm wide, sweet, maturing in late summer.

Form: Shrub or small tree to 6 m (20 ft), often forming thickets as a result of root sprouting.

Habitat and Ecology: Found along fencerows and in open or disturbed areas.

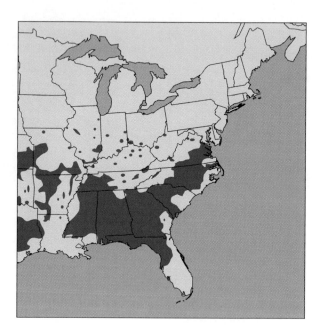

Uses: Fruit used in jellies and preserves. Fruit eaten by white-tailed deer, black bear, gray fox, raccoon and opossum. Thickets form excellent nesting and escape cover for a variety of songbirds and game birds.

Botanical Name: *Prunus* is Latin for "plum tree," *angustifolia* means "narrow leaf."

Similar Species: Flatwoods plum or hog plum (*Prunus umbellata* Elliott) does not form thickets and has the same range but is distinguished by ovate to oval leaves with margin teeth lacking red-tipped glands, a downy leaf midrib, and smaller purple fruit.

Prunus angustifolia leaves

Prunus angustifolia flowers

Prunus angustifolia fruit

Prunus angustifolia bark and thorns

Prunus angustifolia bark

Rosaceae

Prunus caroliniana (P. Mill.) Ait.

Carolina laurel cherry, laurel cherry, cherry laurel

Quick Guide: *Leaves* alternate, simple, evergreen, shiny, elliptical, margin with an occasional hooked red tooth; *Fruit* a blue-black drupe persisting over winter; *Bark* smooth and gray-brown; *Habitat* a variety of sites.

Leaves: Simple, alternate, evergreen, leathery, shiny, elliptical to oblanceolate, 5 to 12 cm long, apex and base acute, margin entire or with several hooked red teeth, underside glabrous, petiole red and lacking glands.

Twigs: Slender, red-brown, glaucous, with lenticels; leaf scar small, heart shaped, raised, with three bundle scars.

Buds: Terminal bud acute, 6 mm long, with red-black overlapping scales; flower buds plump, red-white.

Bark: Gray-brown and mostly smooth with horizontal lenticels.

Flowers: Perfect, small, erect, short-stalked, in yellow-white racemes in spring before the new leaves.

Fruit: Drupe, blue-black, nearly round, 1 to 1.5 cm wide, maturing in late summer or fall and persisting over winter.

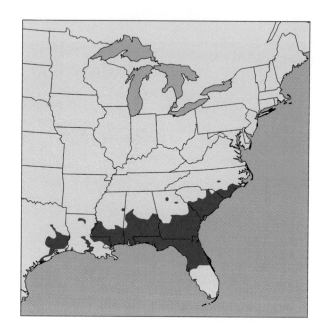

Form: Up to 9 m (30 ft) in height.

Habitat and Ecology: Found in a variety of habitats including upland forests, near rivers and streams, along fencerows, in open disturbed areas, and in coastal forests.

Uses: Leaves and twigs may be poisonous to livestock. Fruit relished by songbirds, particularly American robins and cedar waxwings.

Botanical Name: *Prunus* is Latin for "plum tree," *caroliniana* refers to the geographic range.

Prunus caroliniana fruit

Prunus caroliniana leaves

Prunus caroliniana bark

Prunus caroliniana flowers

Prunus mexicana S. Wats.
Mexican plum, bigtree plum, fall plum

Quick Guide: *Leaves* alternate, simple, ovate, with prominent venation and a sharply serrate margin, underside densely pubescent; *Flowers* in white umbels; *Fruit* a large, sweet, red drupe; *Bark* with rough, scaly ridges; *Habitat* a variety of sites.

Leaves: Simple, alternate, deciduous, ovate to elliptical, 5 to 13 cm long, veins on upper surface deeply sunken, apex acute or acuminate, base acute to rounded, margin sharply serrate, densely pubescent below, petiole red and pubescent with and without glands at the leaf base; autumn color yellow.

Twigs: Pubescent when young, becoming stiff, maroon-gray, glaucous or glossy, with lenticels and spur branches, found with and without thorns; leaf scar semicircular with hair on the top margin and three bundle scars.

Buds: Acute, 6 mm long, with overlapping brown-maroon scales.

Bark: Maroon to dark gray with horizontal lenticels and scaly; large trees furrowed with rough, loose ridges.

Flowers: Perfect, in showy white umbels before or with the leaves.

Fruit: Drupe, 2.5 to 4 cm wide, nearly round, sweet, glaucous, red to purplish when mature in late summer.

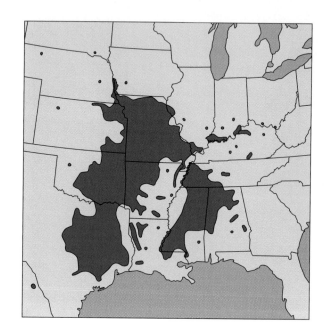

Form: Up to 9 m (30 ft) in height, does not form thickets.

Habitat and Ecology: Shade intolerant; found on a variety of sites including fencelines, woodlands, and bottomlands.

Uses: Fruits used in jams and preserves. Flowers visited by bees and fruit eaten by birds and mammals. This tree can be planted as an ornamental for the flowers and tolerance of dry soils.

Botanical Name: *Prunus* is Latin for "plum tree"; *mexicana* refers to the geographic range.

Prunus mexicana immature fruit

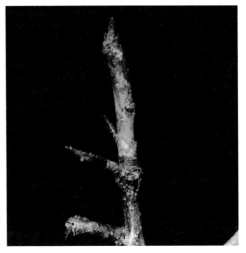

Prunus mexicana twig

Prunus mexicana leaves

Prunus mexicana bark

Prunus mexicana flowers

Rosaceae 435

Prunus pensylvanica L. f.

pin cherry, fire cherry, wild red cherry, bird cherry

Quick Guide: *Leaves* alternate, simple, lanceolate, apex long tapering, margin finely serrate, petiole bright red with one or two glands near the blade; *Flowers* in white umbels after the leaves; *Fruit* a bright red drupe in summer; *Bark* red to red-brown with lenticels, shiny on small trees and branches; *Habitat* a variety of sites.

Leaves: Simple, alternate, deciduous, lanceolate, 8 to 15 cm long, apex acuminate and long tapering, base acute to rounded, margin finely serrate, underside glabrous, petiole bright red with one or two glands near the blade; autumn color red or orange.

Twigs: Slender, bright red to red-brown, shiny or glaucous, with lenticels; leaf scar small, semicircular, with three bundle scars.

Buds: Terminal buds ovoid to rounded, clustered, 2 mm long; scales red-brown, overlapping.

Bark: Small trees shiny, satiny, red to maroon, with horizontal lenticels; large trees red-brown with rough lenticels, sometimes peeling.

Flowers: Perfect, in white umbels late spring after the leaves.

Fruit: Drupe, bright red, round, 7 mm wide, maturing in summer.

Form: Up to 12 m (40 ft) in height.

Habitat and Ecology: Shade intolerant; found on a variety of sites, frequently forming thickets on burned or disturbed areas. Forest associates include *Abies fraseri, Acer pensylvanicum, Acer rubrum, Acer saccharum, Betula alleghaniensis, Betula lenta, Picea rubens, Pinus strobus,* and *Tsuga canadensis* in the southern Appalachians and farther north also *Betula papyrifera, Fagus grandifolia, Populus tremuloides,* and *Quercus rubra.*

Uses: Wood not commercially important. Same wildlife uses as for *Prunus angustifolia* plus fruit eaten by ruffed grouse and stems used by beaver.

Botanical Name: *Prunus* is Latin for "plum tree"; *pensylvanica* refers to the geographic range.

Prunus pensylvanica leaves

Prunus pensylvanica bark

Prunus pensylvanica bark

Prunus pensylvanica fruit

Prunus serotina Ehrh.

black cherry, rum cherry, wild black cherry

Quick Guide: *Leaves* alternate, simple, oval, oblong or lanceolate, margin serrate, midrib with tawny pubescence, petiole with one or two glands near the blade; *Flowers* in white racemes after the leaves; *Fruit* a purple-black juicy drupe; *Bark* with black scales; *Habitat* a variety of soils but preferring cool, moist sites.

Leaves: Simple, alternate, deciduous, shiny, oval, oblong or lanceolate, 5 to 15 cm long, apex acuminate, base acute to rounded, margin serrate, midrib with tawny pubescence, petiole with one or two glands near the blade; autumn color yellow.

Twigs: Slender, red-brown, shiny or glaucous, with lenticels and an almondlike smell when cut; leaf scar small, semicircular, with three bundle scars.

Buds: Terminal bud acute, 4 mm long; scales overlapping, varying in color from brown to red-brown, red-green, or pinkish (see page 415).

Bark: On small trees red-brown, smooth, lustrous, and with horizontal lenticels; large trees with gray-black scales or plates.

Flowers: Perfect, in long white racemes in spring with the young leaves.

Fruit: Drupe, red-purple to black, round, 0.8 to 1.3 cm wide, maturing from summer to fall.

Form: Up to 24 m (80 ft) in height and 1 m (3 ft) in diameter.

Habitat and Ecology: Shade intolerant; found on a variety of soils but best growth is on moist, fertile soils. Forest associates are numerous and include *Abies balsamea, Acer rubrum, Acer saccharum, Betula lenta, Fagus grandifolia, Liriodendron tulipifera, Picea rubens, Pinus strobus, Prunus pensylvanica, Quercus alba, Quercus rubra, Quercus velutina,* and *Tsuga canadensis.*

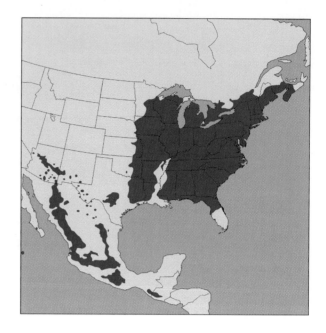

Uses: Wood rich red-brown, lusterous when finished, moderately heavy, moderately hard, straight grained; commercially valuable, used for furniture, veneer, and woodenware. Fruit was added to cordials, brandies, and rum as a flavoring. A beautiful tree for landscaping but susceptible to fungal disease and infestation by the forest tent caterpillar. Wilted foliage may be poisonous to livestock because of the release of hydrocyanic acid. A valuable wildlife fruit because of the consistency of production and maturation when few other fruits are available. Fruit eaten by a variety of birds including ruffed grouse, northern bobwhite, and numerous songbirds, as well as black bear, red and gray foxes, raccoon, fox and gray squirrels, opossum, and numerous small mammals. Occasionally browsed by white-tailed deer.

Botanical Name: *Prunus* is Latin for "plum tree"; *serotina* means "late ripening or late opening."

Prunus serotina leaves

Prunus serotina flowers

Prunus serotina fruit

Prunus serotina bark

Prunus serotina bark

Prunus virginiana L.

chokecherry, eastern chokecherry, common chokecherry

Quick Guide: *Leaves* alternate, simple, oval to obovate, lacking tawny pubescence on the midrib, margin with small sharp teeth, petiole with glands; *Twigs* with a strong odor when cut; *Fruit* bright red to purplish; *Bark* dark with lenticels; *Habitat* a variety of open sites.

Leaves: Simple, alternate, deciduous, oval to obovate, 4 to 10 cm long, apex abruptly acuminate, base cuneate to rounded, margin sharply and finely serrate, underside with small tufts of hair in vein axils, petiole with small glands; autumn color yellow.

Twigs: Slender, gray or red-brown, glabrous, with lenticels, emitting a strong unpleasant odor when cut; leaf scar semicircular with three bundle scars.

Buds: Acute, up to 1 cm long; scales overlapping, brown, glabrous.

Bark: Red-brown to gray-black with lenticels on small trees; on large trees becoming shallowly fissured; releasing an unpleasant odor when cut.

Flowers: Perfect, white, in long terminal racemes, appearing late spring and early summer.

Fruit: Drupe, bright red to purplish, about 9 mm wide, in drooping clusters late summer and fall, very astringent when immature.

Form: A shrub or small tree up to 9 m (30 ft) tall with a low crown, often forming thickets.

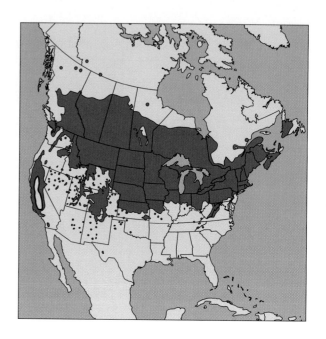

Habitat and Ecology: Shade intolerant; found on a variety of open and disturbed areas.

Uses: Fruit eaten by many game and songbirds and small mammals; browsed by ruffed grouse, rabbits, white-tailed deer, and moose. The unripe fruit is very bitter (hence the name), but the ripe fruit is used in jams, jellies, and wines. Cultivars are available for landscaping.

Botanical Name: *Prunus* is Latin for "plum tree"; *virginiana* refers to the geographic range.

Prunus virginiana leaf

Prunus virginiana fruit

Prunus virginiana bark

Prunus virginiana flowers

Sorbus americana Marsh.

mountain-ash, American mountain-ash

Quick Guide: *Leaves* alternate, pinnately compound, leaflets 9 to 17 and lanceolate with a sharply serrate margin, rachis bright red; *Flowers* in white flat-topped clusters after the leaves; *Fruit* a pome in orange-red clusters; *Bark* gray-brown, smooth or scaly; *Habitat* a variety of sites.

Leaves: Pinnately compound, alternate, deciduous, up to 25 cm long, often arching, rachis bright red and grooved. Leaflets 9 to 17, lanceolate, 5 to 10 cm long, sessile, apex acuminate, margin sharply serrate, underside pale; autumn color red.

Twigs: Stout, red-brown, glabrous, with lenticels; leaf scar crescent shaped with at least five bundle scars.

Buds: Terminal bud up to 1.3 cm long, acute, with red-brown sticky scales.

Bark: Gray-brown, smooth with lenticels, becoming scaly.

Flowers: Perfect, white, five-petalled, in large flat-topped clusters in late spring.

Fruit: Pome, orange-red to scarlet, round, 6 to 10 mm wide, in bright clusters in fall.

Form: Up to 9 m (30 ft) in height.

Habitat and Ecology: Found at high elevations in the southern Appalachians in association with *Abies fraseri* and *Picea rubens,* and in the North on edges of swamps, bogs, and lakes as well as on dry sites.

Uses: Can be planted as an ornamental but susceptible to insect and disease problems. Fruit eaten by ruffed grouse and numerous songbirds, in particular evening grosbeaks, American robin, and cedar waxwings.

Botanical Name: *Sorbus* is the Latin name for mountain-ash, *americana* refers to the New World.

Similar Species: Showy or northern mountain-ash (*Sorbus decora* (Sarg.) C. K. Schneid.) is found in the northeastern United States and distinguished by leaves usually not arching, larger flowers in more

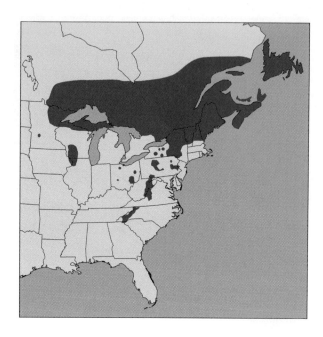

open clusters (versus the dense clusters of *Sorbus americana*), and larger fruits (1.3 cm wide). European mountain-ash (*Sorbus aucuparia* L.) has naturalized in some areas and is identified by coarsely toothed leaflets, wooly buds, and orange fruit.

Sorbus americana flowers

Sorbus americana fruit

Sorbus americana leaves

Sorbus americana bark

Cephalanthus occidentalis L.

buttonbush, common buttonbush, honey-balls

Quick Guide: *Leaves* opposite or whorled, simple, elliptical, margin entire; *Leaf scar* with one bundle scar; *Flowers* and *Fruit* in compact round heads; *Bark* red-brown with loose ridges; *Habitat* wet sites.

Leaves: Simple, opposite or whorled, deciduous, elliptical to ovate, 7 to 18 cm long, apex acute to acuminate, base acute to truncate, margin entire, underside pale and pubescent, petiole grooved and sometimes purplish; autumn color yellow.

Twigs: Red-brown, pubescent, with corky lenticels and stipular scars; leaf scar circular with one crescent shaped bundle scar, stipular scars connect the leaf scars. The tips of shoots die back.

Buds: True terminal bud lacking; lateral buds minute, sunken.

Bark: Red-brown, with loose or scaly ridges.

Flowers: Perfect, white, in round long-stalked heads during summer.

Fruit: Nutlets, in a compact round head, 2 to 3 cm wide, maturing in late summer and early fall, persistent over winter.

Form: Shrub or tree to 6 m (20 ft).

Habitat and Ecology: Shade intolerant; found in wet areas such as edges of swamps and ponds,

next to streams, and in marshes and drainage ditches.

Uses: Fruit eaten by a variety of songbirds and some species of waterfowl, such as wood duck and mallard. Flowers popular with bees. Thickets provide excellent roosting and brood rearing habitat for wood ducks.

Botanical Name: *Cephalanthus* means "head" in Latin, referring to the flowers and fruit; *occidentalis* refers to the Western Hemisphere.

Cephalanthus occidentalis fruit

Cephalanthus occidentalis flowers

Cephalanthus occidentalis twig

Cephalanthus occidentalis bark

Cephalanthus occidentalis leaves

Rubiaceae

Pinckneya bracteata (Bartr.) Raf.
pinckneya, fever-tree, Georgia-bark

Quick Guide: *Leaves* opposite, simple, elliptical, densely pubescent, margin entire, petiole red and pubescent; *Twig* very hairy with corky lenticels; *Leaf scar* with one bundle scar; *Sepals* large, creamy white to rose red; *Fruit* a bumpy, flat-topped, round capsule; *Habitat* wet sites.

Leaves: Simple, opposite, deciduous, elliptical to ovate, 7 to 20 cm long, margin entire, apex acute, base cuneate, densely pubescent above and below, petiole red and pubescent.

Twigs: Red-brown, downy pubescent or tomentose, with corky lenticels and stipular scars; leaf scar nearly round with one crescent shaped bundle scar and stipular scars connecting the leaf scars.

Buds: Terminal bud 7 mm long, sharp-pointed, with red-brown scales; lateral buds minute, round.

Bark: Brown, shallowly grooved or lightly scaly.

Flowers: Perfect, tubular; sepals leafy, ovate, white to pale pink or rose-red, some large (7 cm long); petals yellow-green with maroon stripes; blooming in late spring or early summer.

Fruit: Capsule, 1 to 2 cm wide, woody, brown, nearly round, flat-topped, bumpy, splitting in half, seeds flat and winged, maturing in fall.

Form: Shrub or tree to 6 m (20 ft).

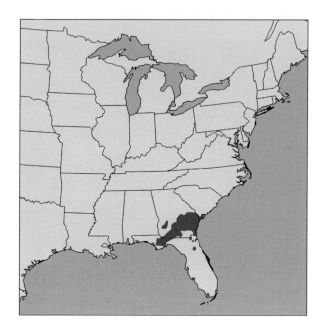

Habitat and Ecology: Uncommon, found in swamps, bays, and bogs of the Coastal Plain, often in association with *Toxicodendron vernix* (Godfrey and Wooten 1981).

Uses: Inner bark was used to treat fevers and malaria.

Botanical Name: *Pinckneya* is for Charles Pinckney, a South Carolina statesman and botany enthusiast; *bracteata* refers to the bractlike sepals.

Pinckneya bracteata flowers

Pinckneya bracteata twig

Pinckneya bracteata bark

Pinckneya bracteata fruit

Pinckneya bracteata leaves and fruit

Rubiaceae

Rutaceae Guide

1. Leaves trifoliately compound. *Ptelea trifoliata*
2. Leaves pinnately compound; twigs with paired spines. *Zanthoxylum americanum*
3. Leaves pinnately compound; twigs with scattered prickles. *Zanthoxylum clava-herculis*

Zanthoxylum americanum terminal bud

Zanthoxylum clava-herculis twig

Zanthoxylum americanum leaves

Zanthoxylum clava-herculis leaf

Zanthoxylum clava-herculis flowers

Zanthoxylum clava-herculis fruit

Ptelea trifoliata L.

common hoptree, wafer ash, stinking ash

Quick Guide: *Leaves* alternate, trifoliately compound, glandular dots above and below, with a rank odor when crushed; *Buds* yellow-brown, pubescent, and sunken within a U-shaped leaf scar; *Fruit* a round samara; *Bark* red-brown with warty lenticels; *Habitat* rocky uplands.

Leaves: Trifoliately compound, alternate, deciduous, up to 21 cm long. Leaflets ovate to elliptical, 5 to 15 cm long, apex acuminate to rounded, base cuneate to rounded, margin shallowly serrate, dotted with glands above and below, emitting an unpleasant odor when crushed; autumn color yellow.

Twigs: Red-brown, pubescent, with lenticels and a rank odor when cut; leaf scar U shaped with three bundle scars.

Buds: True terminal bud lacking; lateral buds, yellow-brown, densely hairy, buried within the leaf scar.

Bark: Red-brown to gray-brown with warty lenticels on small trees; large trees becoming ridged and scaly.

Flowers: Perfect or functionally imperfect, with green-white petals, in malodorous terminal clusters in spring after the leaves.

Fruit: A waferlike samara, round, flat, with two fused wings, 2 cm wide, maturing in fall, clusters persistent over winter.

Form: Shrub or small tree up to 8 m (25 ft) in height.

Habitat and Ecology: An occasional tree found in the understory of rocky, upland forests.

Uses: Fruits reported to have been used as a substitute for hops in beer.

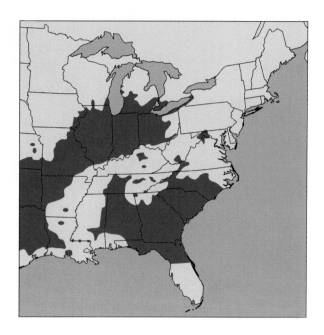

Botanical Name: *Ptelea* is from Greek for "elm tree," referring to the similarities in fruit; *trifoliata* means "three leaved."

Ptelea trifoliata fruit

Ptelea trifoliata flowers

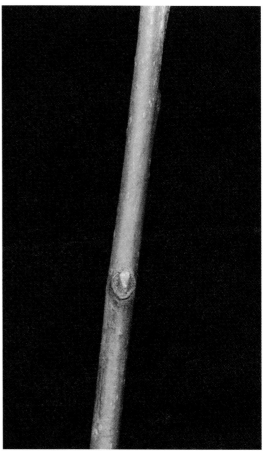
Ptelea trifoliata twig

Ptelea trifoliata bark

Ptelea trifoliata leaf

Zanthoxylum americanum P. Mill.

common prickly-ash, northern prickly ash, toothache tree

Quick Guide: *Leaves* alternate, pinnately compound, leaflets five to eleven, rachis with spines; *Twigs* with paired spines; *Bark* gray and mostly smooth; *Habitat* a variety of sites.

Leaves: Pinnately compound, alternate, deciduous, up to 25 cm long, rachis often red and with small spines. Leaflets five to eleven, ovate to lanceolate or elliptical, 3 to 6 cm long, with a bitter taste, apex acute, base slightly inequilateral, margin serrate or crenate with pale glands, underside pubescent; autumn color yellow. See page 448 also.

Twigs: Slender, brown-gray to maroon, pubescent, with paired spines and a lemon smell when cut; leaf scar semicircular with three bundle scars.

Buds: Terminal bud small, round; scales red and overlapping with wooly, red-brown hairs; lateral buds superposed (see page 448).

Bark: Brown-gray, mostly smooth.

Flowers: Imperfect, species is dioecious; in small yellow-green clusters with the leaves.

Fruit: Follicles, nearly round, red-brown, 5 mm wide, containing shiny black seeds; maturing in late summer.

Form: A shrub or small tree up to 8 m (25 ft) in height, often forming thickets.

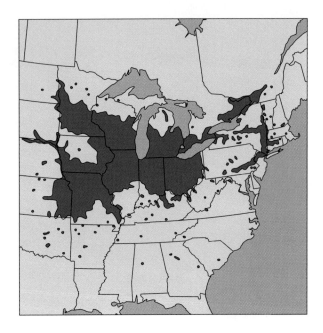

Habitat and Ecology: Found on a wide range of sites including stream edges, fencerows, and dry upland forests.

Uses: Thickets provide cover for wildlife, birds eat the seeds.

Botanical Name: *Zanthoxylum* means "golden wood"; *americanum* refers to the New World.

Zanthoxylum americanum twig

Zanthoxylum americanum leaves

Zanthoxylum americanum bark

Zanthoxylum americanum fruit

Zanthoxylum clava-herculis L.

Hercules'-club, toothache-tree, southern prickly-ash

Quick Guide: *Leaves* alternate, pinnately compound, leaflets shiny, margins crenate with yellow glands on margin teeth; *Spines* often on the rachis, petiole, and young branches; *Bark* with sharp or corky pyramidal growths; *Habitat* coastal forests.

Leaves: Pinnately compound, alternate, deciduous, up to 38 cm long, rachis with or without spines. Leaflets 5 to 19, usually 7 to 9, lanceolate to ovate or falcate, 2.5 to 7 cm long, shiny, margin crenate with yellow glands between teeth, apex acute or acuminate, base inequilateral or rounded, petiole notably red or purplish, underside glandular, with a rank odor when crushed; autumn color yellow. See page 449 also.

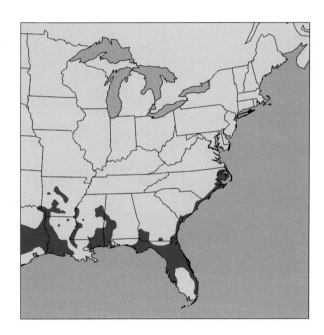

Twigs: Stout, green-brown, glabrous, with scattered sharp prickles; leaf scar shield shaped with three bundle scars. See page 449 also.

Buds: Terminal bud small, round; scales pubescent, green-brown, with red-brown dots.

Bark: Brown-gray and smooth with lenticels and sharp prickles on small trees that become dull, corky, and pyramidal (resembling the mythical club of Hercules) on large trees.

Flowers: Perfect and imperfect, species often dioecious, petals yellow-green and small in terminal clusters in spring or early summer (see page 449).

Fruit: Follicle, nearly round, red-brown, 6 mm wide, containing a shiny black seed, maturing in late summer. See page 449 also.

Form: Up to 12 m (40 ft) in height and 46 cm (1.5 ft) in diameter.

Habitat and Ecology: Found near streams and rivers, in sand dunes and hammocks, and along fences.

Uses: The bark was chewed to numb toothaches. Birds eat the fruit.

Botanical Name: *Zanthoxylum* means "golden wood"; *clava* means "a club or knotty wood."

Rutaceae

Zanthoxylum clava-herculis fruit

Zanthoxylum clava-herculis bark

Zanthoxylum clava-herculis twig

Zanthoxylum clava-herculis leaf

Zanthoxylum clava-herculis bark

Salicaceae Guide

1. Leaves triangular, apex acuminate, margin with coarse rounded teeth, underside glabrous, petiole long and flattened. *Populus deltoides*
2. Similar to *Populus deltoides* but leaves ovate with smaller margin teeth and a round petiole. *Populus heterophylla*
3. Leaves ovate, margin crenate or finely serrate, underside silvery and brown blotched, petiole round. *Populus balsamifera*
4. Leaves ovate to orbicular, margin with large coarse teeth, petiole long and flattened. *Populus grandidentata*
5. Leaves nearly round, margin crenate, underside mostly glabrous, petiole long and flattened. *Populus tremuloides*
6. Leaves lanceolate, margin with small red teeth, underside mostly glabrous. *Salix nigra*
7. Similar to *Salix nigra* but leaf margin with yellow teeth and usually a shrub. *Salix caroliniana*
8. Leaves lanceolate, margin serrate, underside glaucous, petiole up to 3 cm long and often twisted. *Salix amygdaloides*
9. Leaves lanceolate and leathery, margin serrate, underside glaucous; twigs crack when broken. *Salix fragilis*
10. Leaves lanceolate, margin serrate, shiny above and below. *Salix lucida*
11. Leaves lanceolate, margin serrate, underside prominently white and silky-pubescent. *Salix alba*
12. Leaves oblong to lanceolate, margin entire or irregularly serrate, underside white or bluish. *Salix discolor*
13. Leaves elliptical to obovate, margin entire or irregularly serrate, appearing wrinkled. *Salix bebbiana*

Salix nigra staminate flowers

Populus deltoides staminate flowers

Salix nigra pistillate flowers

Populus deltoides pistillate flowers

Salicaceae

Populus balsamifera L.

balsam poplar, eastern balsam poplar, balm of Gilead, Tacamahac

Quick Guide: *Leaves* alternate, simple, ovate, margin crenate, petiole round with glands, underside silvery with brown spots; *Buds* waxy and fragrant; *Bark* gray and smooth or with flattened ridges; *Habitat* a variety of sites.

Leaves: Simple, alternate, deciduous, ovate, 8 to 15 cm long, apex acute or acuminate, base obtuse to rounded or cordate, margin crenate or finely serrate, petiole round with glands at the blade, dark green above, underside silvery and brown-blotched; autumn color yellow.

Twigs: Somewhat stout, orange-brown to maroon, glabrous, with lenticels; leaf scar cordate with three bundle scars.

Buds: Acute, up to 2.5 cm long, very resinous, waxy, fragrant; scales overlapping, green to red-brown, glabrous.

Bark: Green-gray and smooth on young stems; large trees white to dark gray and furrowed with broad ridges.

Flowers: Imperfect, species is dioecious, staminate and pistillate catkins before the leaves.

Fruit: Capsule, ovoid, two valved, 7 mm long, glabrous, releasing cottonlike seeds in late spring.

Form: Up to 25 m (85 ft) in height and 61 cm (2 ft) in diameter but sometimes larger.

Habitat and Ecology: Shade intolerant; found on a variety of sites including old fields and woodland openings but common on moist or wet soils. Forest associates include *Acer rubrum, Betula*

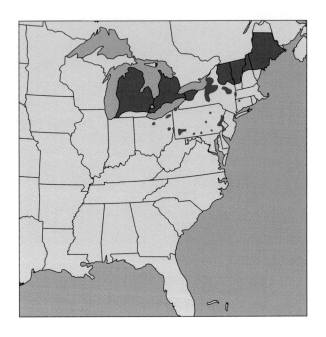

papyrifera, Fraxinus nigra, Picea glauca, Picea mariana, Populus tremuloides, Salix spp., *Thuja occidentalis,* and *Ulmus americana.*

Uses: Wood similar to *Populus tremuloides;* used for pulp, boxes, crates, and construction; hybrids used in short rotation plantations. Also used in windbreaks. A browse for grouse, deer, beaver and moose, and young stands provide cover for wildlife. The waxy coating on the buds reported to be used by bees to seal cracks in the hive; the sap was used by Native Americans to seal canoes (Rogers 1920).

Botanical Name: *Populus* is Latin for "poplar tree"; *balsamifera* refers to the fragrant resin from the buds. Also named *Populus balsamifera* spp. *balsamifera* L.

Populus balsamifera bark

Populus balsamifera leaf

Populus balsamifera twig

Populus balsamifera leaves

Populus balsamifera bark

Populus balsamifera fruit

Salicaceae

Populus deltoides Bartr. ex Marsh.

eastern cottonwood, southern cottonwood, eastern poplar

Quick Guide: *Leaves* alternate, simple, large, triangular, apex acuminate, petiole long and flattened; *Twig* and terminal *Bud* stout; *Bark* ash gray and deeply grooved; *Habitat* bottomlands.

Leaves: Simple, alternate, deciduous, triangular, blade 8 to 18 cm long, apex acute or acuminate, base truncate, margin with coarse rounded teeth, underside glabrous, petiole long (10 cm) with glands near the blade and flattened causing the leaves to shake in the wind; autumn color yellow.

Twigs: Stout, yellow-brown, glabrous, angled; leaf scar heart shaped with three bundle scars.

Buds: Terminal bud stout, acute, angled, up to 2 cm long; scales overlapping, yellow-green, shiny, sticky.

Bark: Yellow-green to light gray-brown and smooth or shallowly grooved on small trees; large trees light brown to ash gray and deeply grooved with thick ridges.

Flowers: Imperfect, species is dioecious, staminate and pistillate flowers in catkins before the leaves (see page 457).

Fruit: Capsule, conical, two- to four-valved, 6 to 12 mm long, releasing cottonlike seeds in late spring.

Form: Up to 30 m (100 ft) in height and 1 m (4 ft) in diameter.

Habitat and Ecology: Shade intolerant; found in bottomlands and near rivers and streams with *Acer negundo, Acer rubrum, Acer saccharinum, Betula nigra, Fraxinus* spp., *Liquidambar styraciflua, Platanus occidentalis, Quercus macrocarpa, Quercus phellos, Salix nigra,* and *Ulmus americana*.

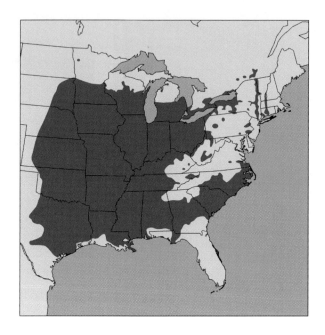

Uses: Wood white to gray, light, soft; used for pulpwood, veneer, crates, and fuel. Because of very early rapid growth it is used in windbreaks, waste site reclamation, naturalizing wet areas, and short rotation woody crops. Browsed somewhat by various deer and rabbit species, and catkins and seeds used by various grouse and songbirds. Bark eaten by beaver.

Botanical Name: *Populus* is Latin for "poplar tree"; *deltoides* refers to the Greek letter delta and the triangular leaves.

Similar Species: swamp cottonwood (*Populus heterophylla* L.) is found primarily in the Coastal Plain in wet clay bottoms and swamps, and is distinguished by ovate leaves with smaller teeth on the margin, a mostly acute apex, a more rounded or cordate leaf base, and a round petiole, and by a smaller bud.

Populus deltoides leaves

Populus deltoides twig

Populus deltoides fruit

Populus deltoides bark

Populus heterophylla leaves

Salicaceae

Populus grandidentata Michx.

bigtooth aspen, largetooth aspen

Quick Guide: *Leaves* alternate, simple, widely ovate, margin with large teeth; *Bark* green-gray and smooth, becoming gray and grooved with flattened ridges; *Habitat* a variety of sites.

Leaves: Simple, alternate, deciduous, widely ovate to orbicular, 8 to 12 cm long, apex acute to acuminate, base truncate or rounded, margin with large coarse teeth, underside with white pubescence becoming glabrous, petiole long and flattened; autumn color orange to yellow.

Twigs: Stout, brown-gray, pubescent; leaf scar shield or heart shaped with three bundle scars.

Buds: Terminal bud stout, acute, up to 1 cm long; scales overlapping, maroon, with pubescence.

Bark: Green-gray or white-gray with warty lenticels on small trees; becoming gray-brown and grooved with flattened ridges.

Flowers: Imperfect, species is dioecious, staminate and pistillate flowers in catkins before the leaves.

Fruit: Capsule, conical, two-valved, 6 mm long, releasing cottonlike seeds in late spring.

Form: Up to 21 m (70 ft) in height and 61 cm (2 ft) in diameter.

Habitat and Ecology: Shade intolerant; found on a wide range of sites including sandy, dry upland soils but best growth is on moist, rich soils. Rapidly colonizes disturbed sites by seed or sprouting.

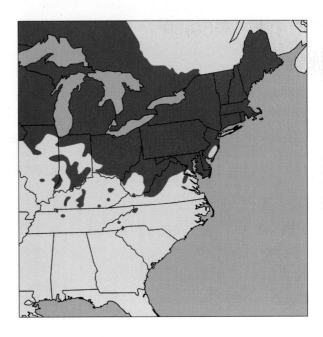

Forest associates include *Abies balsamea, Acer rubrum, Acer saccharum, Betula alleghaniensis, Picea glauca, Pinus banksiana, Pinus resinosa, Pinus strobus, Populus tremuloides, Prunus pensylvanica, Prunus serotina, Quercus alba, Quercus macrocarpa, Quercus rubra,* and *Tilia americana*.

Uses: wood white to gray, light, soft; used for pulpwood, veneer, crates, and fuel. A browse for grouse, deer, and beaver.

Botanical Name: *Populus* is Latin for "poplar tree"; *grandidentata* refers to the large teeth on the leaves.

Populus grandidentata twig

Populus grandidentata bark

Populus grandidentata bark

Populus grandidentata leaf

Salicaceae

Populus tremuloides Michx.

trembling aspen, quaking aspen, golden aspen, trembling poplar

Quick Guide: *Leaves* alternate, simple, nearly round, margin with small rounded teeth, petiole very long and flattened; *Bark* green-gray to white and smooth, becoming darker and grooved with flattened ridges; *Habitat* a variety of sites, common on disturbed areas.

Leaves: Simple, alternate, deciduous, ovate to orbicular, 5 to 15 cm long, apex acute, base rounded to cordate, margin crenate, petiole long (up to 8 cm) and flattened causing the leaves to shake in the wind, underside mostly glabrous; autumn color vivid yellow.

Twigs: Slender, maroon to green, shiny; leaf scar crescent or heart shaped with three bundle scars.

Buds: Acute, appressed, 6 mm long; scales overlapping, green-red-brown, shiny.

Bark: Gray, green-gray, or white and smooth on small trees; on large trees becoming darker and grooved with flattened ridges.

Flowers: Imperfect, species is dioecious, staminate and pistillate flowers in catkins before the leaves.

Fruit: Capsule, conical, two-valved, 6 mm long, releasing cottonlike seeds in late spring.

Form: Usually up to 18 m (60 ft) in height and 61 cm (2 ft) in diameter but can be larger. Pure stands may originate from a single clone, which may be thousands of years old.

Habitat and Ecology: Very shade intolerant; the most widely distributed tree in North America found on a wide variety of sites and an aggressive

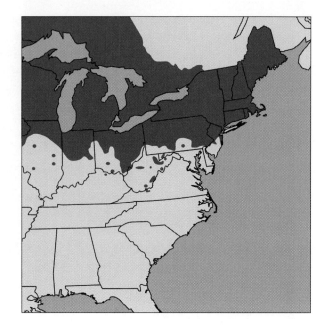

colonizer from seed or sprouts on disturbed or abandoned sites. Forest associates are numerous and include *Acer saccharum, Betula alleghaniensis, Picea glauca, Picea mariana, Pinus strobus, Populus grandidentata,* and *Tsuga canadensis.*

Uses: Wood white to gray, light, soft; used for pulpwood, lumber, plywood, veneer, shingles, paper, particleboard, crates, matches, and fuel. Native Americans brewed a medicinal tea from the bark. Planted as an ornamental for the bark and fall color. A browse for grouse, deer, snowshoe hare, and beaver; young stands provide cover for many small and large mammals, and numerous species of birds.

Botanical Name: *Populus* is Latin for "poplar tree"; *tremuloides* refers to the leaves that tremble in the wind.

Populus tremuloides leaves

Populus tremuloides bark

Populus tremuloides bark

Populus tremuloides twig

Salix nigra Marsh.
black willow, swamp willow

Quick Guide: *Leaves* alternate, lanceolate, margin with small glandular teeth; *Branches* slender, brittle, and yellow-orange; *Bark* gray-brown to red-brown with scaly ridges or loose plates; *Habitat* water margins.

Leaves: Simple, alternate, deciduous, lanceolate, 8 to 15 cm long, apex acuminate, base acute to rounded, margin finely serrate and teeth tipped with red glands, underside mostly glabrous, nearly sessile; autumn color yellow.

Twigs: Slender, brittle, yellow-orange to maroon, pubescent or glabrous, with stipular scars; leaf scar V shaped with three bundle scars.

Buds: True terminal bud lacking; lateral buds 2 to 5 mm long, conical, with one yellow-red glabrous scale.

Bark: Gray-brown to dark red-brown, scaly or with loose plates, inner bark chocolate brown.

Flowers: Imperfect, species is dioecious, staminate and pistillate flowers in catkins with the leaves (see page 457).

Fruit: Capsule, conical, two-valved, up to 6 mm long, releasing cottonlike seeds in late spring or summer.

Form: Usually only up to 15 m (50 ft) in height but larger on good sites.

Habitat and Ecology: Shade intolerant; found in floodplains and on margins of streams, rivers, and swamps with *Acer negundo*, *Acer rubrum*, *Acer saccharinum*, *Betula nigra*, *Fraxinus* spp., *Juglans nigra*, *Magnolia virginiana*, *Nyssa* spp., *Platanus occidentalis*, and *Populus deltoides*.

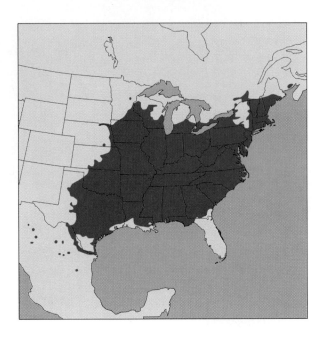

Uses: Wood white to red-brown, light, soft; used for pulpwood, boxes, baskets, wicker furniture, and crates. Also planted for soil stabilization. Bark and leaves were used to make aspirin. Browsed by deer, elk, and moose. Buds and twigs eaten by various grouse species. Bark eaten by beaver and snowshow hare.

Botanical Name: *Salix* is Latin for "willow"; *nigra* means "dark," referring to the bark.

Similar Species: Coastal Plain willow (*Salix caroliniana* Michx.) is usually a shrub that has similar leaves but with yellow rather than red glands on margin teeth.

Salix nigra bark

Salix nigra fruit

Salix nigra twig

Salix nigra leaves

Salix nigra bark

Salicaceae

Other willows

Salix alba L. White willow, a tree up to 30 m (100 ft) tall, has naturalized in the eastern United States on moist soils. Leaves are alternate, simple, lanceolate, 5 to 13 cm long with a serrate margin and white, silky-pubescent underside. Young twigs are golden and buds have one yellow scale. Bark is dark gray with scaly ridges on mature trees. Planted as an ornamental.

Salix amygdaloides Anderss. Peachleaf willow or almond-leaved willow is a shrub or tree up to 25 m (80 ft) in height found on edges of rivers and lakes and in swamps from New York to the Great Lakes region. Leaves are alternate, simple, lanceolate to ovate-lanceolate, and 5 to 15 cm long with an acuminate apex, finely serrate margin, and a pale and waxy underside. The petiole is comparatively long for a willow (up to 3 cm) and often twisted. Buds have one shiny, yellow-brown scale and mature bark is gray-brown with scaly ridges. The inferior wood may be used for cheap lumber and the tree is used for erosion prevention and in wetland reclamation. The bark is chewed by beaver and the tree provides cover for waterfowl.

Salix bebbiana Sarg. Bebb willow is a shrub or small multistemmed tree up to 6 m (20 ft) in height found on a variety of sites including bottomlands, limestone flats, rocky slopes, open hillsides, and sand plains throughout the Northeast and Great Lakes region. Leaves are alternate, simple, elliptical, oblong or obovate, 3 to 10 cm long, wrinkled above, with an entire or irregularly serrate margin, and underside with white-gray pubescence. Buds have one shiny brown scale and bark is red-brown to gray and furrowed when mature. Used for erosion prevention and in land reclamation.

Salix discolor Muhl. Pussy willow is a shrub usually in clumps or small multistemmed tree up to 6 m (20 ft) in height found on stream banks and in bogs, low meadows, and swamps throughout the Northeast and Great Lakes region and as far south as Tennessee. Leaves are alternate, simple, oblong to elliptical or lanceolate, 4 to 11 cm long, bright green above, white and pubescent or glaucous below, with an entire or irregularly serrate margin. Buds have one red-purple, caplike scale and mature bark is gray with flattened or scaly ridges. Flowers are popular with bees and the plant is browsed by moose, muskrat, beaver, rabbits, hares, grouse, and ducks, and provides cover for many birds and mammals. The silky-soft catkins are used in flower arrangements.

Salix fragilis L. Crack willow is a native of Europe and Asia and has naturalized along streams and rivers in the eastern United States. Recognized by alternate, simple, lanceolate, dark green, leathery leaves up to 15 cm long that are glaucous below and with a finely serrate margin. The brittle twigs crack when broken.

Salix lucida Muhl. Shining willow is a shrub or small tree up to 10 m (30 ft) in height with a round crown found in wet areas such as stream banks, lake shores, and swamps throughout the Northeast and Great Lakes region. Leaves are 7 to 15 cm long, alternate, simple, lanceolate, shiny above and below, with an acuminate apex, finely serrate margin, and petiole with glands at the leaf base. Buds have one yellow-brown scale and mature bark is brown and furrowed. The tree is planted as an ornamental and used in erosion prevention. The bark is chewed by beaver and the tree provides cover for waterfowl.

Salix fragilis leaves

Salix fragilis bark

Salix discolor catkins

Salicaceae

Sapotaceae Guide

1. Leaf underside with velvety, rusty or gray pubescence; twigs with rusty or gray pubescence. *Bumelia lanuginosa*
2. Leaf underside mostly glabrous when mature; twigs mostly glabrous. *Bumelia lycioides*
3. Leaf underside with long silver or copper hairs running parallel to lateral veins; twigs with pale pubescence. *Bumelia tenax*

Bumelia lanuginosa leaves

Bumelia lanuginosa twig

Bumelia lanuginosa twig

Bumelia tenax twig

Sapotaceae

Bumelia lanuginosa (Michx.) Pers.

gum bumelia, gum bully, woolly buckthorn

Quick Guide: *Leaves* alternate, simple, oblanceolate, margin entire, underside with rusty dense pubescence; *Twigs* with rusty pubescence, thorns, and milky sap; *Bark* gray-brown and grooved with scaly ridges and red inner bark; *Habitat* stream edges and sandy uplands.

Leaves: Simple, alternate, tardily deciduous, oblanceolate, elliptical, or obovate, 2.5 to 9 cm long, apex mostly rounded, base cuneate to rounded, margin entire, underside with velvety rusty-gray pubescence, petiole densely pubescent. See also page 470.

Twigs: Gray-brown with rusty or gray pubescence and thorns (see page 471). Exuding milky sap when cut. Leaf scar small, semicircular with three bundle scars.

Buds: True terminal bud lacking; lateral buds embedded, scales overlapping with rusty pubescence.

Bark: Gray-brown and shallowly grooved on small trees; large trees with scaly ridges and red inner bark.

Flowers: Perfect, white, in small clusters on maroon pubescent stalks from leaf axils in midsummer (see page 471).

Fruit: Berry, black, obovoid, pubescent, about 8 mm long, maturing in fall.

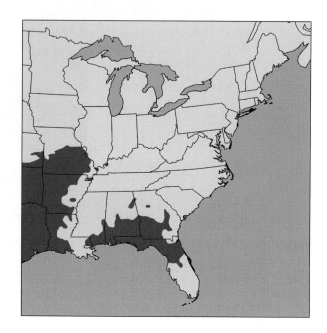

Form: Shrub or small tree to 12 m (40 ft) in height.

Habitat and Ecology: Found on stream edges and dry, sandy or rocky sites.

Uses: Flowers popular with bees and fruit eaten by birds.

Botanical Name: *Bumelia* is from an ancient Greek name for European ash; *lanuginosa* means "with soft hair," referring to the leaves and twigs. Also named *Sideroxylon lanuginosum* Michx.

Bumelia lanuginosa leaves

Bumelia lanuginosa bark

Bumelia lanuginosa bark

Bumelia lanuginosa immature fruit

Bumelia lycioides (L.) Pers.
buckthorn bumelia, buckthorn bully

Quick Guide: *Leaves* alternate, simple, oblanceolate, margin entire; *Twigs* glabrous with thorns and milky sap; *Bark* gray-brown and scaly revealing red inner bark; *Habitat* a variety of sites.

Leaves: Simple, alternate, tardily deciduous, shiny, elliptical to oblanceolate, 5 to 15 cm long, apex acute, base cuneate, margin entire, underside mostly glabrous when mature.

Twigs: Brown-gray, mostly glabrous, with thorns. Exuding milky sap when cut. Leaf scar small, semicircular with three bundle scars.

Buds: True terminal bud lacking; lateral buds embedded in spur shoots; scales overlapping, yellow-green.

Bark: Small trees gray-brown and ridged; large trees scaly with red inner bark.

Flowers: Perfect, white, in small clusters from leaf axils in midsummer.

Fruit: Berry, black, ovoid or ellipsoid, glabrous, 1 to 1.5 cm long, maturing in fall.

Form: Shrub or small tree to 15 m (50 ft) in height.

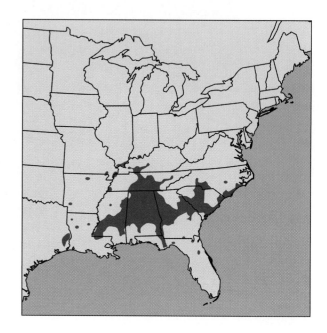

Habitat and Ecology: Found on streambanks and in upland forests.

Uses: Flowers popular with bees and fruit eaten by birds.

Botanical Name: *Bumelia* is from an ancient Greek name for European ash; *lycioides* means "box-thorn-like." Also named *Sideroxylon lycioides* L.

Bumelia lycioides flowers

Bumelia lycioides leaves

Bumelia lycioides twig

Bumelia lycioides bark

Bumelia lycioides fruit

Sapotaceae

Bumelia tenax (L.) Willd.

tough bumelia, tough bully, tough buckthorn, ironwood

Quick Guide: *Leaves* alternate, simple, oblanceolate, underside with long hairs lying parallel to lateral veins; *Twigs* pale pubescent with thorns, and milky sap; *Habitat* scrub forests.

Leaves: Simple, alternate, tardily deciduous or evergreen, shiny, oblanceolate or obovate, 2 to 7 cm long, apex obtuse to rounded, base cuneate, margin entire; underside covered with long, silver, copper, or maroon hairs that run parallel to the lateral veins.

Twigs: Slender, flexible but tough, maroon or gray, with pale pubescence (see page 471), numerous thorns, and milky sap when cut; leaf scar small, semicircular with three bundle scars.

Buds: True terminal bud lacking; lateral buds small, rounded, with rusty pubescence.

Bark: Gray to red-brown and smooth or grooved, becoming scaly.

Flowers: Perfect, white, in small clusters from leaf axils in midsummer, similar to *Bumelia lycioides*.

Fruit: Berry, black, obovoid, 1 to 1.5 cm long, maturing in fall.

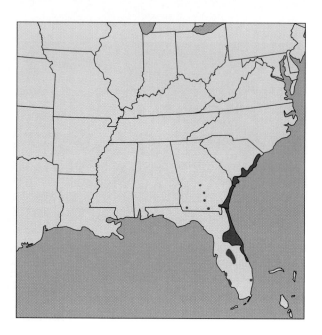

Form: Shrub or small tree to 8 m (26 ft) in height.

Habitat and Ecology: Dry, oak-pine forests.

Uses: Flowers visited by bees; birds eat the fruits.

Botanical Name: *Bumelia* is from an ancient Greek name for European ash; *tenax* means "tough, persistent, tenacious," referring to the branches. Also named *Sideroxylon tenax* L.

Bumelia tenax leaves

Bumelia tenax bark

Bumelia tenax bark

Bumelia tenax immature fruit

Sapotaceae

Paulownia tomentosa (Thumb.) Sieb. & Zucc. ex Steud.

Royal paulownia, princess-tree, empress-tree

Quick Guide: *Leaves* opposite, simple, large, heart shaped, tomentose below; *Flowers* in large, purple, terminal clusters before leaf out; *Fruit* a nutlike capsule persistent over winter; *Bark* gray-brown and smooth; *Habitat* open and disturbed sites.

Leaves: Simple, opposite or occasionally whorled, deciduous, heart shaped, 13 to 40 cm long, apex acuminate, base cordate, margin entire, underside tomentose, petiole long and stout; autumn color yellow. Seedlings often showing lobes or coarse teeth.

Twigs: Stout, yellow-brown to green-brown, pubescent, with corky white lenticels; leaf scar circular, notched at the top, raised, with bundle scars forming a circle; pith chambered or hollow.

Buds: True terminal bud lacking; lateral buds often superposed, blunt, embedded; flower buds large, drooping, velvety brown, conspicuous in late summer through winter.

Bark: Gray-brown and mostly smooth with lenticels, becoming lightly ridged.

Flowers: Perfect, 5 cm long, tubular, pale purple, fragrant, in erect terminal clusters before the leaves.

Fruit: Capsule, ovoid, nutlike, woody, two-valved, about 3 cm long, releasing small winged seeds in summer, clusters of old capsules persistent.

Form: Up to 15 m (50 ft) in height, very fast growing.

Habitat and Ecology: Originally from Asia, naturalized in open and disturbed areas throughout the East. Considered an invasive exotic pest plant in Tennessee.

Uses: Wood light and strong, commercially important in Asia and used for cabinets, furniture, and tea chests. Planted as an ornamental for the attractive flowers but suffers from poor form and messy leaves and fruit in fall, and produces lots of seed.

Botanical Name: *Paulownia* is for Anna Paulowna; *tomentosa* refers to the hair on the leaf underside.

Paulownia tomentosa leaf

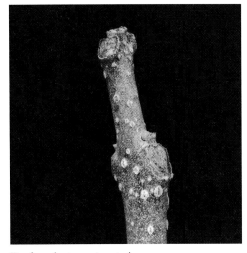
Paulownia tomentosa twig

Paulownia tomentosa flowers

Paulownia tomentosa bark

Paulownia tomentosa fruit

Scrophulariaceae

Ailanthus altissima (P. Mill.) Swingle
tree-of-heaven, ailanthus

Quick Guide: *Leaves* alternate, pinnately compound, leaflets lanceolate and up to 41 in number, margin entire except for coarse basal teeth tipped with black glands; *Branches* stout and curving upward; *Fruit* a twisted samara in clusters; *Bark* mostly smooth and gray-brown; *Habitat* open, disturbed areas.

Leaves: Pinnately compound, alternate, deciduous, up to 1m long. Leaflets 11–41, lanceolate, 8 to 15 cm long, margin entire except for coarse basal teeth tipped with black glands on the underside, apex acuminate, base cordate or inequilateral, underside lightly pubescent; autumn color yellow or red. Crushed leaves with an unpleasant odor.

Twigs: Very stout, tan, downy, with white lenticels; leaf scar large, broadly U shaped, raised, pale, with many bundle scars; with a strong odor when cut.

Buds: True terminal bud lacking; lateral buds round, embedded, pubescent.

Bark: Gray-brown and smooth with vertical striations, becoming lightly ridged.

Flowers: Imperfect, species is dioecious; pistillate and staminate flowers small, yellow-white, in terminal clusters in late spring after the leaves; staminate flowers with an unpleasant odor.

Fruit: Samara, yellow to brown, twisted, about 3 cm long, with the seed in the center, clusters maturing in summer or fall and persistent over winter.

Form: Up to 15 m (50 ft) in height, very fast growing.

Habitat and Ecology: Originally from Asia, naturalized in open and disturbed areas throughout the East and an aggressive colonizer. This plant is considered an invasive species in many states.

Uses: First planted as an ornamental in the early 1800s because of excellent tolerance of poor sites, but can take over an area due to excessive seed production and root sprouting. Fruit eaten by birds.

Botanical Name: *Ailanthus* means "reaching to heaven"; *altissima* means "high or tall."

Ailanthus altissima flowers

Ailanthus altissima leaf

Ailanthus altissima twig

Ailanthus altissima bark

Ailanthus altissima fruit

Staphylea trifolia L.
American bladdernut, bladdernut

Quick Guide: *Leaves* opposite, trifoliately compound; *Twigs* red-green mottled with a pair of buds at the tip; *Flowers* white and bell shaped; *Fruit* a bladderlike capsule; *Bark* gray-brown, smooth, and streaked; *Habitat* moist, fertile soils.

Leaves: Trifoliately compound, opposite, deciduous, up to 23 cm long. Leaflets three or rarely five, ovate, 4 to 10 cm long, apex acuminate, base rounded, margin finely serrate, underside pubescent, petiole long; autumn color yellow.

Twigs: Red-green, mottled or striped, glaucous, with lenticels; leaf scar triangular with three bundle scars and a stipular scar on either side of the leaf scar.

Buds: True terminal bud lacking; lateral buds usually in pairs at the terminal, ovoid, with overlapping, green-brown, glabrous scales.

Bark: Gray-brown to black and smooth with vertical white stripes, becoming lightly ridged.

Flowers: Perfect, white, bell shaped, in drooping clusters in late spring after the leaves.

Fruit: Capsule, three lobed, bladderlike, thin walled, inflated, papery, 4 to 6 cm long, maturing in early fall.

Form: Up to 9 m (30 ft) in height, can form thickets.

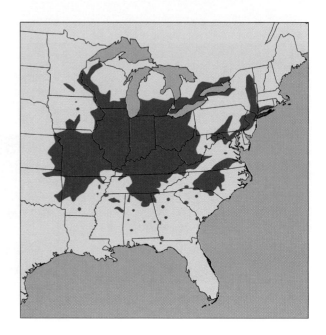

Habitat and Ecology: Found on moist, fertile soils of floodplains, stream banks, and north-facing slopes.

Uses: Sometimes planted as an ornamental.

Botanical Name: *Staphylea* comes from Greek for "clusters of grapes," referring to the flowers; *trifolia* means "three leaved."

Staphylea trifolia twig

Staphylea trifolia fruit

Staphylea trifolia leaf

Staphylea trifolia bark

Staphylea trifolia flowers

Staphyleaceae

Styracaceae Guide

1. Leaf margin with irregular coarse teeth; flower petals fused only at the base; drupe two-winged. *Halesia diptera*
2. Leaf margin with small sharp teeth; flower petals fused; drupe four-winged. *Halesia tetraptera*
3. Leaf margin wavy and irregularly toothed; flower petals fused only at the base and recurved; drupe unwinged and pubescent. *Styrax americanus*
4. Leaf oval to orbicular and margin entire or with irregular fine or dentate teeth; flower petals fused only at the base and not recurved; drupe unwinged and pubescent. *Styrax grandifolius*

Styrax americanus twig

Styrax grandifolius twig

Halesia diptera twig

Halesia tetraptera bark

Halesia tetraptera twig

Halesia diptera Ellis
two-wing silverbell, snowdrop tree

Quick Guide: Similar to *Halesia tetraptera*, but distinguished by *Leaves* with irregular, coarse teeth; *Flowers* with petals fused only at the base; *Fruit* a two-winged drupe.

Leaves: Simple, alternate, deciduous, obovate, ovate or elliptical, 5 to 16 cm long, apex acute or acuminate, base obtuse to rounded, margin with coarse and irregular teeth, underside with white pubescence; autumn color yellow.

Twigs: Red-brown and pubescent with lenticels; leaf scar heart shaped to nearly round with one crescent shaped bundle scar (see page 485).

Buds: True terminal bud lacking; lateral buds small, pubescent, superposed.

Bark: Gray-brown to red-brown, striped, and smooth on small trees; large trees furrowed with scaly ridges.

Flowers: Perfect, white, bell shaped, petals fused only at the base, in showy clusters before and with the leaves.

Fruit: Drupe, two-winged, papery, about 4 cm long, maturing in early fall.

Form: Up to 9 m (30 ft) in height.

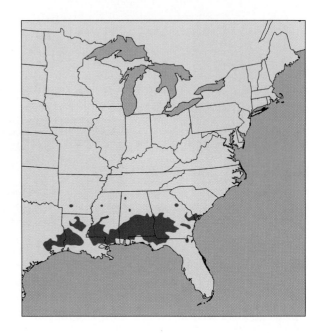

Habitat and Ecology: Shade tolerant; found on moist soils of ravines, stream edges, bottomlands, and swamp margins.

Uses: An attractive specimen for planting on moist soils. Seeds eaten by birds and small mammals; flowers popular with bees.

Botanical Name: *Halesia* is for the plant physiologist Stephen Hales; *diptera* means "two winged."

Halesia diptera flowers

Halesia diptera leaf

Halesia diptera fruit

Styracaceae

Halesia tetraptera Ellis
Carolina silverbell, snowdrop tree, mountain silver bell

Quick Guide: *Leaves* alternate, simple, margin with small sharp teeth; *Flowers* white and bell shaped; *Fruit* a persistent four-winged drupe; *Bark* on small stems white striped, becoming red-brown to black and scaly on large trees; *Habitat* moist soils.

Leaves: Simple, alternate, deciduous, ovate, obovate or elliptical, 5 to 15 cm long, apex acuminate, base acute to rounded, margin with small sharp teeth or entire, underside with white pubescence; autumn color yellow.

Twigs: Green-brown possibly with white streaks, shreddy when older (see page 485); leaf scar shield shaped and raised with one crescent shaped bundle scar.

Buds: True terminal bud lacking; lateral buds small, reddish, pubescent, superposed.

Bark: Red-brown to black and smooth with pale stripes on small trees; large trees with loose, oblong, brown-gray to black scales and red-brown inner bark.

Flowers: Perfect, white, bell shaped, petals fused, in showy clusters before and with the leaves.

Fruit: Drupe, four-winged, papery, about 4 cm long, maturing in early fall.

Form: Up to 30 m (100 ft) in height and 1 m (3 ft) in diameter in the Great Smoky Mountains. Very large trees often with burls.

Habitat and Ecology: Shade tolerant; found on moist soils in the Piedmont and Appalachians with *Acer rubrum, Acer saccharum, Aesculus flava,*

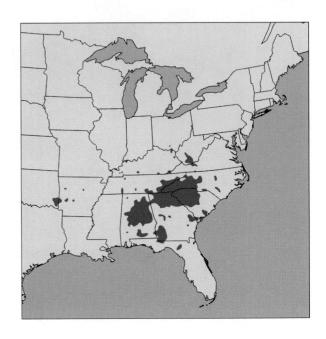

Betula lenta, Fagus grandifolia, Liriodendron tulipifera, Magnolia spp., *Pinus strobus, Prunus serotina, Quercus rubra,* and *Tsuga canadensis.* In the Coastal Plain, associated with *Acer negundo, Acer rubrum, Fagus grandifolia, Quercus alba, Quercus michauxii,* and *Quercus pagoda.*

Uses: Wood white to red-brown, soft, light, close grained; used for pulpwood, veneer, cabinetry, and woodenware. Planted as an ornamental and cultivars with pink flowers are available. Seeds eaten by birds and small mammals; flowers popular with bees.

Botanical Name: *Halesia* is for the plant physiologist Stephen Hales; *tetraptera* means "four winged." Some consider trees in the southern range as *Halesia carolina* L.

Halesia tetraptera flowers and fruit

Halesia tetraptera bark

Halesia tetraptera bark

Halesia tetraptera bark

Halesia tetraptera leaf

Styrax americanus Lam.
American snowbell

Quick Guide: *Leaves* alternate, simple, elliptical, margin wavy or irregularly toothed; *Flowers* white with petals fused only at the base and recurved; *Fruit* an unwinged pubescent drupe; *Buds* tan, scurfy, and superposed; *Bark* gray-brown and mostly smooth; *Habitat* moist to wet soils.

Leaves: Simple, alternate, deciduous, obovate or elliptical, 3 to 8 cm long, apex tapering to a blunt point, base acute, margin wavy or irregularly toothed, underside glabrous or with pubescence, petiole pubescent.

Twigs: Red-brown, pubescent or glabrous; leaf scar shield shaped with one crescent shaped bundle scar.

Buds: True terminal bud lacking; lateral buds stalked, naked, tan, scurfy pubescent, 3 mm long, superposed (see page 485).

Bark: Gray-brown, mostly smooth.

Flowers: Perfect, white, petals fused only at the base and recurved, in showy clusters in spring after the leaves.

Fruit: Drupe, round, pubescent, 6 to 13 mm wide, maturing in early fall.

Form: A shrub or small tree up to 6 m (20 ft) in height.

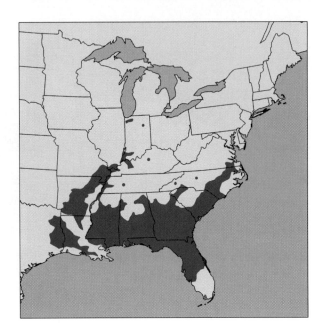

Habitat and Ecology: Shade tolerant; found on moist to wet soils of marshes, swamps, ditches, and stream margins.

Uses: Can be planted as an ornamental on moist sites. Seeds eaten by birds and small mammals; flowers popular with bees.

Botanical Name: *Styrax* means "resinous gum tree"; *americanus* refers to the New World.

Styrax americanus flowers

Styrax americanus leaves

Styrax americanus bark

Styrax americanus fruit

Styrax grandifolius Ait.
bigleaf snowbell

Quick Guide: *Leaves* alternate, simple, round, margin entire or irregularly toothed; *Flowers* white and bell shaped, petals fused only at the base; *Fruit* an unwinged pubescent drupe; *Buds* golden brown, fuzzy; *Bark* gray-brown-orange striped; *Habitat* mesic woods.

Leaves: Simple, alternate, deciduous, oval to nearly round, 8 to 16 cm long, apex quickly tapering to a sharp point or rounded, base rounded, margin entire or with irregular fine or dentate teeth, underside pubescent.

Twigs: Red-brown, pubescent or glabrous; leaf scar shield shaped with one crescent shaped bundle scar (see page 485).

Buds: True terminal bud lacking; lateral buds stalked, naked, golden brown, fuzzy, 6 mm long, superposed.

Bark: Gray-brown-orange striped and smooth on small trees; large trees shallowly grooved with somewhat scaly ridges.

Flowers: Perfect, white, bell shaped, petals fused only at the base, in showy clusters in spring after the leaves.

Fruit: Drupe, nearly round or ellipsoid, pubescent, 7–10 mm wide, maturing in early fall.

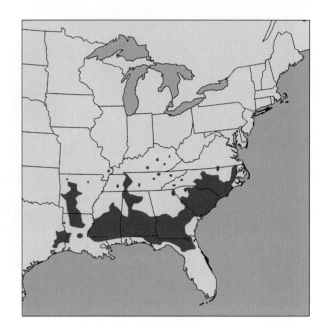

Form: Shrub or small tree up to 6 m (20 ft) in height.

Habitat and Ecology: Shade tolerant; found on moist soils next to streams, in floodplains, and in mesic woods.

Uses: Can be planted as an ornamental on moist sites. Seeds eaten by birds and small mammals; flowers popular with bees.

Botanical Name: *Styrax* means "resinous gum tree"; *grandifolius* means "big leaf."

Styrax grandifolius leaf

Styrax grandifolius fruit

Styrax grandifolius bark

Styrax grandifolius flowers

Symplocos tinctoria (L.) L'Her.

sweetleaf, horse-sugar, dye bush, yellow wood

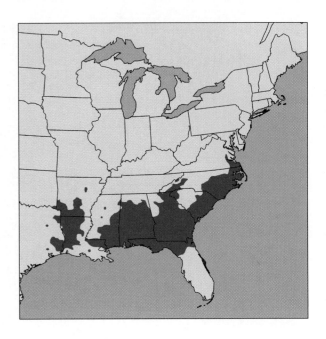

Quick Guide: *Leaves* alternate, simple, leathery, sweet tasting, margin mostly entire, petiole and midrib bright yellow; *Flowers* in yellow-white clusters in spring; *Bark* gray-brown, striped, smooth or shallowly grooved; *Habitat* moist soils.

Leaves: Simple, alternate, evergreen or tardily deciduous, leathery, elliptical to oblong, 5 to 15 cm long, apex acute, base cuneate, margin mostly entire or with fine teeth, underside pubescent or glabrous, petiole and midrib bright yellow, leaf sweet tasting, especially near the midrib.

Twigs: Red-brown, glabrous or pubescent, with a chambered pith; leaf scar shield shaped with one crescent shaped bundle scar.

Buds: Terminal bud ovoid, 6 mm long, with green-brown, overlapping, pubescent scales; flower buds plump.

Bark: Gray-brown to green-brown, striped, smooth or lightly grooved, with warts.

Flowers: Perfect, short stalked, in yellow-white clusters in spring before the new leaves.

Fruit: Drupe, oblong, green to brown, 1 to 1.3 cm long, maturing in early fall.

Form: Up to 9 m (30 ft) in height.

Habitat and Ecology: Shade tolerant; found on moist soils of upland forests, swamp borders, floodplains, and stream banks, and occasionally on dry sites.

Uses: A yellow dye for wool was made from the leaves and bark in colonial times. Leaves enjoyed by livestock and moderately browsed by white-tailed deer.

Botanical Name: *Symplocos* refers to the united stamens; *tinctoria* refers to tincture or dye made from the leaves and bark.

Symplocos tinctoria leaves

Symplocos tinctoria fruit

Symplocos tinctoria bark

Symplocos tinctoria flowers

Gordonia lasianthus (L.) Ellis

loblolly-bay, gordonia

Quick Guide: *Leaves* alternate, simple, evergreen, elliptical, margin with fine teeth; *Flowers* with five white, silky petals in summer; *Bark* gray-brown and deeply grooved on large trees; *Habitat* acidic, wet soils.

Leaves: Simple, alternate, evergreen, leathery, shiny, elliptical, 8 to 16 cm long, apex acute, base cuneate, margin with fine teeth, underside mostly glabrous.

Twigs: Red-brown to gray, glabrous or pubescent; leaf scar shield shaped with bundle scars forming a U shape.

Buds: Terminal bud ovoid and pubescent.

Bark: Red-brown to gray and shallowly grooved on small trees; large trees gray-brown, deeply grooved with thick ridges.

Flowers: Perfect, white, 7 cm wide, fragrant, with five fringed and silky hairy petals; flowers blooming during summer.

Fruit: Capsule, ovoid, 1 to 2 cm long, with white downy pubescence, splitting along five seams, containing winged seeds, maturing in early fall.

Form: Up to 24 m (80 ft) in height.

Habitat and Ecology: Shade tolerant; found on acidic wet soils of shrub bogs, evergreen bay forests, and pocosins in association with *Chameacyparis thyoides*, *Cyrilla racemiflora*, *Ilex cassine*, *Magnolia virginiana*, *Nyssa biflora*, *Persea palustris*, *Pinus serotina*, and *Taxodium distichum* var. *imbricarium*.

Uses: Reddish wood occasionally used for specialty items; planted as an ornamental due to shiny foliage and flowers but difficult to cultivate.

Botanical Name: *Gordonia* is for James Gordon, a British nurseryman; *lasianthus* means "flower with hair."

Gordonia lasianthus fruit

Gordonia lasianthus leaf

Gordonia lasianthus bark

Gordonia lasianthus bark

Gordonia lasianthus flower

Theaceae 497

Stewartia ovata (Cav.) Weatherby
mountain-camellia, mountain stewartia

Quick Guide: *Leaves* alternate, simple, deciduous, 2 ranked, margin ciliate and finely toothed, underside pubescent; *Flowers* showy with five white petals; *Fruit* a five celled, egg-shaped capsule; *Habitat* understory of mesic woods.

Leaves: Simple, alternate, deciduous, thin, elliptical, ovate or obovate to nearly oval, appearing 2 ranked, 5 to 13 cm long, margin ciliate with fine and widely spaced teeth, underside and petiole pubescent; autumn color yellow to red.

Twigs: Red-green, glabrous or pubescent; leaf scar with one bundle scar.

Buds: 6 mm long, spindlelike, with one or two silvery-white, pubescent scales.

Bark: Gray-brown and shallowly grooved or scaly.

Flowers: Perfect, 5 to 10 cm wide, with five or six white petals, stamens with orange anthers, blooming in summer.

Fruit: Capsule, ovoid, 2.0 cm long, woody, pubescent, splitting along five sutures, maturing in fall.

Form: A shrub or small tree up to 8 m (26 ft) tall.

Habitat and Ecology: Found in the understory on moist soils in the mountains and upper Piedmont.

Uses: Slow-growing and difficult to transplant, cultivars occasionally used as ornamentals.

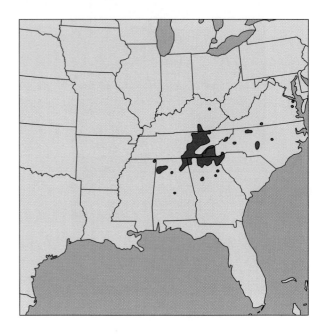

Botanical Name: *Stewartia* is after John Stewart, a 1700s British politician and patron of botany; *ovata* refers to the shape of the leaf or fruit.

Similar Species: Silky camellia or Virginia stewartia (*Stewartia malacodendron* L.) is found in the understory on moist, fertile soils in the southeastern Coastal Plain and lower Piedmont. It is distinguished from *Stewartia ovata* by geographic range, flowers with five petals, stamens with purple filaments and bluish anthers, and more globose capsules.

Stewartia ovata immature fruit

Stewartia ovata leaf

Stewartia malacodendron twig

Stewartia ovata bark

Stewartia malacodendron leaves and flower

Theaceae

Tilia americana L.

basswood, American linden, linden, bee tree

Quick Guide: *Leaves* alternate, simple, heart shaped, margin sharply serrate; *Fruit* a nutlet attached to a leafy bract; *Bark* gray-brown with flattened ridges; *Habitat* mesic sites.

Leaves: Simple, alternate, deciduous, ovate to heart shaped, 10 to 20 cm long, apex acuminate, base cordate and inequilateral, margin sharply serrate, underside glabrous or pubescent; autumn color yellow.

Twigs: Moderately stout, zig-zag, green to red-brown; leaf scar semicircular with numerous bundle scars.

Buds: True terminal bud lacking; lateral buds divergent, ovoid, plump, 6 mm long; scales green or maroon, overlapping.

Bark: Gray-brown and smooth or shallowly grooved on small trees; large trees more deeply furrowed with flattened or somewhat scaly fibrous ridges.

Flowers: Perfect, yellowish, fragrant, in long-stalked drooping clusters attached to a leafy bract, in late spring or summer.

Fruit: Nutlet, 5 to 8 mm wide, pubescent, long-stalked and attached to a leafy bract, in drooping clusters in late summer or early fall.

Form: Up to 36 m (120 ft) tall and 1.2 m (4 ft) in diameter. Sprouting results in clumps of trees.

Habitat and Ecology: Shade tolerant; found on moist, fertile soils but also on drier sites in association with a wide variety of species. The map indicates the range of *Tilia* spp.

Uses: Wood white to yellow-brown, light, soft, fine textured; used for plywood, furniture, crates, boxes, turnery, guitars, and woodenware. The fibrous inner bark was used for making ropes, fishnets, and mats. Can be planted as an ornamental with sufficient space and cultivars are available. Flowers very popular with bees and the source of basswood honey. Lightly browsed by white-tailed deer and cottontail rabbits. Light use of seeds by squirrels and small mammals. Dried leaves were used as winter fodder for cattle.

Botanical Name: *Tilia* is Latin for "linden tree"; *americana* refers to the New World. Classification of the basswoods is confused. Some designate numerous varieties based on range and degree of pubescence on leaves and twigs.

Tilia americana flowers

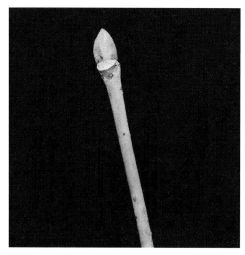
Tilia americana twig

Tilia americana leaves

Tilia americana bark

Tilia americana fruit

Tiliaceae

Ulmaceae Guide

1. Leaves ovate, apex acuminate, margin entire or irregularly serrate, three veins arising from the leaf base; twigs and buds pubescent; drupe orange-red to purple; bark smooth with corky warts. *Celtis laevigata*
2. Leaves ovate, apex acute or acuminate, margin serrate, three veins arising from the leaf base; twigs and buds mostly glabrous; drupe red-purple; bark with warty ridges. *Celtis occidentalis*
3. Leaves widely ovate, apex acute or acuminate, margin irregularly serrate, three veins arising from the leaf base, scabrous above; drupe orange-red; bark with corky warts. *Celtis tenuifolia*
4. Leaves ovate, 2 ranked, apex acute, margin serrate; fruit with fleshy projections; bark scaly or shreddy with red inner bark. *Planera aquatica*
5. Leaves lanceolate, apex acute, margin doubly serrate; twigs with corky wings; samara elliptical, pubescent on the margin and notched; bark with oblong scales. *Ulmus alata*
6. Similar to *Ulmus alata* but leaves very scabrous on the surface; fruits in autumn. *Ulmus serotina*
7. Similar to *Ulmus alata* but fruits in autumn; leaves smaller and scabrous. *Ulmus crassifolia*
8. Leaves obovate, base greatly inequilateral, margin doubly serrate; twigs and buds mostly glabrous; samara round, notched, and pubescent on the margin; bark with flattened ridges. *Ulmus americana*
9. Similar to *Ulmus americana* but the leaf base only slightly inequilateral; twigs with corky wings; fruit more obovate and covered with hair. *Ulmus thomasii*
10. Leaves obovate or elliptical, margin doubly serrate, scabrous above and below; buds with maroon pubescence; samara round, margin glabrous and shallowly notched; bark red-brown with corky or flattened ridges. *Ulmus rubra*

Ulmus alata flowers

Celtis occidentalis flowers

Ulmus americana flowers

Ulmus rubra flowers

Ulmaceae

Celtis laevigata Willd.

sugarberry, southern hackberry, sugar hackberry

Quick Guide: *Leaves* alternate, simple, ovate to lanceolate, apex acuminate, three veins arising from base, margin entire or irregularly serrate; *Twigs* and *Buds* pubescent; *Fruit* an orange-red to purple drupe; *Bark* gray-brown with corky warts or smooth; *Habitat* bottomlands.

Leaves: Simple, alternate, deciduous, ovate to lanceolate, 5 to 13 cm long, three main veins from the base, apex acuminate and long pointed, base rounded sometimes inequilateral, margin entire or irregularly serrate above the middle, mostly smooth but sometimes scabrous above; autumn color yellow.

Twigs: Slender, zig-zag, red-brown to green, with fine pubescence and lenticels; leaf scar crescent shaped with three bundle scars.

Buds: True terminal bud lacking; lateral buds triangular, 3 mm long, appressed; scales overlapping, red-black, pubescent.

Bark: Gray-brown to blue-gray with corky or woody warts; large trees smooth.

Flowers: Perfect and imperfect, small, yellow-green or green-white, without petals; in spring with the leaves.

Fruit: Drupe, orange-red to purplish, round, about 7 mm wide, sweet, maturing in fall.

Form: Usually to 15 m (50 ft) in height but up to 30 m (100 ft) in height on good sites.

Habitat and Ecology: Shade tolerant; found sometimes on uplands but primarily in bottomlands and on edges of streams and swamps with *Carya*

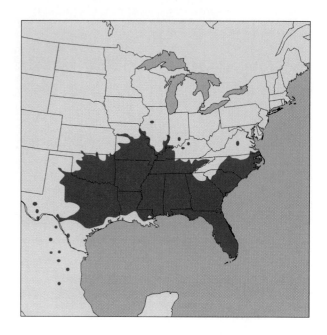

cordiformis, Celtis occidentalis, Fraxinus spp., *Liquidambar styraciflua, Populus deltoides, Quercus lyrata, Quercus pagoda, Quercus phellos, Salix nigra,* and *Ulmus americana.*

Uses: Wood yellow-gray, moderately hard, moderately heavy; used for furniture, veneer, and boxes. Can be planted as a shade tree and used as street trees in the South and Midwest. A favorite fruit eaten by numerous song- and game birds including American robin, yellow-bellied sapsucker, mockingbird, mourning dove, and various quails; fruit also used by a variety of small to midsize mammals. Lightly browsed by deer species.

Botanical Name: *Celtis* was a name for African lotus, which has sweet fruit; *laevigata* means "smooth," referring to the leaves or bark. Also named *Celtis laevigata* var. *laevigata* Willd.

Celtis laevigata flowers

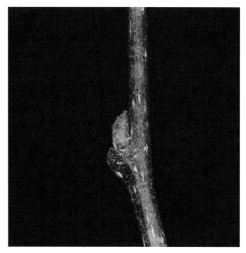
Celtis laevigata twig

Celtis laevigata fruit

Celtis laevigata bark

Celtis laevigata leaves

Celtis occidentalis L.

hackberry, northern hackberry, American hackberry, common hackberry

Quick Guide: *Leaves* alternate, simple, ovate, apex acute or acuminate, base cordate, three veins arising from base, margin serrate; *Twigs* and *Buds* mostly glabrous; *Fruit* a red-purple drupe; *Bark* gray-brown with warty corky ridges; *Habitat* a variety of sites.

Leaves: Simple, alternate, deciduous, ovate, 5 to 13 cm long, 3 main veins from the base, apex acute to acuminate, base cordate and inequilateral, margin serrate, smooth or scabrous above; autumn color yellow.

Twigs: Slender, zig-zag, green-brown to red-brown, mostly glabrous, with lenticels; leaf scar crescent shaped with three bundle scars.

Buds: True terminal bud lacking; lateral buds triangular, 6 mm long, appressed; scales overlapping, red-black, mostly glabrous.

Bark: Gray-brown to ash gray with warty corky ridges; very large trees with scaly ridges.

Flowers: Perfect and imperfect, small, yellow-green or green-white, without petals; in spring with the leaves (see page 503).

Fruit: Drupe, red-purple, round, about 8 mm wide, maturing in fall.

Form: Usually only up to 15 m (50 ft) tall.

Habitat and Ecology: Intermediate shade tolerance; found mostly in bottomlands but also on

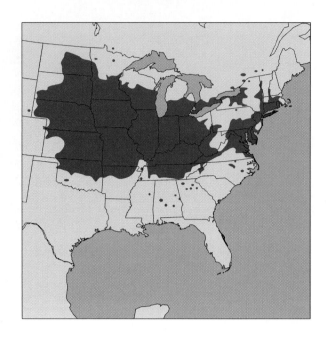

limestone soils, slopes, bluffs, and in upland forests. Forest associates include *Acer saccharum, Fagus grandifolia, Liquidambar styraciflua, Platanus occidentalis, Tilia americana,* and *Ulmus americana*.

Uses: Wood yellow-gray, moderately hard, moderately heavy; used for furniture, veneer, and boxes. Wildlife uses as for *Celtis laevigata*. Planted as an ornamental but use cultivars that do not have leaf gall and witches'-broom.

Botanical Name: *Celtis* was a name for African lotus, which has sweet fruit; *occidentalis* refers to the Western Hemisphere.

Ulmaceae

Celtis occidentalis leaves

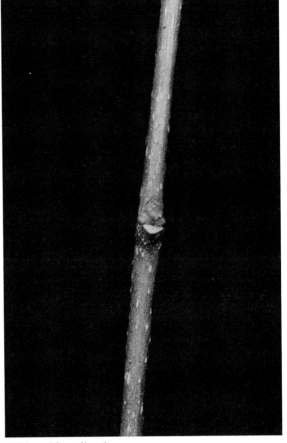
Celtis occidentalis twig

Celtis occidentalis bark

Celtis occidentalis fruit

Celtis tenuifolia Nutt.
Georgia hackberry, dwarf hackberry

Quick Guide: *Leaves* alternate, simple, ovate, apex acute or acuminate, base cordate, three veins arising from base, margin irregularly serrate above the middle, very scabrous above; *Fruit* an orange-red drupe; *Bark* gray-brown with corky warts; *Habitat* well-drained soils.

Leaves: Simple, alternate, deciduous, ovate, 4 to 7 cm long, three main veins from the base, apex acute or acuminate, base cordate and inequilateral, margin irregularly serrate above the middle or entire, very scabrous above; autumn color yellow.

Twigs: Slender, zig-zag, green-brown, mostly glabrous, with lenticels; leaf scar crescent shaped with three bundle scars.

Buds: True terminal bud lacking, lateral buds triangular, 3 mm long, appressed; scales red-black, overlapping, glabrous or pubescent.

Bark: Gray-brown with corky warts, becoming fissured on large stems.

Flowers: Perfect and imperfect, small, yellow-green or green-white, without petals; in spring with the leaves.

Fruit: Drupe, orange-red to red-brown, round, 7 mm wide, maturing in fall.

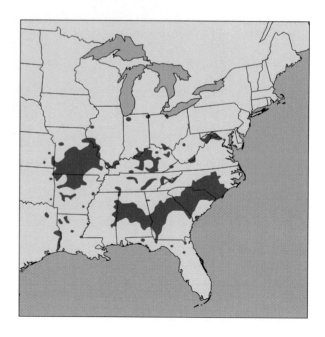

Form: Shrub or small tree up to 9 m (30 ft) in height.

Habitat and Ecology: Shade tolerant; found in the understory of upland forests.

Uses: Wood not commercially important. See *Celtis laevigata* for wildlife uses.

Botanical Name: *Celtis* was a name for African lotus, which has sweet fruit; *tenuifolia* means "thin leaved."

Ulmaceae

Celtis tenuifolia flowers

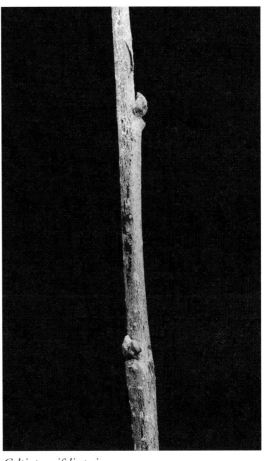
Celtis tenuifolia twig

Celtis tenuifolia bark

Celtis tenuifolia leaves and immature fruit

Ulmaceae 509

Planera aquatica J.F. Gmel.
water-elm, planertree

Quick Guide: *Leaves* alternate, simple, ovate, 2 ranked, apex acute, margin serrate; *Fruit* with fleshy projections; *Bark* gray-brown and scaly with red inner bark; *Habitat* swamps and bottoms.

Leaves: Simple, alternate, deciduous, 2 ranked, ovate to lanceolate, 3 to 7 cm long, apex acute, base rounded, margin serrate.

Twigs: Slender, zig-zag, red-brown, pubescent, with lenticels; leaf scar triangular and minute with three bundle scars.

Buds: True terminal bud lacking; lateral buds ovoid to round, 2 mm long or smaller; scales red-brown, overlapping, with pubescence.

Bark: Gray-brown, thin, scaly, shreddy or shaggy, with red inner bark.

Flowers: Perfect and imperfect, yellow-green or white-green, without petals, from leaf axils in spring with the leaves.

Fruit: Drupe, misshapen, with fleshy projections, 8 to 13 mm long, maturing in spring.

Form: Up to 15 m (50 ft) in height with a low crown and a forked, vase-shaped trunk.

Habitat and Ecology: An uncommon tree, found in the Coastal Plain in wet areas such as

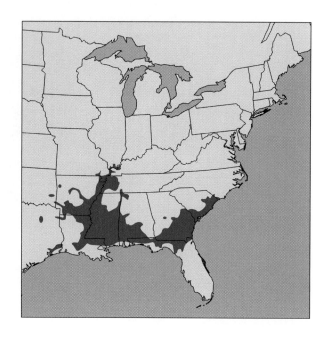

swamps, bottomlands, and stream edges, often associated with *Taxodium* and *Nyssa* spp.

Uses: Wood not commercially important, sometimes used for boats. Fruit eaten by waterfowl.

Botanical Name: *Planera* is for Johann Planer, a German botanist in the eighteenth century; *aquatica* refers to the habitat.

Planera aquatica leaves

Planera aquatica fruit

Planera aquatica flowers

Planera aquatica twig

Planera aquatica bark

Ulmus alata Michx.

winged elm, cork elm, wahoo, small-leaved elm, hard elm

Quick Guide: *Leaves* alternate, simple, elliptical, apex acute, base slightly inequilateral, margin doubly serrate; *Twigs* often with corky wings; *Fruit* an elliptical, pubescent, notched samara; *Bark* gray-brown with flattened or scaly ridges; *Habitat* a variety of sites.

Leaves: Simple, alternate, deciduous, elliptical or lanceolate, 3 to 8 cm long, apex acute, base slightly inequilateral, margin doubly serrate, smooth or sometimes scabrous above; autumn color yellow.

Twigs: Slender, slightly zig-zag, red-brown, mostly glabrous, with lenticels, often but not always with corky wings; leaf scar semicircular with three or more bundle scars.

Buds: True terminal bud lacking; lateral buds acute, 3 mm long, divergent; scales red-brown, overlapping, mostly glabrous.

Bark: Small trees with gray corky ridges or scaly, large trees gray-brown to red-brown with flattened possibly scaly ridges.

Flowers: Perfect, in yellow-red to purplish clusters in spring before the leaves (see page 503).

Fruit: Samara, elliptical, 0.5 to 1 cm long, deeply notched at the apex, margin with dense white pubescence, maturing before or with the leaves.

Form: Up to 15 m (50 ft) in height and 61 cm (2 ft) in diameter with drooping branches.

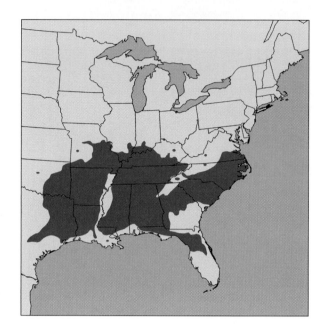

Habitat and Ecology: Shade tolerant; found on a variety of sites such as stream edges, floodplains, open fields, and sandy uplands. Forest associates include *Celtis laevigata, Fraxinus pennsylvanica, Quercus alba, Quercus marilandica, Quercus rubra, Quercus stellata, Quercus velutina, Quercus pagoda,* and *Ulmus americana.*

Uses: Wood brown or red-brown, heavy, hard, shock resistant; used for boxes, crates, furniture, posts, fuel, and hockey sticks. Seed eaten by birds and small mammals.

Botanical Name: *Ulmus* is Latin for "elm tree"; *alata* means "winged," referring to the twigs.

Ulmus alata bark

Ulmus alata fruit

Ulmus alata twig

Ulmus alata leaves and winged twig

Ulmus alata bark

Ulmus americana L.
American elm, white elm, water elm

Quick Guide: *Leaves* alternate, simple, obovate, apex acuminate, base greatly inequilateral, margin doubly serrate; *Twigs* and *Buds* mostly glabrous; *Fruit* a round, notched, pubescent samara; *Bark* gray-brown with scaly ridges and brown-white inner layers; *Habitat* moist soils.

Leaves: Simple, alternate, deciduous, obovate to ovate and elliptical, 8 to 15 cm long, apex acuminate, base greatly inequilateral, margin doubly serrate, smooth or slightly scabrous above (usually very scabrous on saplings), pubescent below; autumn color yellow.

Twigs: Slender, slightly zig-zag, red-brown, mostly glabrous; leaf scar semicircular with three or more bundle scars.

Buds: True terminal bud lacking; lateral buds ovoid to acute, 6 mm long, divergent, with red-black mostly glabrous scales.

Bark: Gray-brown with scaly ridges and when cut with alternating brown and white layers. Old trees with raised interlacing ridges.

Flowers: Perfect, long stalked, in red-brown clusters in spring before the leaves (see page 503).

Fruit: Samara, nearly round, 1 cm wide, deeply notched at the apex, margin with dense white pubescence, maturing in spring with the developing leaves.

Form: Up to 38 m (125 ft) in height and 1.5 m (5 ft) in diameter, with drooping branches and a forked or vase-shaped trunk.

Habitat and Ecology: Intermediate shade tolerance; found on upland sites but more common on bottomlands, terraces, and stream or swamp

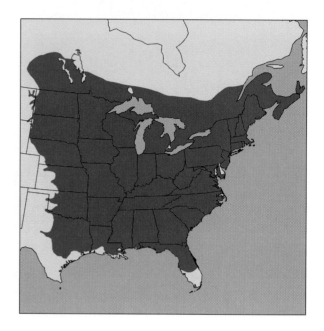

edges with *Acer negundo, Acer rubrum, Acer saccharinum, Betula lenta, Betula papyrifera, Celtis laevigata, Fagus grandifolia, Fraxinus americana, Fraxinus nigra, Fraxinus pennsylvanica, Liquidambar styraciflua, Picea mariana, Quercus bicolor, Quercus macrocarpa, Platanus occidentalis, Populus deltoides,* and *Ulmus rubra*.

Uses: Wood whitish gray to brown, moderately heavy, moderately hard; used for boxes, crates, pallets, and furniture. Was used in shipbuilding and for wheel hubs. Bark was used by Native Americans for rope and canoes. Seeds eaten by wood ducks, wild turkey, and a variety of songbirds; buds eaten by cottontail rabbit and gray, red, and fox squirrels. Before Dutch elm disease, a wilt fungus, this species was a favorite urban shade tree.

Botanical Name: *Ulmus* is Latin for "elm tree"; *americana* refers to the New World.

Ulmus americana bark

Ulmus americana twig

Ulmus americana bark

Ulmus americana fruit

Ulmus americana leaf

Ulmaceae

Ulmus crassifolia Nutt.
cedar elm, basket elm

Quick Guide: Similar to *Ulmus alata* but *Leaves* smaller, coarsely serrate and scabrous; *Flowers* appearing in late summer; *Samara* maturing in autumn.

Leaves: Simple, alternate, deciduous, elliptical, oblong or ovate, usually under 5 cm long, thick, apex acute to rounded, base inequilateral, margin coarsely serrate or doubly serrate, scabrous above, underside pubescent; autumn color yellow.

Twigs: Very slender, red-brown to gray, some zig-zag, with lenticels and corky wings; leaf scar semicircular with three or more bundle scars.

Buds: True terminal bud lacking, lateral buds ovoid-acute, 3 mm long, scales red-brown.

Bark: Gray to gray-brown with scaly ridges.

Flowers: Perfect, appearing in late summer or early fall.

Fruit: Samara, elliptical 10 to 13 mm long, notched at the apex, margin with short white hairs, maturing in late autumn.

Form: Up to 30 m (100 ft) in height with a vaselike form and drooping branches.

Habitat and Ecology: Intermediate shade tolerance; found on moist limestone soils, poorly drained clay soils, streambanks, bottomlands and dry limestone hills. Forest associates in floodplains

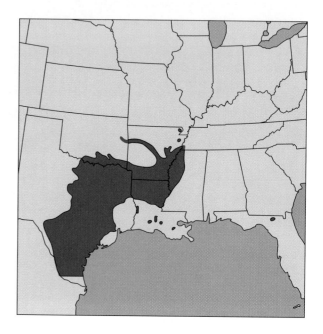

include *Acer rubrum, Carya aquatica, Celtis laevigata, Fraxinus pennsylvanica, Liquidambar styraciflua, Quercus lyrata, Quercus phellos, Ulmus alata,* and *Ulmus americana.*

Uses: Wood light brown to reddish, heavy, hard, shock resistant, similar to *Ulmus thomasii*. Fruit eaten by a variety of birds, small mammals, and deer. Planted as an ornamental but susceptible to Dutch elm disease.

Botanical Name: *Ulmus* is Latin for "elm tree"; *crassifolia* means "thick leaf."

Ulmus crassifolia leaves

Ulmus crassifolia young bark

Ulmus crassifolia twig

Ulmus rubra Muhl.
slippery elm, red elm, soft elm

Quick Guide: *Leaves* alternate, simple, obovate, apex acuminate, base inequilateral, margin doubly serrate, scabrous above and below; *Buds* with maroon pubescence; *Fruit* a round, shallowly notched samara with a glabrous margin; *Bark* gray-brown, with flat ridges and brown inner layers; *Habitat* moist soils.

Leaves: Simple, alternate, deciduous, obovate to oval and elliptical, 10 to 18 cm long, apex acuminate, base inequilateral, margin doubly serrate, scabrous above and below; autumn color yellow.

Twigs: Slender, slightly zig-zag, red-brown to gray, with pubescence, scabrous; leaf scar semicircular with three or more bundle scars. Inner bark slippery and mucilaginous.

Buds: True terminal bud lacking; lateral buds ovoid, 6 mm long, divergent; scales red-black with red-maroon pubescence.

Bark: Gray-brown or red-brown with corky or flattened sometimes interlacing ridges and brown layers when cut.

Flowers: Perfect, short stalked, in red-brown to purplish clusters before the leaves. Reported by Rogers (1965) to turn to purple only if flowers are wet by rain and the pigments can diffuse out (see page 503).

Fruit: Samara, round, 1 to 2 cm wide, only shallowly notched at the apex, seed portion pubescent but margin glabrous, maturing in spring as the leaves develop.

Form: Up to 21 m (70 ft) in height and 1 m (3 ft) in diameter with a vase shape.

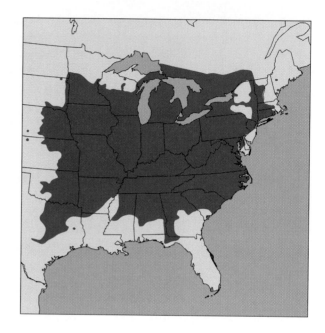

Habitat and Ecology: Shade tolerant; found on a variety of sites including limestone soils, but primarily in bottomlands and moist uplands. Forest associates are numerous and include *Acer rubrum, Acer saccharum, Betula nigra, Carya* spp., *Platanus occidentalis, Quercus alba, Quercus macrocarpa, Quercus muehlenbergii, Quercus rubra, Quercus velutina,* and *Ulmus americana*.

Uses: Wood gray-white to red-brown, moderately heavy, moderately hard; used for boxes, crates, pallets, and furniture. Inner bark used as a home remedy. Seed eaten by birds and small mammals. Susceptible to Dutch elm disease.

Botanical Name: *Ulmus* is Latin for "elm tree"; *rubra* means "red," referring to red-maroon hair on the buds or to the reddish wood.

Ulmus rubra fruit and twig

Ulmus rubra bark

Ulmus rubra bark

Ulmus rubra leaves

Ulmaceae

Ulmus serotina Sarg.
September elm, red elm

Quick Guide: Similar to *Ulmus alata* but *Leaves* scabrous on top; *Samara* with long, silvery hairs on the margin and maturing in autumn.

Leaves: Simple, alternate, deciduous, elliptical to oval, 5 to 9 cm long, yellow-green, apex acute to acuminate, base inequilateral and cordate, margin doubly serrate, coarsely textured or scabrous above, underside pale pubescent; autumn color yellow.

Twigs: Slender, zig-zag, red-brown to gray, with lenticels, glabrous or finely pubescent, older twigs with corky wings; leaf scar semicircular with three or more bundle scars.

Buds: True terminal bud lacking, lateral buds ovoid-acute, up to 6 mm long, divergent, scales red-black and mostly glabrous.

Bark: Gray-brown with thick, corky ridges on small trees; large trees red-brown or gray with scaly, flat ridges.

Flowers: Perfect, long-stalked, green, appearing in September.

Fruit: Samara, elliptical to oblong, 1.0 to 1.3 cm long, deeply notched at the apex, margin with long silvery hairs, maturing in late autumn.

Form: Up to 24 m (80 ft) in height with a vaselike form.

Habitat and Ecology: Shade tolerant; found mostly in mesic woods and along streams but also on limestone outcroppings, uncommon. Forest associates include *Acer barbatum, Betula nigra, Fraxinus americana, Fraxinus quadrangulata, Liquidambar styraciflua, Quercus rubra,* and *Ulmus americana.*

Uses: Wood light brown to reddish, heavy, hard, shock resistant, similar to *Ulmus thomasii.* Fruit eaten by a variety of birds, small mammals, and deer. Planted as an ornamental but susceptible to Dutch elm disease.

Botanical Name: *Ulmus* is Latin for "elm tree"; *serotina* means "late," referring to the late-blooming flowers.

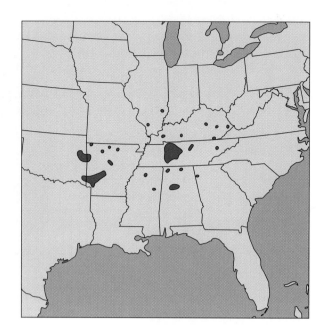

Ulmus serotina twig

Ulmus serotina young stem

Ulmus serotina young bark

Ulmus serotina leaves

Ulmus thomasii Sarg.

rock elm, cork elm, hickory elm

Quick Guide: Similar to *Ulmus americana* but *Leaves* less inequilateral at the base; *Twigs* with corky wings; *Fruit* hairy all over.

Leaves: Simple, alternate, deciduous, obovate to oval and elliptical, 5 to 11 cm long, apex acuminate, base inequilateral, margin doubly serrate, smooth above, underside pubescent; autumn color yellow.

Twigs: Slender, zig-zag, red-brown to gray, pubescent to glabrous, small branches with corky wings; leaf scar semicircular with three or more bundle scars.

Buds: True terminal bud lacking, lateral buds ovoid-acute, 6 to 10 mm long, divergent; scales red-brown, pubescent. More pointed than for *ulmus americana*.

Bark: Gray-brown, furrowed, with flattened scaly ridges.

Flowers: Perfect, in red-brown clusters in spring before the leaves.

Fruit: Samara, ovate or obovate, 1 to 2 cm long, shallowly notched at the apex, pubescent all over; maturing in spring with the developing leaves.

Form: Up to 30 m (100 ft) in height with a narrow crown, lacking the vase shape.

Habitat and Ecology: Intermediate shade tolerance; found on moist well-drained soils, limestone outcroppings, heavy clay soils, and rocky ridges. Forest associates include *Acer rubrum, Acer

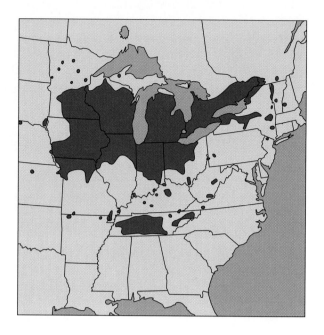

saccharum, Betula alleghaniensis, Fagus sylvatica, Fraxinus nigra, Prunus serotina, Tsuga canadensis,* and *Ulmus americana.*

Uses: Wood light brown to reddish, heavy, hard, shock resistant; used for furniture, boxes, crates, plywood, veneer, pianos, hockey sticks, boat frames, and tool handles. Considered the best wood of all the elms and was used for wheels, bridge timbers, agricultural implements, axe handles, railroad ties, automobile bodies, and refrigerators. Fruit eaten by a variety of birds, small mammals, and deer.

Botanical Name: *Ulmus* is Latin for "elm tree"; *thomasii* is for the horticulturist David Thomas.

Ulmus thomasii leaves

Ulmus thomasii twig and fruit

Ulmus thomasii young bark

Ulmus thomasii branch

Ulmaceae

Glossary

2 ranked: in two vertical rows on an axis.
4 ranked: in four vertical rows on an axis.
Achene: a dry, unwinged, one-seeded, indehiscent fruit; sycamore.
Acuminate: gradually tapering to an acute angle; attenuated.
Acute: forming an angle less than 90 degrees, pointed.
Alternate: one leaf at a node.
Angiosperm: the flowering plants, immature seeds enclosed in an ovary.
Anther: pollen-bearing structure of the stamen.
Apex: the tip of the leaf.
Apophysis: the exposed portion of the cone scale when the cone is closed, the "lip" of the cone scale that bears the prickle.
Appressed: pressed against.
Arcuate: archlike, referring to lateral venation that parallels the leaf margin and curves toward the apex.
Armed: possessing a sharp spine, prickle, or thorn.
Asymmetrical: uneven, unequal in size.
Auriculate: with an earlike appendage, usually referring to the leaf base.
Awl-like: like a pointed tool, tapering to a slender point.
Axil: angle between the branch and petiole, or midrib and lateral vein.
Berry: a fleshy fruit with one or more seeds; persimmon, pawpaw.
Bipinnately compound: twice pinnately compound.
Blade: usually referring to the leaf minus the petiole.
Bog: naturally waterlogged spongy ground; usually consists of a peat and sphagnum moss substrate.
Botanical name: the Latin binomial name of a plant consisting of the generic name (genus) followed by the specific epithet, which is often descriptive. Developed by Linnaeus in 1753 in *Species Plantarum*.
Bottomland: low alluvial land.
Bract: a leaflike structure.
Bundle scar: scars in the leaf scar left by the xylem and phloem.
Capsule: a dehiscent fruit with more than one pistil; sweetgum, poplars, willow.

Catkins: apetalous, sessile, unisexual flowers clustered on a single long axis; oaks, birches.
Chambered pith: pith with empty cells.
Compound leaf: a leaf with at least two leaflets.
Cordate: heart shaped.
Crenate: with rounded teeth on the leaf margin.
Cruciform: crosslike.
Cuneate: the base of a leaf that gradually tapers to an acute angle, wedge-shaped.
Cyme: a flat-topped cluster of flowers blooming at the center first.
Dehiscent: splitting along defined sutures or seams.
Deltoid: triangular.
Dentate: leaf margin with teeth pointing outward.
Diaphragmed pith: barred pith with closed cells.
Dimorphic: having two forms.
Dioecious: having flowers of only one sex on a tree.
Divergent: wide spreading, spread apart.
Doubly serrate: leaf margin with a smaller tooth within a larger tooth.
Drupe: a fleshy fruit with an inner stony wall enclosing the seed; blackgum, cherry.
Elliptical: shaped like an ellipse with the widest point in the middle.
Emarginate: notched at the tip.
Embedded: buried, enclosed, or fixed.
Entire: leaf margin without teeth, continuous, smooth.
Erose: eroded, roughly toothed, jagged.
Exserted: visible, extending past the surrounding parts.
Falcate: sickle shaped.
Fascicle: a group of needles usually bundled by a sheath.
Filament: threadlike structure of the stamen supporting the anther.
Follicle: a dry, dehiscent fruit splitting along one suture; magnolia.
Furrowed: grooved, plowed.
Glabrous: smooth, without hair.
Glandular: with glands.
Glaucous: with a waxy coating.
Globose: globelike, spherical.
Gymnosperm: plants with seeds not borne in an ovary.

Hammock: an elevated, well-drained site surrounded by a marsh or bog.
Head: a round or flat-topped cluster of sessile flowers.
Imbricate: overlapping.
Imperfect flower: flower of only one sex; with stamens or pistils; unisexual.
Indehiscent: not splitting along defined sutures or seams.
Inequilateral: unequal in shape or length.
Inflorescence: flower cluster.
Keeled: with a prominent longitudinal ridge.
Lanceolate: lance shaped, longer than wide, widest below the middle.
Lateral bud: on the side of the twig, usually smaller than the terminal bud.
Leaf scar: scar made by the petiole where it attached to the branch.
Legume: a dry, podlike fruit splitting along two opposite seams; eastern redbud, mimosa, honeylocust.
Lenticel: a lense-shaped bump.
Linear: long and narrow with parallel sides.
Lobe: rounded part of an organ.
Lustrous: shiny.
Margin: the edge.
Mesic: intermediate between wet and dry, moist.
Midrib: the central vein on the leaf.
Monoecious: having flowers of both sexes on the same tree.
Mucilaginous: slimy.
Mucronate: a sharp or rigid point.
Naked bud: bud lacking visible scales.
Node: position on the twig where the petiole(s) or branch(es) arises.
Nut: a one-seeded fruit with a hard shell; oak, hickory, walnut.
Nutlet: a small nut.
Oblanceolate: reverse lanceolate, widest above the middle.
Oblong: longer than wide with parallel sides.
Obovate: reverse ovate, widest at the apex.
Obtuse: forming an angle greater than 90 degrees, blunt.
Opposite: two leaves per node on opposite sides of the twig.
Orbicular: nearly circular.
Oval: broadly elliptical.
Ovary: in the flower, the portion of the pistil that encloses the ovules or seeds.
Ovate: egg-shaped outline, widest at the base.
Ovoid: egg shaped, three dimensional.
Ovule: immature seed.
Palmate: arising from a common point, like fingers on a hand.
Panicle: a branched flower structure; sumac, buckeye.
Peltate: shield shaped, attached from the center.
Pendent: hanging down.
Perfect flower: a flower with both sexes, with stamens and pistils; bisexual.
Petal: part of the flower surrounding the reproduc- tive organs, usually white or brightly colored.
Petiole: the leaf stalk.
Pinnately compound: a leaf with leaflets arranged along a central axis, like a feather.
Pistil: female part of a flower made up of an ovary, style, and stigma.
Pistillate: female or seed flower.
Pith: the spongy center of a twig or root.
Pocosin: an evergreen shrub-tree bog in the Atlantic Coastal Plain.
Pome: a fleshy fruit with a papery interior wall enclosing the seed; crabapple, serviceberry.
Prickle: a sharp or corky outgrowth from the bark, or sharp structure on a cone scale.
Pseudoterminal bud: a lateral bud assuming the terminal position due to an aborted twig tip. The bud has a leaf scar and the twig is zig-zag.
Pubescent: with soft hairs.
Raceme: an unbranched inflorescence, the flowers on a single axis.
Rachis: the main axis of a compound leaf.
Reflexed: bent downward or backward.
Repand: irregularly or slightly wavy.
Revolute: with the leaf margin rolled under.
Rhombic: diamond shaped.
Rugose: wrinkled.
Samara: a dry, winged, indehiscent fruit; maples, elms.
Scabrous: sandpapery to the touch.
Scurfy: with small scales.
Sepal: the outer whorl of a flower, usually green but can appear petal-like as in flowering dogwood and fevertree.
Serotinous: late in flowering or in cone opening.
Serrate: sharp sawlike teeth on the leaf margin pointing toward the apex.
Sessile: lacking a petiole or stalk.
Shade tolerance: refers to a plant's vigor in the shaded forest understory and ability to grow

from a seedling to an adult under dense cover; may vary with site quality, tree age, and climate.
Simple: a leaf with one leaflet, unbranched.
Sinuate: wavy.
Sinus: depression or gap between lobes.
Spatulate: spatulalike with the widest point at the apex, gradually widening towards the tip.
Spine: a modified leaf or stipule, usually sharp.
Stalked: with a supporting structure, like a neck.
Stamen: the male reproductive flower consisting of an anther and filament.
Staminate: the male or pollen flower bearing the stamens.
Stellate: star shaped.
Stigma: the top of the style that receives the pollen.
Stipular scar: scar left on the twig by the stipule.
Stipule: a leafy appendage at the petiole, usually smaller than the leaves and appearing before the leaves.
Striations: parallel streaks, stripes, or fine grooves.

Style: between the ovary and stigma.
Subopposite: nearly opposite.
Superposed buds: a bud above another.
Swamp: a seasonally flooded bottomland that is better drained than a bog.
Terminal bud: bud at the tip of the shoot, usually larger than the lateral buds, lacking a leaf scar.
Thorn: a sharp, modified branch.
Tomentose: with matted soft or wooly hairs.
Trifoliately compound: with three leaflets.
Tripinnately compound: thrice pinnately compound.
Truncate: squarelike.
Umbel: a flat-topped flower, with flowers arising from the same point.
Valvate: not overlapping, edge to edge.
Whorled: three or more leaves per node.
Winged: with wings or extensions from the seed, flower, rachis, or twig.
Wooly: with long tangled or matted hairs.

Bibliography

Baldwin, H.I. 1993. *Forest leaves. How to identify trees and shrubs of northern New England.* Portsmouth, NH: Peter E. Randall Publisher.

Barnes, B.V., and W.H. Wagner, Jr. 2004. *Michigan trees: A guide to the trees of the Great Lakes Region.* Ann Arbor, MI: The University of Michigan Press.

Bell, C.R., and A.H. Lindsey. 1990. *Fall color and woodland harvests.* Chapel Hill, N.C.: Laurel Hill Press.

Bishop, G.N. 2000. *Native trees of Georgia.* Athens, GA: Georgia Forestry Commission.

Brown, C.L., and L.K. Kirkman. 1990. *Trees of Georgia and adjacent states.* Portland, Ore.: Timber Press.

Brown, H.P. 1938. *Trees of the northeastern United States native and naturalized.* Boston, MA: The Christopher Publishing House.

Burns, R.M., and B.H. Honkala, eds. 1990. *Silvics of North America.* Vol. 1, *Conifers.* Agriculture Handbook 654. Washington, D.C.: USDA Forest Service.

Burns, R.M., and B.H. Honkala, eds. 1990. *Silvics of North America.* Vol. 2, *Hardwoods.* Agriculture Handbook 654. Washington, D.C.: USDA Forest Service.

Core, E.L., and N.P. Ammons. 1977. *Woody plants in winter.* Pacific Grove, Calif.: The Boxwood Press.

Davis, D.E., N.D. Davis, and L.J. Samuelson. 1999. *Guide and key to Alabama trees.* Dubuque, Iowa: Kendall/Hunt Publishing.

Dean, B.E. 1988. *Trees and shrubs of the southeast.* Birmingham, Ala.: Audubon Society Press.

Dickson, J.G., ed. 2001. *Wildlife of southern forests.* Blaine, Wash.: Hancock House Publishers.

Dirr, M.A. 1997. *Dirr's hardy trees and shrubs.* Portland, Ore.: Timber Press.

Dorr, L.J., and K.C. Nixon. 1985. Typification of the oak (Quercus) taxa described by B.S. Buckley (1809–1884). *Taxon* 34:211–228.

Duncan, W.H., and M.B. Duncan. 1988. *Trees of the southeastern United States.* Athens, GA: The University of Georgia Press.

Dwelley, M.J. 2000. *Trees and shrubs of New England.* Camden, ME: Down East Books.

Farrar, J.L. 1995. *Trees of the northern United States and Canada.* Ames, Iowa: Iowa State University Press.

Fralish, J.S., and S.B. Franklin. 2002. *Taxonomy and ecology of woody plants in North American forests.* New York: John Wiley & Sons, Inc.

Gill, J.D., and W.M. Healy, eds. 1974. *Shrubs and vines for northeastern wildlife.* USDA Forest Service General Technical Report NE-9. Upper Darby, Penn.: Northeastern Forest Experiment Station.

Gledhill, D. 2002. *The names of plants.* Cambridge, U.K.: Cambridge University Press.

Godfrey, R.K. 1988. *Trees, shrubs, and woody vines of northern Florida and adjacent Georgia and Alabama.* Athens, GA: The University of Georgia Press.

Godfrey, R.K., and J.W. Wooten. 1981. *Aquatic and wetland plants of the southeastern United States.* Athens, GA: The University of Georgia Press.

Grimm, W.C. revised by J. Kartesz. 2002. *The illustrated book of trees.* Mechanicsburg, PA: Stackpole Books.

Halls, L.K., ed. 1977. *Southern fruit-producing woody plants used by wildlife.* USDA Forest Service General Technical Report SO-16. Upper Darby, Penn.: Southern Forest Experiment Station.

Halls, L.K., and T.H. Ripley. 1961. *Deer browse plants of southern forests.* U.S. Forest Service Report. Upper Darby, Penn.: South and Southeast Forest Experiment Station.

Hardin, J.W., D.J. Leopold, and F.M. White. 2001. *Harlow & Harrar's textbook of dendrology.* New York: McGraw-Hill Book Company, Inc.

Harlow, W.M. 1946. *Fruit key and twig key to trees and shrubs.* New York: Dover Publications.

Harrar, E.S., and J.G. Harrar. 1962. *Guide to southern trees.* New York: Dover Publications.

Harris, J.G., and M.W. Harris. 1999. *Plant identification terminology.* Spring Lake, Utah: Spring Lake Publishing.

Hosie, R.C. 1969. *Native trees of Canada.* Ottawa, Canada: The Queen's Printer.

Hunter, C.G. 2004. *Trees, shrubs, & vines of Arkansas.* Little Rock, AR: The Ozark Society Foundation.

Hutnick, R.J., and H.W. Yawney. 1961. *Silvical characteristics of red maple (Acer rubrum).* USDA Forest Service Station Paper 142. Upper Darby, Penn.: Northeastern Forest Experiment Station.

Leopold, D.J. 2003. *Trees of New York State native and naturalized.* Syracuse, NY: Syracuse University Press.

Leopold, D.J., W.C. McComb, and R.N. Muller. 1998. *Trees of the central hardwood forests of North America.* Portland, Ore.: Timber Press.

Little, E.L., Jr. 1971. *Atlas of United States trees.* Vol. 1, *Conifers and important hardwoods.* Miscellaneous Publication No. 1146. Washington D.C.: USDA Forest Service.

Little, E.L., Jr. 1977. *Atlas of United States trees.* Vol. 4, *Minor eastern hardwoods.* Miscellaneous Publication No. 1342. Washington D.C.: USDA Forest Service.

Little, E.L., Jr. 1979. *Checklist of United States trees (native and naturalized).* Agriculture Handbook No. 541. Washington, D.C.: USDA Forest Service.

Martin, A.C., H.S. Zinn, and A.L. Nelson. 1951. *American wildlife and plants: A guide to wildlife food habits.* New York: McGraw-Hill Book Company, Inc.

Matoon, W.R. 1948. *Common forest trees of Florida.* Tallahassee, FLA.: Florida Board of Forestry and Parks.

Miller, H.A., and S.H. Lamb. 1985. *Oaks of North America.* Happy Camp, Calif.: Naturegraph Publishers, Inc.

Miller, J.H., and K.V. Miller. 1999. *Forest plants of the southeast and their wildlife uses.* Champagne, Ill.: Southern Weed Society.

Panshin, A.J., and C. de Zeeuw. 1980. *Textbook of wood technology.* New York: McGraw-Hill Book Company, Inc.

Preston, R.J., and R.R. Braham. 2002. *North American Trees.* Ames, Iowa: Iowa State Press.

Rogers, J.E. 1920. *The tree book: A popular guide to a knowledge of the trees of North America and to their uses and cultivation.* New York: Doubleday, Page & Company.

Rogers, W.E. 1965. *Tree flowers of forest, park and street.* New York: Dover Publications.

Samuelson, L.J., and M.E. Hogan. 2003. *Forest trees: A guide to the southeastern and mid-atlantic regions of the United States.* Upper Saddle River, NJ: Prentice Hall.

Settergren, C., and R.E. McDermott. 1995. *Trees of Missouri.* University Extension Publication SB 767. Columbia, MO: University of Missouri Press.

Sharitz, R.R., and J.W. Gibbons. 1982. *The ecology of southeastern shrub bogs (pocosins) and Carolina bays: A community profile.* Publication No. FWS/OBS-82/04. Washington, D.C.: U.S. Fish and Wildlife Service, Division of Biological Services.

Smith, N.F. 1995. *Trees of Michigan and the upper Great Lakes.* Holt, MI: Thunder Bay Press.

Sternberg, G., and J. Wilson. 2004. *Native trees for North American landscapes.* Portland, OR: Timber Press.

Stupka, A. 1993. *Trees, shrubs and woody vines of Great Smoky Mountains National Park.* Knoxville, Tenn.: The University of Tennessee Press.

Vitale, A.T. 1997. *Leaves in myth, magic and medicine.* New York: Stewart, Tabori & Chang.

Wagner, W.H. 1979. *Modern carpentry.* South Holland, Ill.: The Goodheart-Willcox Co., Inc.

Wharton, M.E., and R.. Barbour. 1973. *Trees & shrubs of Kentucky.* Lexington, KY: The University Press of Kentucky.

York, H.H. 1999. *100 forest trees of Alabama.* Auburn, Ala.: Alabama Forestry Commission and State Department of Education, Office of Career/Technical Education, Agriscience Technology Education.

Index

Abies balsamea, 16–17
Abies fraseri, 18–19
Acer barbatum, 72–73
Acer barbatum pistillate flowers, 69
Acer leucoderme, 74–75
Acer negundo, 76–77
Acer negundo pistillate flowers, 71
Acer negundo staminate flowers, 71
Acer nigrum, 78–79
Acer pensylvanicum, 80–81
Acer platanoides, 86
Acer rubrum, 82–83
Acer rubrum pistillate flowers, 71
Acer rubrum staminate flowers, 71
Acer rubrum var. *drummondii*, 82
Acer rubrum var. *trilobum*, 82
Acer saccharinum, 84–85
Acer saccharum, 86–87
Acer spicatum, 88–89
Aceraceae, 70–89
    Acer barbatum, 72–73
    Acer leucoderme, 74–75
    Acer negundo, 76–77
    Acer nigrum, 78–79
    Acer pensylvanicum, 80–81
    Acer platanoides, 86
    Acer rubrum, 82–83
    Acer saccharinum, 84–85
    Acer saccharum, 86–87
    Acer spicatum, 88–89
    black maple, 78–79
    boxelder, 76–77
    chalk maple, 74–75
    Florida maple, 72–73
    mountain maple, 88–89
    Norway maple, 86
    overview, 70
    red maple, 82–83
    silver maple, 84–85
    striped maple, 80–81
    sugar maple, 86–87
Aesculus flava, 306–307
Aesculus flava leaf and flowers, 305
Aesculus glabra, 308–309
Aesculus glabra fruit, 304
Aesculus parviflora, 310
Aesculus parviflora flowers, 305
Aesculus pavia, 310–311
Aesculus pavia flowers, 305
Aesculus sylvatica, 310
Aesculus sylvatica flowers, 305
Ailanthus, 480–481
Ailanthus altissima, 480–481
Albizia julibrissin flowers, 69
Albizzi, Filippo, 368
Allegheny chinkapin, 224–225
Allegheny serviceberry, 418–419
Almond-leaved willow, 468

Alnus incana spp. *rugosa*, 126–127
Alnus rugosa, 126–127
Alnus serrulata, 128–129
Alternate-leaf dogwood, 176–177
Amelanchier arborea, 416–417
Amelanchier arborea twig, 415
Amelanchier laevis twig, 415
Amelanchier laevis, 418–419
American arborvitae, 12–13
American beech, 218, 226–227
American bladdernut, 482–483
American chestnut, 222–223
American crabapple, 424–425
American elder, 160–161
American elm, 514–515
American hackberry, 506–507
American holly, 118–119
American hophornbeam, 142–143
American hornbeam, 140–141
American larch, 20
American linden, 500–501
American mountain-ash, 442–443
American planetree, 410–411
American plum, 428–429
American smoketree, 92–93
American snowbell, 490–491
American sycamore, 410–411
American walnut, 340–341
American witch-hazel, 300
American yellowwood, 214–215
Anacardiaceae, 90–101
    Cotinus obovatus, 92–93
    overview, 90
    poison-sumac, 100–101
    Rhus capallina, 94–95
    Rhus glabra, 96–97
    Rhus typhina, 98–99
    smoketree, 92–93
    smooth sumac, 96–97
    staghorn sumac, 98–99
    Toxicodendron vernix, 100–101
    winged sumac, 94–95
Angelica-tree, 122–123
Angiosperm families, 65–523
    Aceraceae, 70–89. *See also* Aceraceae
    Anacardiaceae, 90–101. *See also* Anacardiaceae
    anise tree family, 312–313
    Annonaceae, 66, 102–103
    Aquifoliaceae, 104–121. *See also* Aquifoliaceae
    Araliaceae, 67, 122–123
    beech, 218–299
    Betulaceae, 124–143. *See also* Betulaceae
    Bignoniaceae, 144–147
    birch, 124–143
    bitterwood family, 68, 480–481

bladdernut family, 65, 482–483
buckeye, 304–311
buckthorn, 67, 412–413
Caesalpiniaceae, 148–157. *See also* Caesalpiniaceae
camellia family, 67, 496–499
Caprifoliaceae, 158–171. *See also* Caprifoliaceae
cashew, 90–101
Celastraceae, 65, 172–173
Cornacea, 174–195. *See also* Cornacea
custard apple, 66, 102–103
cyrilla, 196–199
Cyrillaceae, 196–199
dogwood, 174–195
Ebenaceae, 66, 200–201
ebony, 66, 200–201
elm family, 502–523
Ericaceae, 202–211. *See also* Ericaceae
Euphorbiaceae, 66, 212–213
Fabaceae, 214–217
Fagaceae, 218–299. *See also* Fagaceae
figwort family, 65, 478–479
ginseng, 67, 122–123
Hamamelidaceae, 66, 300–303
heath, 202–211
hickory, 314–331
Hippocastanaceae, 304–311
holly, 104–121
honeysuckle, 158–171
Illiciaceae, 312–313
Juglandaceae, 314–331. *See also* Juglandaceae
Lauraceae, 342–349. *See also* Lauraceae
laurel, 342–349
leaves alternate, simple, and lobed, 65
leaves alternate, simple, and unlobed (fragrant), 65–66
leaves alternate, simple, and unlobed (leaf margin without teeth), 66
leaves alternate, simple, and unlobed (margin toothed), 66–67
leaves alternate, simple, and unlobed (twigs armed), 66
leaves alternate and compound (twigs armed), 67
leaves alternate and compound (twigs unarmed), 67–68
leaves opposite and compound, 65
leaves opposite or whorled and simple, 65
legume, 148–157, 214–217
linden family, 67, 500–501
madder family, 65, 444–447
mahogany, 68, 366–367

maple, 70–89
Meliaceae, 68, 366–367
mimosa, 67, 368–369
Mimosaceae, 67, 368–369
Moraceae, 370–379. *See also Moraceae*
mulberry, 370–379
Myricaceae, 380–387. *See also Myricaceae*
oak, 218–299
Oleaceae, 388–409. *See also Oleaceae*
olive, 388–409
overview, 65–68
pawpaw, 66, 102–103
Platanaceae, 65, 410–411
poplar family, 456–469
Rhamnaceae, 67, 412–413
Rosaceae, 414–443. *See also Rosaceae*
rose family, 414–443
Rubiaceae, 65, 444–447
rue family, 448–455
Rutaceae, 448–455. *See also Rutaceae*
Salicaceae, 456–469. *See also Salicaceae*
sapodilla, 470–477
Sapotaceae, 470–477. *See also Sapotaceae*
Scrophulariaceae, 65, 478–479
Simaroubaceae, 68, 480–481
spurge, 66, 212–213
staff-tree, 172–173
Staphyleaceae, 65, 482–483
storax family, 484–493
Styracaceae, 484–493. *See also Styracaceae*
sweetleaf family, 66, 494–495
sycamore, 65, 410–411
Symplocaceae, 66, 494–495
Theaceae, 67, 496–499
Tiliaceae, 67, 500–501
trumpet creeper, 144–147
Ulmaceae, 502–523. *See also Ulmaceae*
walnut, 314–331
wax myrtle, 380–387
willow family, 456–469
witch hazel, 66, 300–303
Anise tree family, 312–313
Annonaceae, 66, 102–103
Apple, 426–427
Aquifoliaceae, 104–121
    American holly, 118–119
    Carolina holly, 114
    common winterberry, 114
    dahoon, 108–109
    Ilex ambigua, 114
    Ilex amelanchier, 106–107
    Ilex cassine, 108–109
    Ilex coriacea, 110–111
    Ilex decidua, 112–113

    Ilex glabra, 110
    Ilex montana, 114–115
    Ilex myrtifolia, 116–117
    Ilex opaca, 118–119
    Ilex verticillata, 114, 115
    Ilex vomitoria, 120–121
    inkberry, 110
    large gallberry, 110–111
    mountain winterberry, 114–115
    myrtle-leaved holly, 116–117
    overview, 104
    possumhaw, 112–113
    sarvis holly, 106–107
    yaupon, 120–121
Arkansas oak, 268
Arkansas pine, 32–33
Arrowwood, 170, 171
Ashe, Margaret, 266
Ashe juniper, 6
Ashe magnolia, 360, 361
Ashleaf maple, 76–77
Asimina parviflora, 102–103
Asimina triloba, 102–103
Aspen-leaved birch, 138–139
Atlantic white-cedar, 4–5

Baldcypress, 8–9
Balm of Gilead, 458
Balsam fir, 16–17
Balsam poplar, 458
Banks, Joseph, 28
Banksian pine, 28–29
Barren oak, 268–269
Basket ash, 398–399
Basket elm, 516–517
Basket oak, 270–271
Basswood, 500–501
Bastard white oak, 232–233
Bay pine, 46–47
Bear oak, 252–253
Bebb willow, 468
Bee tree, 500–501
Bee-tupelo, 192–193
Beech, 226–227
Beech family, 218–299. *See also Fagaceae*
Beechnut tree, 226–227
Betula alleghaniensis, 130–131
Betula lenta, 132–133
Betula nigra, 134–135
Betula papyrifera, 136–137
Betula populifolia, 138–139
Betulaceae, 124–143
    Alnus rugosa, 126–127
    Alnus serrulata, 128–129
    Betula alleghaniensis, 130–131
    Betula lenta, 132–133
    Betula nigra, 134–135
    Betula papyrifera, 136–137
    Betula populifolia, 138–139
    black birch, 132–133

    Carpinus caroliniana, 140–141
    gray birch, 138–139
    hazel alder, 128–129
    hophornbeam, 142–143
    hornbeam, 140–141
    Ostrya virginiana, 142–143
    overview, 124
    paper birch, 136–137
    river birch, 134–135
    speckled alder, 126–127
    yellow birch, 130–131
Big buckeye, 306–307
Big bud hickory, 336–337
Big shagbark hickory, 324–325
Bigleaf magnolia, 360–361
Bigleaf shagbark hickory, 324–325
Bigleaf snowbell, 492–493
Bignoniaceae, 144–147
    Catalpa bignonioides, 144–145, 146
    Catalpa speciosa, 146–147
    northern catalpa, 146–147
    southern catalpa, 144–145, 146
Bigtooth aspen, 462
Bigtree plum, 434–435
Birch family, 124–143. *See also Betulaceae*
Bird cherry, 436–437
Bitter pecan, 316–317
Bitternut hickory, 318–319
Bitterwood family, 68, 480–481
Black-alder, 114
Black ash, 398–399
Black birch, 132–133
Black cherry, 438–439
Black-cypress, 10–11
Black hickory, 334–335
Black maple, 78–79
Black oak, 242–243, 268–269, 296–297
Black spruce, 24–25
Black sugar maple, 78–79
Black titi, 196–197, 198–199
Black tupelo, 194–195
Black walnut, 340–341
Black willow, 457, 466–467
Blackgum, 190–191, 194–195
Blackhaw, 166–167
Blackjack oak, 268–269
Bladdernut, 482–483
Bladdernut family, 65, 482–483
Blue ash, 404–405
Blue beech, 140–141
Bluehaw, 168–169
Bluejack oak, 256–257
Bluff oak, 232–233
Bodark, 374–375
Bog spruce, 24–25
Bottlebrush buckeye, 310
Bottom white pine, 36–37
Bottomland post oak, 292
Bottomland red oak, 280–281
Bowwood, 374–375

Boxelder, 76–77
Boxwood, 118–119, 182–183
Broussonet, Auguste, 372
Buckeye family, 304–311. *See also*
    Hippocastanaceae
Buckthorn, 67, 412–413
Buckthorn bully, 474–475
Buckthorn bumelia, 474–475
Buckwheat-bush, 196–197
Bull bay, 358–359
Bumelia lanuginosa, 471, 472–473
Bumelia lycioides, 474–475
Bumelia tenax, 471, 476–477
Bur oak, 264–265
Burningbush, 172–173
Butternut, 338–339, 340
Buttonball-tree, 410–411
Buttonbush, 444
Buttonwood, 410–411

Caesalpiniaceae, 148–157
    Cercis canadensis, 150–151
    eastern redbud, 150–151
    Gleditsia aquatica, 152–153
    Gleditsia triacanthos, 154–155
    Gymnocladus dioicus, 156–157
    honeylocust, 154–155
    Kentucky coffeetree, 156–157
    overview, 148
    waterlocust, 152–153
Camellia family, 67, 496–499
Canada balsam, 16–17
Canada hemlock, 56–57
Canada plum, 428
Canadian spruce, 22–23
Candleberry, 381–383, 386–387
Canoe birch, 136–137
Caprifoliaceae, 158–171
    arrowwood, 170, 171
    blackhaw, 166–167
    elderberry, 160–161
    highbush cranberry, 170
    hobblebush, 170
    maple-leaved virburnum, 170, 171
    nannyberry, 162–163
    overview, 158
    possumhaw, 164–165
    rusty blackhaw, 168–169
    Sambucus canadensis, 160–161
    Sambucus pubens, 160
    scarlet elder, 160
    small-leaf viburnum, 170
    small viburnum, 170
    Viburnum abovatum, 170
    Viburnum acerifolium, 170, 171
    Viburnum alnifolium, 170
    Viburnum cassinoides, 170
    Viburnum dentatum, 170, 171
    Viburnum lantanoides, 170
    Viburnum lentago, 162–163
    Viburnum nudum, 164–165
    Viburnum nudum var. *cassinoides*, 170
    Viburnum opulus var.
        americanum, 170
    Viburnum prunifolium, 166–167
    Viburnum rufidulum, 168–169
    Viburnum trilobum, 170
    wild raisin, 170
    witch hobble, 170
    witherod, 170
Carolina ash, 396–397
Carolina buckthorn, 412–413
Carolina hemlock, 56, 58–59
Carolina holly, 114
Carolina laurel cherry, 432–433
Carolina red maple, 82
Carolina silverbell, 488–489
Carpinus caroliniana, 140–141
Carpinus caroliniana pistillate
    flowers, 125
Carpinus caroliniana staminate
    flowers, 125
Carya alba, 336
Carya aquatica, 316–317
Carya carolinae-septentrionalis, 330
Carya cordiformis, 318–319
Carya glabra, 320–321
Carya illinoinensis, 322–323
Carya illinoinensis staminate flowers, 315
Carya laciniosa, 324–325
Carya myristiciformis, 326–327
Carya myristiciformis staminate
    flowers, 315
Carya ovalis, 328–329
Carya ovata, 330–331
Carya pallida, 332–333
Carya texana, 334–335
Carya tomentosa, 336–337
Carya tomentosa staminate flowers, 315
Cashew family, 90–101. *See also*
    Anacardiaceae
Castanea dentata, 222–223
Castanea mollissima, 224
Castanea pumila, 224–225
Cat spruce, 22–23
Catalpa bignonioides, 144–145, 146
Catalpa speciosa, 146–147
Cedar elm, 516–517
Cedar family, 2–13. *See also Cupressaceae*
Cedar pine, 36–37
Celastraceae, 65, 172–173
Celtis laevigata, 504–505
Celtis laevigata var. *laevigata*, 504–505
Celtis occidentalis, 506–507
Celtis occidentalis flowers, 503
Celtis tenuifolia, 508–509
Cephalanthus occidentalis, 444
Cercis canadensis, 150–151
Cercis canadensis flowers, 149
Chalk maple, 74–75
Chamaecyparis thyoides, 4–5
Chameacyparis thyoides bark, 3
Chapman, Alan, 236
Chapman oak, 236–237
Cherry birch, 132–133
Cherry laurel, 432–433
Cherrybark oak, 280–281
Chestnut oak, 286–287
Chestnuts, 218
Chickasaw plum, 430
Chinese chestnut, 224
Chinese mulberry, 376–377
Chinese privet, 406–407
Chinese umbrella, 366–367
Chinkapin oak, 272–273
Chinquapin, 224–225
Chionanthus virginicus, 390–391
Chittamwood, 92–93
Chokecherry, 440–441
Christmas holly, 118–119, 120–121
Cigar-tree, 144–145, 146–147
Cinnamon wood, 343, 348–349
Cirillo, Domenico, 198
Clifton, William, 196
Coastal Plain willow, 466
Common alder, 128–129
Common apple, 426–427
Common buttonbush, 444–445
Common catalpa, 144–145, 146–147
Common chokecherry, 440–441
Common elder, 160–161
Common hackberry, 506–507
Common hoptree, 450–451
Common juniper, 6
Common mulberry, 376–377
Common pawpaw, 102–103
Common prickly-ash, 452–453
Common smoketree, 92–93
Common winterberry, 114
Cork elm, 512–513, 522–523
Cornacea, 174–195
    alternate-leaf dogwood, 176–177
    blackgum, 194–195
    Cornus alternifolia, 176–177
    Cornus amomum, 178–179
    Cornus drummondii, 180–181
    Cornus florida, 182–183
    Cornus racemosa, 184–185
    Cornus rugosa, 184
    Cornus stolonifera, 184, 185
    Cornus stricta, 186–187
    flowering dogwood, 182–183
    gray dogwood, 184–185
    Nyssa aquatica, 188–189
    Nyssa biflora, 190–191
    Nyssa ogeche, 192–193
    Nyssa sylvatica, 194–195
    Ogeeche tupelo, 192–193
    overview, 174
    red-osier dogwood, 184, 185
    roundleaf dogwood, 184
    silky dogwood, 178–179

swamp dogwood, 186–187
swamp tupelo, 190–191
water tupelo, 188–189
Cornus alternifolia, 176–177
Cornus amomum, 178–179
Cornus drummondii, 180–181
Cornus florida, 182–183
Cornus florida pistillate flowers, 69
Cornus foemina, 186
Cornus racemosa, 184–185
Cornus rugosa, 184
Cornus sericea, 184
Cornus stolonifera, 184, 185
Cornus stricta, 186–187
Cotinus obovatus, 92–93
Cotinus obovatus twig, 91
Cotton-gum, 188–189
Cow oak, 270–271
Crack willow, 468
Crataegus spp., 420–421
Cuban pine, 34–35
Cucumber magnolia, 354–355
Cucumbertree, 354–355
Cupressaceae, 2–13
    ashe juniper, 6
    Atlantic white-cedar, 4–5
    baldcypress, 8–9
    Chamaecyparis thyoides, 4–5
    common juniper, 6
    eastern redcedar, 6–7
    Juniperus ashei, 6
    Juniperus communis, 6
    Juniperus virginiana, 6–7
    Juniperus virginiana var. *silicicola*, 6
    northern white-cedar, 12–13
    overview, 2
    pondcypress, 10–11
    southern red-cedar, 6
    Taxodium distichum var. *distichum*, 8–9
    Taxodium distichum var. *imbricarium*, 10–11
    Thuja occidentalis, 12–13
Custard apple, 66, 102–103
Cypress family, 2–13. *See also Cupressaceae*
Cyrilla, 196–199
Cyrillaceae, 196–199

Dahoon, 108–109
Darlington oak, 250–251, 260–261
Deciduous holly, 112–113
Definitions (glossary), 524–526
Delta post oak, 292
Devilwood, 408–409
Diamond leaf oak, 250–251, 260–261
Dogwood family, 174–195. *See also Cornacea*
Downy serviceberry, 416–417
Drummond, Thomas, 180
Drummond red maple, 82

Durand, Elias, 240
Durand oak, 240–241
Durand white oak, 240–241
Dwarf black oak, 252–253
Dwarf chestnut, 224–225
Dwarf hackberry, 508–509
Dwarf pawpaw, 102–103
Dwarf post oak, 266–267
Dwarf sumac, 94–95
Dye bush, 494–495

Ear-leaved magnolia, 356–357
Eastern balsam poplar, 458
Eastern chokecherry, 440–441
Eastern cottonwood, 457, 460–461
Eastern fir, 16–17
Eastern hemlock, 56–57
Eastern hophornbeam, 142–143
Eastern larch, 20
Eastern red oak, 288–289
Eastern redbud, 150–151
Eastern redcedar, 6–7
Eastern spruce, 26–27
Eastern swampprivet, 392–397
Eastern white-cedar, 12–13
Eastern white pine, 48–49
Ebenaceae, 66, 200–201
Ebony, 66, 200–201
Elderberry, 160–161
Elliott, Stephen, 34
Elm family, 502–523. *See also Ulmaceae*
Empress-tree, 478–479
Ericaceae, 202–211
    Kalmia latifolia, 204–205
    mountain laurel, 204–205
    overview, 202
    Oxydendrum arboreum, 206–207
    purple rhododendron, 208, 209
    Rhododendron catawbiense, 208, 209
    Rhododendron maximum, 208–209
    rosebay rhododendron, 208–209
    sourwood, 206–207
    sparkleberry, 210–211
    Vaccinium arboreum, 210–211
Euonymus atropurpurea, 172–173
Euphorbiaceae, 66, 212–213
European mountain-ash, 442
Evergreen bayberry, 384–385
Evergreen holly, 118–119, 120–121
Evergreen magnolia, 358–359

Fabaceae, 214–217
Fagaceae, 218–299
    Allegheny chinkapin, 224–225
    American beech, 218, 226–227
    American chestnut, 222–223
    bear oak, 252–253
    black oak, 296–297
    blackjack oak, 268–269
    bluejack oak, 256–257

bluff oak, 232–233
bur oak, 264–265
Castanea dentata, 222–223
Castanea mollissima, 224
Castanea pumila, 224–225
Chapman oak, 236–237
cherrybark oak, 280–281
chestnut oak, 286–287
chestnuts, 218
Chinese chestnut, 224
chinkapin oak, 272–273
Delta post oak, 292
Durand oak, 240–241
Fagus grandifolia, 226–227
Georgia oak, 248–249
laurel oak, 250–251
live oak, 298–299
lobed red oaks, 218
lobed white oaks, 219
myrtle oak, 274–275
northern pin oak, 242–243
northern red oak, 288–289
Nuttall oak, 294–295
Oglethorpe oak, 278–279
overcup oak, 262–263
overview, 218–220
pin oak, 282–283
post oak, 292–293
Quercus acutissima, 228–229
Quercus alba, 230–231
Quercus austrina, 232–233
Quercus bicolor, 234–235
Quercus buckleyi, 294
Quercus chapmanii, 236–237
Quercus coccinea, 238–239
Quercus durandii, 240–241
Quercus ellipsoidalis, 242–243
Quercus falcata, 244–245
Quercus geminata, 246–247
Quercus georgiana, 248–249
Quercus hemisphaerica, 250–251
Quercus ilicifolia, 252–253
Quercus imbricaria, 254–255
Quercus incana, 256–257
Quercus laevis, 258–259
Quercus laurifolia, 260–261
Quercus lyrata, 262–263
Quercus macrocarpa, 264–265
Quercus margaretta, 266–267
Quercus marilandica, 268–269
Quercus michauxxi, 270–271
Quercus muehlenbergii, 272–273
Quercus myrtifolia, 274–275
Quercus nigra, 276–277
Quercus oglethorpensis, 278–279
Quercus pagoda, 280–281
Quercus palustris, 282–283
Quercus phellos, 284–285
Quercus prinus, 286–287
Quercus rubra, 288–289
Quercus shumardii, 290–291
Quercus similis, 292

Quercus stellata, 292–293
Quercus texana, 294–295
Quercus velutina, 296–297
Quercus virginiana, 298–299
  sand live oak, 246–247
  sand post oak, 266–267
  sawtooth oak, 228–229
  scarlet oak, 238–239
  shingle oak, 254–255
  Shumard oak, 290–291
  southern red oak, 244–245
  swamp chestnut oak, 270–271
  swamp laurel oak, 260–261
  swamp white oak, 234–235
  Texas Shumard oak, 294
  turkey oak, 258–259
  unlobed red oaks, 219
  unlobed white oaks, 219
  water oak, 276–277
  white oak, 230–231
  willow oak, 284–285
Fagus grandifolia, 226–227
Fall plum, 434–435
False acacia, 216–217
False pignut hickory, 328–329
Farkleberry, 210–211
Fetid buckeye, 308–309
Fever-tree, 446–447
Fiddle oak, 276–277
Figwort family, 65, 478–479
Fire cherry, 436–437
Fish-bait tree, 144–145, 146
Flameleaf sumac, 94–95
Flatwoods plum, 430
Florida anise, 312–313
Florida aspen, 212–213
Florida maple, 72–73
Florida slash pine, 34
Florida torreya, 62–63
Florida yew, 60–61
Flowering ash, 390–391
Flowering cornel, 182–183
Flowering dogwood, 182–183
Forestiera acuminata, 392–397
Fragrant olive, 408–409
Fraser, John, 18
Fraser fir, 18–19
Fraser magnolia, 356–357
Fraxinus americana, 389, 394–395
Fraxinus caroliniana, 396–397
Fraxinus nigra, 398–399
Fraxinus pennsylvanica, 389, 400–401
Fraxinus profunda, 402–403
Fraxinus quadrangulata, 404–405
Fringe-tree, 390–391

Georgia-bark, 446–447
Georgia hackberry, 508–509
Georgia oak, 248–249
Gleditsch, Johann, 152, 154
Gleditsia aquatica, 152–153

Gleditsia triacanthos, 154–155
Gleditsia triacanthos flowers, 149
Gleditsia triacanthos fruit, 149
Gleditsia triacanthos leaves, 149
Glossary, 524–526
Golden aspen, 464
Golden elm, 343, 348–349
Gopherwood, 62–63, 214–215
Gordon, James, 496
Gordonia, 496–497
Gordonia lasianthus, 496–497
Gray alder, 126–127
Gray-beard, 390–391
Gray birch, 130–131, 138–139
Gray dogwood, 184–185
Gray oak, 288–289
Gray pine, 28–29
Gray-stemmed dogwood, 184–185
Great laurel, 208–209
Great laurel magnolia, 358–359
Great rhododendron, 208–209
Green ash, 389, 400–401
Gulf-cypress, 8–9
Gum bully, 471, 472–473
Gum bumelia, 471, 472–473
Gymnocladus dioicus, 156–157
Gymnocladus dioicus flowers, 149
Gymnosperm families, 1–63
  cedar, 2–13
  Cupressaceae, 2–13. *See also Cupressaceae*
  cypress, 2–13
  overview, 1
  Pinaceae, 14–59. *See also Pinaceae*
  pine, 14–59
  Taxaceae, 60–63. *See also Taxaceae*
  yew, 60–63

Hackberry, 506–507
Hales, Stephen, 486, 488
Halesia carolina, 488
Halesia diptera, 486–487
Halesia diptera twig, 485
Halesia tetraptera, 488–489
Halesia tetraptera twig, 485
Hamamelidaceae, 66, 300–303
Hamamelis virginiana, 300
Hard elm, 512–513
Hard maple, 78–79, 86–87
Hard pine, 38–39
Haws, 420–421
Hawthorns, 420–421
Hazel alder, 128–129
Heart pine, 38–39
Heath family, 202–211. *See also Ericaceae*
Hedge-apple, 374–375
Hemlock spruce, 56–57
Hercules'-club, 454–455
Hercules-club, 122–123
Hickory elm, 522–523

Hickory family, 314–341. *See also Juglandaceae*
Hickory pine, 40–41
Highbush cranberry, 170
Hill's oak, 242–243
Hippocastanaceae, 304–311
  Aesculus flava, 306–307
  Aesculus glabra, 308–309
  Aesculus parviflora, 310
  Aesculus pavia, 310–311
  Aesculus sylvatica, 310
  bottlebrush buckeye, 310
  Ohio buckeye, 308–309
  overview, 304
  painted buckeye, 310
  red buckeye, 310–311
  yellow buckeye, 306–307
Hoary alder, 126–127
Hobblebush, 170
Hog plum, 430
Holly family, 104–121. *See also Aquifoliaceae*
Honey-balls, 444–445
Honeylocust, 154–155
Honeysuckle family, 158–171. *See also Caprifoliaceae*
Hoop ash, 398–399
Hophornbeam, 142–143
Hornbeam, 140–141
Horse-sugar, 494–495

Ilex ambigua, 114
Ilex amelanchier, 106–107
Ilex cassine, 108–109
Ilex coriacea, 110–111
Ilex decidua, 112–113
Ilex glabra, 110
Ilex glabra pistillate flowers, 105
Ilex montana, 114–115
Ilex myrtifolia, 116–117
Ilex opaca, 118–119
Ilex opaca pistillate flowers, 105
Ilex opaca staminate flowers, 105
Ilex verticillata, 114, 115
Ilex vomitoria, 120–121
Illiciaceae, 312–313
Indian-bean, 144–145, 146–147
Indian cherry, 412–413
Inkberry, 110
Iron oak, 292–293
Ironwood, 140–141, 142–143, 471, 476–477

Jack oak, 242–243, 268–269
Jack pine, 28–29
Jersey pine, 54–55
Judas-tree, 150–151
Juglandaceae, 314–341
  bitternut hickory, 318–319
  black hickory, 334–335

black walnut, 340–341
butternut, 338–339, 340
Carya aquatica, 316–317
Carya carolinae-septentrionalis, 330
Carya cordiformis, 318–319
Carya glabra, 320–321
Carya illinoinensis, 322–323
Carya laciniosa, 324–325
Carya myristiciformis, 326–327
Carya ovalis, 328–329
Carya ovata, 330–331
Carya pallida, 332–333
Carya texana, 334–335
Carya tomentosa, 336–337
Juglans cinerea, 338–339, 340
Juglans nigra, 340–341
mockernut hickory, 336–337
nutmeg hickory, 326–327
overview, 314
pecan, 322–323
pignut hickory, 320–321
red hickory, 328–329
sand hickory, 332–333
shagbark hickory, 330–331
shellbark hickory, 324–325
southern shagbark hickory, 330
water hickory, 316–317
Juglans cinerea, 338–339, 340
Juglans nigra, 340–341
Juneberry, 416–417
Juniperus ashei, 6
Juniperus communis, 6
Juniperus communis bark, 3
Juniperus communis leaves, 3
Juniperus virginiana, 6–7
Juniperus virginiana bark, 3
Juniperus virginiana var. *silicicola*, 6

Kalm, Pehr, 204
Kalmia latifolia, 204–205
Kentucky coffeetree, 156–157
Kingnut hickory, 324–325
Knotty pine, 44–45

Large gallberry, 110–111
Large-leaved cucumbertree, 360–361
Largetooth aspen, 462
Larix laricina, 20–21
Lauraceae, 342–349
 overview, 342
 Persea borbonia, 344–345
 Persea palustris, 346–347
 redbay, 344–345
 sassafras, 343, 348–349
 Sassafras albidum, 343, 348–349
 swamp redbay, 346–347
Laurel cherry, 432–433
Laurel family, 342–349. *See also*
 Lauraceae

Laurel oak, 250–251, 254–255, 260–261
Le-Forestier, Charles, 392
Legume, 214–217
Legume family, 148–157. *See also*
 Caesalpiniaceae
Ligustrum sinense, 406–407
Lily-of-the-valley-tree, 206–207
Linden, 500–501
Linden family, 67, 500–501
Liquidambar styraciflua, 302
Liriodendron tulipifera, 352–353
Little shellbark hickory, 330–331
Live oak, 298–299
Lobed red oaks, 218
Lobed white oaks, 219
Loblolly-bay, 496–497
Loblolly pine, 52–53
Longleaf pine, 38–39
Longstraw pine, 38–39
Low maple, 88–89

Maclure, William, 374
Madder family, 65, 444–447
Magnol, Peter, 354, 358, 362, 364
Magnolia acuminata, 354–355
Magnolia acuminata twig, 351
Magnolia ashei, 360, 361
Magnolia family, 350–365. *See also*
 Magnoliaceae
Magnolia fraseri, 356–357
Magnolia fraseri twig, 351
Magnolia grandiflora, 358–359
Magnolia grandiflora twig, 351
Magnolia macrophylla, 360–361
Magnolia macrophylla twig, 351
Magnolia pyramidata, 356, 357
Magnolia tripetala, 362–363
Magnolia virginiana, 364–365
Magnoliaceae, 350–365
 ashe magnolia, 360, 361
 bigleaf magnolia, 360–361
 cucumbertree, 354–355
 Fraser magnolia, 356–357
 Liriodendron tulipifera, 352–353
 Magnolia acuminata, 354–355
 Magnolia ashei, 360, 361
 Magnolia fraseri, 356–357
 Magnolia grandiflora, 358–359
 Magnolia macrophylla, 360–361
 Magnolia pyramidata, 356, 357
 Magnolia tripetala, 362–363
 Magnolia virginiana, 364–365
 overview, 350
 pyramid magnolia, 356, 357
 southern magnolia, 358–359
 sweetbay magnolia, 364–365
 umbrella magnolia, 362–363
 yellow-poplar, 352–353
Mahogany, 68, 366–367

Malus angustifolia, 422–423
Malus coronaria, 424–425
Malus pumila, 426–427
Malus sylvestris, 426
Manitoba maple, 76–77
Maple family, 70–89. *See also* Aceraceae
Maple-leaved viburnum, 170, 171
Marsh pine, 46–47
Meadow holly, 112–113
Meadow pine, 52–53
Meliaceae, 68, 366–367
Mexican plum, 434–435
Michaux, F., 270
Mimosa, 67, 368–369
Mimosaceae, 67, 368–369
Mist tree, 92–93
Mockernut hickory, 336–337
Moose maple, 80–81, 88–89
Moosewood, 80–81
Moraceae, 370–379
 Broussonetia papyrifera, 372–373
 Maclura pomifera, 374–375
 Morus alba, 376–377
 Morus rubra, 371, 378–379
 osage-orange, 374–375
 overview, 370
 paper-mulberry, 372–373
 red mulberry, 371, 378–379
 white mulberry, 376–377
Morella caroliniensis, 384
Morella cerifera, 382
Morella inodora, 384
Morella pensylvanica, 386
Mossy-overcup oak, 264–265
Mossycup oak, 264–265
Mountain-ash, 442–443
Mountain-camellia, 498–499
Mountain holly, 114–115
Mountain ivy, 204–205
Mountain laurel, 204–205
Mountain magnolia, 354–355, 356–357
Mountain maple, 88–89
Mountain pine, 40–41
Mountain rosebay, 208, 209
Mountain silver bell, 488–489
Mountain stewartia, 498–499
Mountain winterberry, 114–115
Mulberry family, 370–379. *See also*
 Moraceae
Musclewood, 140–141
Myrica cerifera, 381–383
Myrica heterophylla, 384–385
Myrica inodora, 384
Myrica pensylvanica, 386–387
Myricaceae, 380–387
 Myrica cerifera, 381–383
 Myrica heterophylla, 384–385
 Myrica inodora, 384
 Myrica pensylvanica, 386–387
 northern bayberry, 386–387
 odorless bayberry, 384

southern bayberry, 381–383
swamp candleberry, 384–385
Myrtle dahoon, 116–117
Myrtle-leaved holly, 116–117
Myrtle oak, 274–275

Nannyberry, 162–163
Nannyplum, 162–163
Narrow-leaved crabapple, 422–423
Negundo maple, 76–77
Northern bayberry, 386–387
Northern catalpa, 146–147
Northern hackberry, 506–507
Northern hemlock, 56–57
Northern mountain-ash, 442
Northern pin oak, 242–243
Northern prickly ash, 452–453
Northern red oak, 288–289
Northern white-cedar, 12–13
Northern white pine, 48–49
Norway maple, 86
Norway pine, 42–43
Norway spruce, 22
Nutmeg hickory, 326–327
Nuttall's oak, 294–295
Nyssa aquatica, 188–189
Nyssa biflora, 190–191
Nyssa biflora twig, 175
Nyssa ogeche, 192–193
Nyssa ogeche leaves, 175
Nyssa ogeche twig, 175
Nyssa sylvatica, 194–195
Nyssa sylvatica twig, 175

Oak family, 218–299. *See also Fagaceae*
Odorless bayberry, 384
Ogeeche tupelo, 192–193
Ogeechee-lime, 192–193
Oglethorpe oak, 278–279
Ohio buckeye, 308–309
Oil nut, 338–339, 340
Old-field birch, 138–139
Old-field pine, 52–53
Old-man's-beard, 390–391
Oleaceae, 388–409
  black ash, 398–399
  blue ash, 404–405
  Carolina ash, 396–397
  Chinese privet, 406–407
  Chionanthus virginicus, 390–391
  devilwood, 408–409
  Forestiera acuminata, 392–397
  Fraxinus americana, 389, 394–395
  Fraxinus caroliniana, 396–397
  Fraxinus nigra, 398–399
  Fraxinus pennsylvanica, 389, 400–401
  Fraxinus profunda, 402–403
  Fraxinus quadrangulata, 404–405

  fringe-tree, 390–391
  green ash, 389, 400–401
  Ligustrum sinense, 406–407
  Osmanthus americanus, 408–409
  overview, 388
  pumpkin ash, 402–403
  swamp-privet, 392–397
  white ash, 389, 394–395
Olive family, 388–409. *See also Oleaceae*
Opossum oak, 276–277
Osmanthus americanus, 408–409
Ostrya virginiana, 142–143
Ostrya virginiana staminate and pistillate flowers, 125
Overcup oak, 262–263
Oxydendrum arboreum, 206–207
Oxydendrum arboreum bark, 203
Oxydendrum arboreum twig, 203

Pagoda dogwood, 176–177
Painted buckeye, 310
Pale dogwood, 178–179
Pale hickory, 332–333
Pale-leaf hickory, 332–333
Panicled dogwood, 184–185
Paper birch, 136–137
Paulownia tomentosa, 478–479
Pawpaw, 66, 102–103
Peach oak, 284–285
Peachleaf willow, 468
Pecan, 322–323
Pecan hickory, 316–317, 318–319, 322–323
Pencil cedar, 6–7
Pepperidge, 194–195
Persea borbonia, 344–345
Persea palustris, 346–347
Picea abies, 22
Picea glauca, 22–23
Picea mariana, 24–25
Picea rubens, 26–27
Pignut hickory, 320–321
Pin cherry, 436–437
Pin oak, 282–283, 284–285, 294–295
Pinaceae, 14–59
  Abies balsamea, 16–17
  Abies fraseri, 18–19
  balsam fir, 16–17
  black spruce, 24–25
  Carolina hemlock, 56, 58–59
  eastern hemlock, 56–57
  eastern white pine, 48–49
  Fraser fir, 18–19
  jack pine, 28–29
  Larix laricina, 20–21
  loblolly pine, 52–53
  longleaf pine, 38–39
  Norway spruce, 22
  Picea abies, 22
  Picea glauca, 22–23

  Picea mariana, 24–25
  Picea rubens, 26–27
  Pinus banksiana, 28–29
  Pinus clausa, 30–31
  Pinus echinata, 32–33
  Pinus elliottii, 34–35
  Pinus glabra, 36–37
  Pinus palustris, 38–39
  Pinus pungens, 40–41
  Pinus resinosa, 42–43
  Pinus rigida, 44–45
  Pinus serotina, 46–47
  Pinus strobus, 48–49
  Pinus sylvestris, 50–51
  Pinus taeda, 52–53
  Pinus virginiana, 54–55
  pitch pine, 44–45
  pond pine, 46–47
  red pine, 42–43
  red spruce, 26–27
  sand pine, 30–31
  Scotch pine, 50–51
  shortleaf pine, 32–33
  slash pine, 34–35
  spruce pine, 36–37
  Table Mountain pine, 40–41
  tamarack, 20–21
  Tsuga canadensis, 56–57
  Tsuga caroliniana, 56, 58–59
  Virginia pine, 54–55
  white spruce, 22–23
Pinckney, Charles, 446
Pinckneya, 446–447
Pinckneya bracteata, 446–447
Pine family, 14–59. *See also Pinaceae*
Pinus banksiana, 28–29
Pinus clausa, 30–31
Pinus echinata, 32–33
Pinus elliottii, 34–35
Pinus elliottii var. *densa*, 34
Pinus glabra, 36–37
Pinus glabra pollen cones, 15
Pinus palustris, 38–39
Pinus palustris pollen cones, 15
Pinus pungens, 40–41
Pinus resinosa, 42–43
Pinus rigida, 44–45
Pinus serotina, 46–47
Pinus strobus, 48–49
Pinus sylvestris, 50–51
Pinus taeda, 52–53
Pinus taeda pollen cones, 15
Pinus virginiana, 54–55
Pinus virginiana pollen cones, 15
Pitch pine, 34–35, 38–39, 44–45
Planer, Johann, 510
Planera aquatica, 510–511
Planertree, 510–511
Platanaceae, 65, 410–411
Platanus occidentalis, 410–411
Pocosin pine, 46–47

Poison-elderberry, 100–101
Poison-sumac, 100–101
Pond pine, 46–47
Pondcypress, 10–11
Pop ash, 396–397
Popcorn tree, 212–213
Poplar family, 456–469. *See also* Salicaceae
Populus balsamifera, 458
Populus balsamifera spp. *balsamifera*, 458
Populus deltoides, 457, 460–461
Populus grandidentata, 462
Populus heterophylla, 460, 461
Populus tremuloides, 464
Possumhaw, 112–113, 164–165
Possumwood, 200
Post oak, 292–293
Poverty pine, 40–41
Prickly-ash, 122–123
Prickly pine, 40–41
Pride of India, 366–367
Princess-tree, 478–479
Prunus americana, 428–429
Prunus angustifolia, 430
Prunus angustifolia twig, 415
Prunus caroliniana, 432–433
Prunus mexicana, 434–435
Prunus nigra, 428
Prunus pensylvanica, 436–437
Prunus serotina, 438–439
Prunus serotina twig, 415
Prunus umbellata, 430
Prunus virginiana, 440–441
Ptelea trifoliata, 450–451
Pumpkin ash, 402–403
Purple rhododendron, 208, 209
Pussy willow, 468
Pyramid magnolia, 356, 357

Quaking aspen, 464
Quercitron oak, 296–297
Quercus acutissima, 228–229
Quercus alba, 230–231
Quercus alba staminate flowers, 221
Quercus arkansana, 268
Quercus austrina, 232–233
Quercus bicolor, 234–235
Quercus buckleyi, 294
Quercus chapmanii, 236–237
Quercus coccinea, 238–239
Quercus durandii, 240–241
Quercus ellipsoidalis, 242–243
Quercus falcata, 244–245
Quercus geminata, 246–247
Quercus georgiana, 248–249
Quercus hemisphaerica, 250–251
Quercus ilicifolia, 252–253
Quercus imbricaria, 254–255
Quercus incana, 256–257

Quercus laevis, 258–259
Quercus laurifolia, 260–261
Quercus lyrata, 262–263
Quercus macrocarpa, 264–265
Quercus margaretta, 266–267
Quercus marilandica, 268–269
Quercus michauxxi, 270–271
Quercus montana, 286
Quercus muehlenbergii, 272–273
Quercus muehlenbergii pistillate flowers, 221
Quercus myrtifolia, 274–275
Quercus nigra, 276–277
Quercus nigra pistillate flowers, 69, 221
Quercus oglethorpensis, 278–279
Quercus pagoda, 280–281
Quercus palustris, 282–283
Quercus phellos, 284–285
Quercus phellos staminate flowers, 221
Quercus prinus, 286–287
Quercus rubra, 288–289
Quercus shumardii, 290–291
Quercus similis, 292
Quercus stellata, 292–293
Quercus texana, 294–295
Quercus velutina, 296–297
Quercus virginiana, 298–299
Quercus virginiana staminate flowers, 221

Red ash, 389, 400–401, 402–403
Red birch, 134–135
Red buckeye, 310–311
Red-cypress, 8–9
Red elm, 518–519, 520–521
Red hickory, 328–329, 334–335
Red juniper, 6–7
Red maple, 82–83
Red oak, 238–239, 280–281, 288–289, 290–291
Red-osier dogwood, 184, 185
Red pine, 42–43
Red plum, 428–429
Red river oak, 294–295
Red spruce, 26–27
Redbay, 344–345
Redbud, 150–151
Redcedar, 6–7
Redgum, 302
Rhamnaceae, 67, 412–413
Rhamnus caroliniana, 412–413
Rhododendron catawbiense, 208, 209
Rhododendron maximum, 208–209
Rhododendron maximum flower bud, 203
Rhus copallina, 94–95
Rhus copallina twig, 91
Rhus copallinum, 94
Rhus glabra, 96–97
Rhus glabra form, 91
Rhus glabra twig, 91

Rhus hirta, 98
Rhus typhina, 98–99
Rhus typhina twig, 91
River birch, 134–135
River maple, 84–85
Robin, Jean and Vespasion, 216
Rock chestnut oak, 286–287
Rock elm, 522–523
Rock maple, 86–87
Rock oak, 272–273, 286–287
Rosaceae, 414–443
  allegheny serviceberry, 418–419
  Amelanchier arborea, 416–417
  Amelanchier laevis, 418–419
  American plum, 428–429
  black cherry, 438–439
  Canada plum, 428
  Carolina laurel cherry, 432–433
  chickasaw plum, 430
  chokecherry, 440–441
  common apple, 426–427
  Crataegus spp., 420–421
  downy serviceberry, 416–417
  European mountain-ash, 442
  flatwoods plum, 430
  hawthorns, 420–421
  Malus angustifolia, 422–423
  Malus coronaria, 424–425
  Malus pumila, 426–427
  Mexican plum, 434–435
  mountain-ash, 442–443
  northern mountain-ash, 442
  overview, 414
  pin cherry, 436–437
  Prunus americana, 428–429
  Prunus angustifolia, 430
  Prunus caroliniana, 432–433
  Prunus mexicana, 434–435
  Prunus nigra, 428
  Prunus pensylvanica, 436–437
  Prunus serotina, 438–439
  Prunus umbellata, 430
  Prunus virginiana, 440–441
  Sorbus americana, 442–443
  Sorbus aucuparia, 442
  Sorbus decora, 442
  southern crabapple, 422–423
  sweet crabapple, 424–425
Rose family, 414–443. *See also Rosaceae*
Rosebay, 208–209
Rosebay rhododendron, 208–209
Roundleaf dogwood, 184
Royal paulownia, 478–479
Rubiaceae, 65, 444–447
Rue family, 448–455. *See also Rutaceae*
Rum cherry, 438–439
Russian mulberry, 376–377
Rusty blackhaw, 168–169
Rusty blackhaw viburnum, 168–169
Rusty nannyberry, 168–169

Rutaceae, 448–455
    common hoptree, 450–451
    common prickly-ash, 452–453
    Hercules'-club, 454–455
    overview, 448
    Ptelea trifoliata, 450–451
    Zanthoxylum americanum, 452–453
    Zanthoxylum clava-herculis, 454–455

Salicaceae, 456–469
    almond-leaved willow, 468
    balsam poplar, 458
    bebb willow, 468
    bigtooth aspen, 462
    black willow, 457
    Coastal Plain willow, 466
    crack willow, 468
    eastern cottonwood, 457
    overview, 456
    peachleaf willow, 468
    Populus balsamifera, 458
    Populus deltoides, 457
    Populus grandidentata, 462
    Populus heterophylla, 460
    Populus tremuloides, 464
    pussy willow, 468
    Salix alba, 468
    Salix amygdaloides, 468
    Salix bebbiana, 468
    Salix caroliniana, 466
    Salix discolor, 468
    Salix fragilis, 468
    Salix lucida, 468
    Salix nigra, 457
    shining willow, 468
    swamp cottonwood, 460
    trembling aspen, 464
    white willow, 468
Salix alba, 468
Salix amygdaloides, 468
Salix bebbiana, 468
Salix caroliniana, 466
Salix discolor, 468, 469
Salix fragilis, 468, 469
Salix lucida, 468, 469
Salix nigra, 457, 466–467
Sambucus canadensis, 160–161
Sambucus nigra spp. *canadensis*, 160
Sambucus pubens, 160
Sambucus pubens fruit, 159
Sand hickory, 332–333
Sand holly, 114
Sand jack oak, 256–257
Sand live oak, 246–247
Sand pine, 30–31
Sand post oak, 266–267
Sapgum, 302
Sapodilla, 470–477. See also *Sapotaceae*
Sapotaceae, 470–477
    buckthorn bumelia, 474–475

Bumelia lanuginosa, 471, 472–473
Bumelia lycioides, 474–475
Bumelia tenax, 471, 476–477
    gum bumelia, 471, 472–473
    overview, 470
    tough bumelia, 471, 476–477
Sarvis holly, 106–107
Sassafras, 343, 348–349
Sassafras albidum, 343, 348–349
Satin walnut, 302
Sawtooth oak, 228–229
Scalybark hickory, 330–331
Scarlet elder, 160
Scarlet maple, 82–83
Scarlet oak, 238–239
Scotch pine, 50–51
Scots pine, 50–51
Scrophulariaceae, 65, 478–479
Scrub chestnut oak, 272–273
Scrub oak, 252–253, 256–257, 258–259
Scrub pine, 28–29, 30–31, 54–55
Scrubby post oak, 266–267
September elm, 520–521
Shadbush, 416–417
Shagbark hickory, 330–331
Sheepberry, 162–163
Shellbark hickory, 324–325
Shingle oak, 254–255
Shining sumac, 94–95
Shinning willow, 468
Shortleaf pine, 32–33
Shortstraw pine, 32–33
Showy mountain-ash, 442
Shumard oak, 290–291
Sideroxylon lanuginosum, 471, 472–473
Sideroxylon lycioides, 474
Sideroxylon tenax, 476
Silk-tree, 368–369
Silkworm mulberry, 376–377
Silky camellia, 498
Silky cornel, 178–179
Silky dogwood, 178–179
Silver birch, 130–131
Silver maple, 84–85
Silverleaf maple, 84–85
Simaroubaceae, 68, 480–481
Simmon, 200
Skunk spruce, 22–23
Slash pine, 34–35
Slippery elm, 518–519
Small-leaf viburnum, 170
Small-leaved elm, 512–513
Small viburnum, 170
Smoketree, 92–93
Smooth alder, 128–129
Smooth hickory, 320–321
Smooth juneberry, 418–419
Smooth serviceberry, 418–419
Smooth sumac, 96–97

Snakebark maple, 80–81
Snowdrop tree, 486–487, 488–489
Soft elm, 518–519
Soft maple, 82–83, 84–85
Soft pine, 48–49
Sorbus americana, 442–443
Sorbus aucuparia, 442
Sorbus decora, 442
Sorrel tree, 206–207
Sour tupelo, 192–193
Sourgum, 188–189, 194–195
Sourwood, 206–207
South Florida slash pine, 34
Southern balsam fir, 18–19
Southern bayberry, 381–383
Southern catalpa, 144–145, 146
Southern cottonwood, 457, 460–461
Southern crab, 422–423
Southern crabapple, 422–423
Southern-cypress, 8–9
Southern hackberry, 504–505
Southern magnolia, 358–359
Southern pine, 34–35
Southern prickly-ash, 454–455
Southern red-cedar, 6
Southern red oak, 244–245
Southern shagbark hickory, 330
Southern sugar maple, 72–73
Southern white-cedar, 4–5
Southern yellow pine, 32–33, 34–35, 38–39
Spanish oak, 238–239, 244–245
Sparkleberry, 210–211
Speckled alder, 126–127
Spindletree, 172–173
Spruce pine, 30–31, 36–37
Spurge, 66, 212–213
Staff-tree, 172–173
Staghorn sumac, 98–99
Staphylea trifolia, 482–483
Staphyleaceae, 65, 482–483
Stave oak, 230–231
Stewart, John, 498
Stewartia malacodendron, 498, 499
Stewartia ovata, 498–499
Stiff cornel, 186–187
Stiff dogwood, 186–187
Stink-bush, 312–313
Stinking ash, 450–451
Stinking buckeye, 308–309
Stinking-cedar, 62–63
Stone Mountain oak, 248–249
Storax family, 484–493. See also *Styracaceae*
Striped maple, 80–81
Striped oak, 294–295
Styracaceae, 484–493
    American snowbell, 490–491
    bigleaf snowbell, 492–493
    Carolina silverbell, 488–489

Halesia diptera, 486–487
Halesia tetraptera, 488–489
overview, 484
Styrax americanus, 490–491
Styrax grandifolius, 492–493
two-wing silverbell, 486–487
Styrax americanus, 490–491
Styrax americanus twig, 485
Styrax grandifolius, 492–493
Styrax grandifolius twig, 485
Sugar hackberry, 504–505
Sugar maple, 72–73, 86–87
Sugarberry, 504–505
Swamp ash, 389, 396–401
Swamp bay, 364–365
Swamp birch, 130–131
Swamp blackgum, 190–191
Swamp candleberry, 384–385
Swamp cedar, 4–5
Swamp-cedar, 12–13
Swamp chestnut oak, 270–271
Swamp cottonwood, 460, 461
Swamp dogwood, 178–179, 180–181, 186–187
Swamp hickory, 318–319, 326–327
Swamp holly, 112–113
Swamp laurel oak, 250–251, 260–261
Swamp magnolia, 364–365
Swamp maple, 82–83
Swamp oak, 282–283
Swamp pine, 34–35, 38–39
Swamp post oak, 262–263
Swamp-privet, 392–397
Swamp red oak, 280–281, 290–291
Swamp redbay, 346–347
Swamp Spanish oak, 282–283
Swamp spruce, 24–25
Swamp-sumac, 100–101
Swamp tupelo, 188–189, 190–191
Swamp white oak, 234–235
Swamp willow, 457, 466–467
Swamp willow oak, 284–285
Swampbay, 346–347
Swamphaw, 164–165
Sweet bay, 364–365
Sweet birch, 132–133
Sweet buckeye, 306–307
Sweet crabapple, 424–425
Sweet gallberry, 110
Sweet-locust, 154
Sweet maple, 86–87
Sweet pecan, 322–323
Sweet pignut, 320–321
Sweet viburnum, 162–163
Sweetbay magnolia, 364–365
Sweetgum, 302
Sweetleaf, 494–495
Sweetleaf family, 66, 494–495
Sycamore, 65, 410–411
Symplocaceae, 66, 494–495
Symplocos tinctoria, 494–495

Table Mountain pine, 40–41
Tacamahac, 458
Tag alder, 126–127, 128–129
Tallowtree, 212–213
Tamarack, 20–21
Tanbark oak, 286–287
Tapa cloth tree, 372
Taxaceae, 60–63
    Florida torreya, 62–63
    Florida yew, 60–61
    Taxus floridiana, 60–61
    Torreya taxifolia, 62–63
Taxodium ascendens, 10
Taxodium distichum, 8
Taxodium distichum var. *distichum*, 8–9
Taxodium distichum var. *distichum* bark, 3
Taxodium distichum var. *imbricarium*, 10–11, 193
Taxodium distichum var. *imbricarium* bark, 3
Taxus floridiana, 60–61
Terminology (glossary), 524–526
Texas hickory, 334–335
Texas red oak, 294–295
Texas Shumard oak, 294
Theaceae, 67, 496–499
Thomas, David, 522
Thorny-locust, 154
Three-leaved maple, 76–77
Three-thorned acacia, 154
Thuja occidentalis, 12–13
Thunderwood, 100–101
Tidewater cypress, 8–9
Tilia americana, 500–501
Tiliaceae, 67, 500–501
Toothache tree, 122–123, 452–453, 454–455
Torch pine, 52–53
Torrey, John, 62
Torreya taxifolia, 62–63
Tough buckthorn, 471, 476–477
Tough bully, 471, 476–477
Tough bumelia, 471, 476–477
Toxicodendron vernix, 100–101
Tree huckleberry, 210–211
Tree-of-heaven, 480–481
Tree sparkleberry, 210–211
Trembling aspen, 464
Trembling poplar, 464
Triadica sebifera, 212
Trident maple, 82
Trumpet creeper, 144–147. *See also* Bignoniaceae
Tsuga canadensis, 56–57
Tsuga caroliniana, 56, 58–59
Tulip-poplar, 352–353
Tulip tree, 352–353
Tupelo, 194–195
Tupelo-gum, 194–195
Turkey oak, 258–259, 292–293

Two-flowered tupelo, 190–191
Two-wing silverbell, 486–487

Ulmaceae, 502–523
    American elm, 514–515
    cedar elm, 516–517
    Celtis laevigata, 504–505
    Celtis occidentalis, 506–507
    Celtis tenuifolia, 508–509
    Georgia hackberry, 508–509
    hackberry, 506–507
    overview, 502
    Planera aquatica, 510–511
    rock elm, 522–523
    September elm, 520–521
    slippery elm, 518–519
    sugarberry, 504–505
    Ulmus alata, 512–513
    Ulmus americana, 514–515
    Ulmus crassifolia, 516–517
    Ulmus rubra, 518–519
    Ulmus serotina, 520–521
    Ulmus thomasii, 522–523
    water-elm, 510–511
    winged elm, 512–513
Ulmus alata, 512–513
Ulmus alata flowers, 503
Ulmus americana, 514–515
Ulmus americana flowers, 503
Ulmus crassifolia, 516–517
Ulmus rubra, 518–519
Ulmus rubra flowers, 503
Ulmus serotina, 520–521
Ulmus thomasii, 522–523
Umbrella magnolia, 362–363
Umbrella tree, 362–363
Unlobed red oaks, 219
Unlobed white oaks, 219

Vaccinium arboreum, 210–211
Vaccinium arboreum bark, 202
Velvet sumac, 98–99
Viburnum obovatum, 170
Viburnum obovatum leaves and fruit, 159
Viburnum acerifolium, 170, 171
Viburnum alnifolium, 170
Viburnum alnifolium leaf, 159
Viburnum cassinoides, 170
Viburnum cassinoides leaves, 159
Viburnum dentatum, 170, 171
Viburnum lantanoides, 170
Viburnum lentago, 162–163
Viburnum nudum, 164–165
Viburnum nudum var. *cassinoides*, 170
Viburnum opulus var. *americanum*, 170
Viburnum prunifolium, 166–167
Viburnum rufidulum, 168–169
Viburnum trilobum, 170
Virgilia, 214–215

Virginia live oak, 298–299
Virginia pine, 54–55
Virginia stewartia, 498

Wafer ash, 450–451
Wahoo, 172–173, 512–513
Walnut family, 314–341. *See also
    Juglandaceae*
Walter pine, 36–37
Water ash, 389, 396–401
Water birch, 134–135
Water elm, 510–511, 514–515
Water-gum, 188–189
Water hickory, 316–317
Water maple, 82–83, 84–85
Water oak, 250–251, 260–261, 276–277
Water tupelo, 188–189
Water white oak, 262–263
Waterlocust, 152–153
Wax-myrtle, 381–383, 386–387
Wax myrtle family, 380–387. *See also
    Myricaceae*
Weymouth pine, 48–49
Whistlewood, 80–81
White ash, 389, 394–395
White-bark maple, 74–75
White bay, 364–365

White birch, 136–137, 138–139
White-cedar, 4–5
White-cypress, 8–9
White elder, 160–161
White elm, 514–515
White fringetree, 390–391
White hickory, 320–321, 336–337
White holly, 118–119
White maple, 82–83, 84–85
White oak, 230–231
White pine, 48–49
White-poplar, 352–353
White spruce, 22–23
White titi, 198–199
White tupelo, 192–193
White walnut, 338–339, 340
White willow, 468
Whitewood, 352–353
Wild banana tree, 102–103
Wild black cherry, 438–439
Wild crabapple, 424–425
Wild-olive, 408–409
Wild plum, 428–429, 430–431
Wild raisin, 170
Wild red cherry, 436–437
Willow family, 456–469. *See also
    Salicaceae*
Willow oak, 284–285
Winged elm, 512–513

Winged sumac, 94–95
Wire birch, 138–139
Witch hazel, 66, 300–303
Witch hobble, 170
Witherod, 170
Woolly buckthorn, 471, 472–473

Yaupon, 120–121
Yellow ash, 216–217
Yellow birch, 130–131
Yellow buckeye, 306–307
Yellow buckthorn, 412–413
Yellow chestnut oak, 272–273
Yellow-cypress, 8–9
Yellow flower magnolia, 354–355
Yellow locust, 216–217
Yellow oak, 272–273, 296–297
Yellow pine, 32–33, 38–39
Yellow plum, 428–429
Yellow-poplar, 352–353
Yellow spruce, 26–27
Yellow wood, 494–495
Yellowbark oak, 296–297
Yew family, 60–63. *See also Taxaceae*

Zanthoxylum americanum, 452–453
Zanthoxylum clava-herculis, 454–455